9781461287988

The Packaging Media

The Packaging Media

General Editor
F. A. PAINE
B.Sc., C.Chem., F.R.I.C., M.Inst.Pkg.

Published under the authority of The Council of the Institute of Packaging

A HALSTED PRESS BOOK

John Wiley and Sons
New York

BLACKIE & SON LIMITED
Bishopbriggs, Glasgow G64 2NZ
450/452 Edgware Road, London W2 1EG

© 1977 Blackie & Son Ltd
First published 1977

Published in the U.S.A. by
Halsted Press,
a Division of John Wiley and Sons Inc.,
New York

All rights reserved.
No part of this publication may be reproduced,
stored in a retrieval system, or transmitted,
in any form or by any means,
electronic, mechanical, recording or otherwise,
without prior permission of the Publishers

International Standard Book Number
0-470-99369-3
Library of Congress Catalog Card Number
77-92138

Printed in Great Britain by
Thomson Litho Ltd., East Kilbride, Scotland

BACKGROUND TO AUTHORS

ADCOCK, Philip, Ing.Chim.(Lausanne), was Head of the Closures and Plastics Research and Development Section at United Glass and, more recently, General Manager of the Lectraseal Department for applying electronically-sealed membranes to containers of U. G. Closures and Plastics.

ALLEN, D. C., C.Eng., M.I.Mech.E., A.F.R.Ae.S., is Chief Research Engineer of E.P.S. (Research & Development) Ltd., a company of the Export Packing Service Group. He is a founder member of the Packaging Group of the Society of Environmental Engineers, and from 1968 to 1972 was the Packaging Group Chairman.

BRAZIER, A. D., B.Sc., C.Chem., F.R.I.C., M.Inst.Pkg., is Technical Manager of National Adhesives and Resins Ltd., Adhesives Division, having joined the company in 1967.

BRISTON, John H., B.Sc., C.Chem., M.R.I.C., F.P.R.I.F.Inst.Pkg., is a past Chairman of The Institute of Packaging, and co-author of its Correspondence Course. He is author of six books on packaging and plastics.

DAVIES, A. O. D., M.Inst.Pkg., is Machinery and Liaison Development Manager, Samuel Jones & Company Ltd.

FLATMAN, Donald J., F.Inst.Pkg., is Marketing Services Manager of Bakelite Xylonite Limited (Flexible Packaging Division), a subsidiary company of the Union Carbide Corporation. He was Chairman of the National Council of The Institute of Packaging, 1973/74.

FORD, Michael J., B.Sc., is Industrial Liaison Officer of the Fibre Building Board Development Organisation Ltd. (FIDOR). During his career he has been involved closely with the manufacture and commercial operations of the pulp, paper and fibreboard industry.

GODDARD, R. R., M.Inst.Pkg., is Head of the Retail Packaging Group of the Packaging Division, Pira. He has been in the field of packaging for about ten years, before which he was involved in technical service works as an industrial chemist in the food processing industry.

GOOCH, Mrs Joan U, M.I.Inf.Sci., has been a member of Pira's Information & Training Division since 1969, and Editor of Packaging Abstracts since March 1970. She was previously with the Inveresk Paper Group as Librarian/Information Officer at their Research Laboratories in Weybridge.

GREEN, J. R., L.I.M.M.Inst.Pkg., is Deputy Technical Manager of Star Aluminium Company Limited, with 40 years' experience in the aluminium foil industry. He represents the foil industry on a number of national and international committees covering the field of packaging.

HINE, Dennis J., M.Sc., M.Inst.P., F.Inst.Pkg., is Head of the Unit Container Group of the Packaging Division, Pira. In 1953 he began studies on the production and use of cartons and, more recently, has extended his work to the use of flexible and other packaging materials on machines.

LINDOP, B., M.Inst.Pkg., M.Inst.M., started his career in packaging as a management trainee with the Metal Box Company in the Paper Division of the Group at Manchester, eventually becoming Assistant Commercial Manager at the Composite Container Factory. He is Sales Manager of Egremont Tubes Limited, part of Mardon Packaging International Limited.

LOGAN, R. M. C., was apprenticed to Paint Manufacture, spent a short time in the Chemical Industry, and joined Van Leer (UK) Ltd. as a Works Chemist. He eventually became Chief Chemist and Finishing Manager, a position he holds today.

LOTT, Alfred R., B.Tech., M.Inst.Pkg., gained experience of packaging materials in the Packaging Division of Pira and in ICI, Plastics Division. He is now Technical Services Manager of one of the divisions of Smith and Nephew Associated Companies Ltd., and is a member of the London Branch Committee of The Institute of Packaging.

MONTRESOR, J. M., M.A., M.Inst.Pkg., is Packaging Services Manager at Pira.

MOSTYN, Harri, B.Sc., M.Inst.Pkg., is Group Head of Industrial Packaging at Pira. Before joining Pira in 1968, he was Chief Forest Products Officer in the Republic of Zambia.

OSWIN, C. R., M.A., D.Sc., C.Chem., F.R.I.C., formerly Packaging Consultant in the Courtaulds Group. He has been with British Cellophane Limited since 1938, where he was Technical Director from 1962 to 1968.

PAINE, Frank A., B.Sc., C.Chem., F.R.I.C., M.Inst.Pkg., is Director of the Packaging Division of Pira. Other positions include: National Chairman, The Institute of Packaging, 1972/73; Board member of The Institute of Packaging; Chairman, Eurostar Jury, 1961–72; Chairman, World Star Jury, 1970 and 1975; Secretary General, IAPRI (The International Association of Packaging Research Institutes).

PORTEOUS, Gordon J. P., is Managing Director of Universal Pulp Containers Limited, specialists in moulded pulp container manufacture.

PRICE, Dennis, M.I.Mech.E., has been involved in packaging for more than 20 years, and has at various times been concerned with design, production, sales and quality control. He is now Director of a company concerned with metal printing, tin box and can making.

RADCLIFFE, M.Inst.Pkg., is Vice President and Director of Signode Limited, manufacturers of tensional strapping and equipment.

RAWSON, Mike, M.Inst.Pkg., is Director of Bennett and Rawson Ltd., Birmingham, an old-established packing case and packing firm. Trained in tropicalization and packing at the Rover Ministry of Supply, he has lectured on The Institute of Packaging's Residential Course on Export Containers for several years.

ROBINSON, Philip, B.A., is Director of a packaging company, with specialist responsibilities for rigid box production.

SIMPSON, Arthur, F.Inst.Pkg., is Marketing Manager of the Aerosols Division of Metal Box Limited, having previously been Export Sales Manager. He is Regulations Committee Chairman for the European Aerosol Federation, and has been Secretary or Treasurer of the Aerosol Group since its inception over 15 years ago.

WEEDEN, Cyril, B.Sc.(Econ.), F.S.G.T., is Assistant Director of the Glass Manufacturers' Federation. Trained as an engineer, he later became an economist, and is now responsible for research and development in glass industry markets.

Note: The chapters attributed to R. F. D'Lemos (Part Two, chapter 6) and W. G. Atkins (Part Four, chapters 6 and 7) are based on the original chapters in *Packaging Materials and Containers*, and have been revised by F. A. Paine.

EDITOR'S PREFACE

Since 1967, when *Packaging Materials and Containers*, the predecessor to this book, was published, considerable changes in the technology and techniques of packaging have occurred in many areas. Increases in the costs of energy and materials have ensured that the days of relatively cheap packaging are over. The selection of the right packaging media for a specific product and market is even more important today than it was eight years ago. This can be achieved only if the facts about the available alternatives are known to the designer.

This volume attempts to collect together all the basic information on the various packaging media, not only for the benefit of candidates for the Institute of Packaging examination, but also for other students of packaging who need such information in the course of their jobs.

In his Foreword to the first publication, Lord Kings Norton referred to the day when the first Professor of Packaging would be appointed. To my knowledge more than half-a-dozen now exist—in the United States, in Europe and in Asia. The United Kingdom has one degree course in the subject, but so far a chair has not been established, although the UK Institute of Packaging correspondence course has been adopted in several overseas countries, and the examination system is being considered as a possible basis for international agreement on qualification. Industrially, however, Membership of the Institute of Packaging is regarded by most companies as a valuable qualification.

Of the 25 chapters in *Packaging Materials and Containers*, two have been combined, and all have been revised and updated. Seven new chapters, an introduction, and an epilogue have been added, and the revision of many of the original chapters is quite substantial. Chapters

6 of Part Two and 3, 6, 7 of Part Four differ very little from their original versions: indeed, the last three are traditional packaging methods of long standing, and are slowly being replaced in almost all instances. All chapters, except these four, have been converted wherever necessary into SI units (cross-referred to Imperial units where possible confusion might still exist). Chapter 7 of Part One and Capter 5 of Part Three have been contributed by the Industrial Division of Bowater Packaging Ltd. and are not attributed to a specific author. The former is essentially a revision of the chapter in *Packaging Materials and Containers* but the latter is an entirely new contribution.

The result is a book substantially larger than the original and, for this reason, it is now presented in five Parts related to the basic materials, rather than into shipping containers, retail units and the rest.

Once again, within the compass of each chapter the author was free to write (or revise) without the imposition of too many restrictions. My thanks are due to all the contributors for their help and forbearance in allowing me to put the separate chapters together to attempt to produce a connected and, I trust, readable whole.

Frank A. Paine

CONTENTS

BACKGROUND TO AUTHORS
EDITOR'S PREFACE
INTRODUCTION

PART ONE   PAPER-BASED PACKAGING
CHAPTER
1   Paper and Board-making   1.3
    F. A. Paine
2   Wrapping and Packaging Papers   1.20
    R. R. Goddard
3   Multiwall Paper Sacks   1.40
    F. A. Paine
4   Folding Box-board Cartons   1.54
    D. J. Hine
5   Rigid Boxes   1.77
    P. B. Robinson
6   Solid and Corrugated Fibreboard Cases   1.88
    A. R. Lott
7   Fibre Drums   1.106
    Bowater Packaging Ltd.
8   Moulded Pulp Containers   1.116
    G. J. P. Porteous

PART TWO METAL-BASED PACKAGING

1. Metal Packaging—The Basic Materials 2.3
   F. A. Paine
2. Metal Cans 2.13
   D. W. Price
3. Composite Containers 2.52
   B. Lindop
4. Aerosol Packaging 2.73
   A. Simpson
5. Metal Drums and Kegs 2.84
   R. M. C. Logan
6. Collapsible Tubes 2.97
   R. F. D'Lemos
7. Metal Foil Packaging 2.102
   J. R. Green

PART THREE GLASS AND PLASTICS-BASED PACKAGING

1. Glass Containers 3.3
   C. Weeden
2. Moulded Plastics Containers 3.25
   J. H. Briston
3. Packaging with Flexible Barriers 3.44
   C. R. Oswin
4. Sacks made from Plastics Film 3.74
   D. J. Flatman
5. Plastics Drums and Crates 3.94
   Bowater Packaging Ltd.
6. Closures and Dispensing Devices for Glass and Plastics Containers 3.107
   E. P. Adcock

PART FOUR WOOD AND TEXTILE-BASED PACKAGING

1. Wood and Wood Products in Packaging   4.3
   F. A. Paine
2. Timber and Plywood Cases and Crates   4.10
   M. Rawson
3. Wooden Casks and Plywood Kegs   4.20
   F. A. Paine
4. Hardboard and Softboard   4.25
   M. J. Ford and H. P. Mostyn
5. Pallets and Unit Loads   4.45
   H. P. Mostyn
6. Textile Sacks and Bags   4.64
   W. G. Atkins
7. Bales and Baling   4.75
   W. G. Atkins

PART FIVE ANCILLARY MATERIALS

1. Paper, Plastics and Fabric Sealing Tapes   5.3
   A. O. D. Davies
2. Strapping and Stapling   5.9
   V. Radcliffe
3. Labels and Labelling   5.20
   J. U. Gooch and J. M. Montresor
4. Adhesives in Packaging   5.31
   A. D. Brazier
5. Package Cushioning Systems   5.44
   D. C. Allen

EPILOGUE
INDEX

INTRODUCTION

The main packaging materials and their value share of the total packaging used are shown in Table 1.

Table 1. Main packaging materials and their value shares (UK 1972)

Material	% of total value
Paper and paperboard	31
Metal (principally steel and aluminium)	24
Plastics, mouldings and film, including cellulose film	13
Glass	8
Wood	5
Textiles	2
Miscellaneous	17

These figures relate to the United Kingdom in 1972, but similar proportions of the market will be applicable in almost all the sophisticated countries in the Western world. It was estimated that the total value of the packaging materials sold in the UK in 1972 amounted to about £1,350,000,000 or just under 3% of the Gross National Product. This includes, of course, all the industrial packaging for machinery, and all the packaging used for exports, as well as the packaging that reaches the general public through retail outlets.

INTRODUCTION

The forms in which the main packaging materials are employed are shown in Table 2 which also gives some idea of the wide variety available.

Table 2. The forms in which the main packaging materials are employed

Material	Package types
Paper and board	Wrapping papers, bags and carrier bags Boxes and cartons Fitments in board Tubes, spiral and straight-wound Fibreboard cases and fittings Multiwall sacks Fibre drums
Metal	Cans and boxes (including aerosols) Aluminium foil, laminates and labels Collapsible tubes Closures Metal strapping and banding Barrels, kegs and drums Crates or boxes
Glass	Bottles, jars Vials and ampoules
Plastics (including cellulose and rubber)	Films, laminates and sheets Bags, pouches, sachets, etc. Sacks (film and woven tape) Moulded bottles, jars, pots, etc. Thermoformed trays, pots, blisters and fitments, etc. Cushioning materials and fittings Caps and closures Drums Crates Boxes
Timber (including plywood)	Boxes and crates Casks and kegs Pallets and containers Baskets and punnets Wood wool
Textiles	Sacks and bags Baling materials

Packaging costs

Up-to-date figures of the expenditure on packaging by various industries is difficult to obtain, but an analysis of the data given in the 1968 Census of Production gives some idea of the main user industries and the types of material used. The total sales of all the 152 industries listed in the Census was about £51,000 million, while the total packaging purchased by those industries amounted to some £700 million or 1·4%.

Table 3 gives some data on the 26 principal users, who account for about 70% of total packaging purchases. Apart from the top 16, some 130 of the 152 industries listed in the Census spend on average about 2% of their sales value on packaging (it varies from 0·1% to 5·9% depending on the industry). Six industries spend less than 0·1%.

The Food, Drink & Tobacco industries, which accounted in 1968 for just over half the total packaging purchased, spent on average about 5% of their sales value on packaging material. Fresh Fruit and Vegetable packaging accounted for the highest figure (17%) while Milk and Bread averaged 4·2% and 4·3% respectively, and beer less than 2%.

Shipping containers

The principal shipping containers are made from wood, fibreboard and metal. There is a certain usage of glass for carboys of corrosive liquids, such as acids, etchants and other chemicals, but the quantities are small, and the glass must always be protected from impact. Wood is generally used whenever the package is large, or even when it is small if the product is of high density. Thus timber cases and crates are used extensively for weights above 100 kg (2 cwt) while below this weight fibreboard, both solid and corrugated, is the favoured material. Timber is also used for casks for wine, beer, etc., but there is considerable movement towards their replacement by metal (stainless steel or aluminium) either alone or with inner liners of plastics.

There is some use of plastics materials for shipping containers but, since the basic material cost is high, the main use is for returnable containers. Plastics crates are established in the dairy industry, and also in the more exacting use for bottled beers, minerals and soft drinks. Beer crates are usually stacked higher and for longer periods than those for milk; polypropylene is often preferred for beer, while high-density polyethylene has proved adequate for milk crates. Polyethylene is also used in carboys for many liquids, thus avoiding the protection required for their glass counterparts. Polyethylene casks or barrels are also used.

INTRODUCTION

Table 3. Expenditure on packaging by main users (1968 Census)

Industry	Estimated value of Total sales of the industry	Packaging purchased	% of packaging to total value
	£000s	£000s	
Toiletries, etc.	117	27	23
Fruit and vegetables	334	57	17
Soft drinks	129	20	16
Polishes	39	$5\frac{1}{2}$	14
Soap and detergents	150	20	13
Biscuits	174	21	12
Pesticides	38	4	11
Bandages, etc.	48	$4\frac{1}{2}$	10
Confectionery	321	30	$9\frac{1}{2}$
Margarine	47	4	9
Lubricants	69	6	8
British wines	32	$2\frac{1}{2}$	$8\frac{1}{2}$
Starch	294	22	$7\frac{1}{2}$
Pharmaceuticals	326	24	$7\frac{1}{2}$
Paint	168	12	7
Spirits	448	30	7
Photographic	78	$4\frac{1}{2}$	6
Meat and fish	508	24	5
Bread	445	19	$4\frac{1}{2}$
Milk	734	31	4
Animal feed	482	19	4
Grain milling	331	13	4
Sugar	221	$5\frac{1}{2}$	$2\frac{1}{2}$
Tobacco	1436	31	2
Synthetic chemical	457	9	2
Brewing	1047	19	2

INTRODUCTION

Expanded polystyrene finds employment in shipping containers for tomatoes and grapes, and also for both cured and wet fish. In the last two uses the heat insulation properties are useful in keeping the product cold with the minimum of solid coolant.

Solid and corrugated fibreboard cases are probably the most widely used shipping containers. They combine convenience with economy and hygiene. The most common type of case is the one-piece (or regular) slotted container, although open-tray and wrap-around styles are used extensively. The normal range of weight which corrugated and solid fibreboard cases carry lies between 5 and 20 kg, but fibreboard cases can be made to hold loads of up to 50 kg without any special fittings being used. If specially reinforced containers are made, they are capable of being produced to carry loads of powdered or granular material up to 500 kg weight.

In the last decade, a quiet revolution in the movement of goods has taken place. Palletization, modular packaging and freight containers have arrived. In some instances this has led to the pallet superseding the wooden case or crate, and in others the almost complete elimination of the shipping container, since the goods are loaded straight into a freight container and secured within. This is particularly applicable to heavy machinery.

The main purposes of a shipping container can be listed under the following headings:

1 It must contain the product efficiently throughout the journey.
2 It must provide protection against external climatic conditions and contaminants.
3 It must be compatible with the product.
4 It must be easily and efficiently filled and closed.
5 It must be easily handled by the appropriate mechanical or other means.
6 It must remain securely closed in transit, but open easily and when required (as for Customs inspection), be capable of efficient and secure reclosure.
7 It must communicate to the customer, the carrier, the wholesaler and the manufacturer all that it is necessary to know about the product and its destination, as well as how to handle and open the package.
8 Where the product is dangerous or potentially harmful (as for chemicals and acids), the package must be virtually unbreakable.
9 The package must be inexpensive and cost no more than is absolutely necessary to do its job.
10 The package must be readily disposable, or be reusable, or have an after-use for some other purpose.

Very few containers will need to satisfy all these requirements. The emphasis as to which particular ones predominate will be dependent very much upon the product packed.

INTRODUCTION

Retail packaging

Four major considerations are involved in retail package design: the product, its market, the package production problem, and the economics of the packaging operation.

Under product considerations come such questions as susceptibility to damage or deterioration, the need for protection against temperature changes, moisture changes, oxygen, light, mould, or insects, and whether the product will interact with the potential container.

Market considerations include the characteristics of the customer, such as age group and income level; the unit quantities required; aspects of the retail and wholesale operations, and probable methods of handling, warehousing, and distribution. The graphics considered include brand name; whether the product is part of a range; whether there are colours that may not, or must be, used; any pertinent government regulations; and the economic effect of all these factors.

Approximately two-thirds of packaging expenditure is on retail packages. Metal cans and boxes account for 32%, paper boxes and cartons for 20%, plastics packages and cellulose for a similar value, glass for 12%, with paper bags and foil accounting for about 6% each.

Retail packages are sometimes criticized, and it is suggested that a good retail package should be considered in relation to the following guide lines:

1 The package must comply with all relevant legal requirements.
2 The package should adequately contain, protect and preserve the product.
3 The package must be made from materials which have no adverse effect on the contents.
4 Packages for hazardous products must contain them safely, incorporate suitable warnings and be fitted with appropriate closures.
5 The package must not mislead the purchaser about the nature, quality, or quantity of the contents.
6 Packaging should be convenient and safe in normal use and have regard to the nature and purpose of both the pack and its contents.
7 The package should contain convenient quantities of the product.
8 The package should properly identify the product and provide instructions for use where appropriate.
9 The package should be designed with regard to its re-use, recycling and ultimate disposal.

INTRODUCTION

All retail packages must contain, protect and preserve, communicate, and be easily filled and closed appropriately. Many must also sell and provide convenience either in terms of opening, closure and/or reclosure, or by providing useful dispensing methods or simplifying the use of the product in other ways. Unless such advances are real and valuable, the package rapidly ceases to be used.

A good retail package can sell a poor product once, but a badly designed or unduly expensive package will much more easily ruin the market for an excellent product. The consumer far more certainly understands *caveat emptor* than many anti-packaging pressure groups believe.

PART ONE
PAPER—BASED PACKAGING

CHAPTER 1

Paper and Board-making

F. A. Paine

Materials

It was not until the middle of the nineteenth century that the paper industry started to use wood pulp for making paper. Before this, timber had been used for many other purposes—building ships, houses, and for making furniture and boxes, but the beginning of the paper-making industry using pulp wood caused great changes in the practices of cutting trees in the forests.

In the early days, trees were cut and used locally, but shortages of timber had led to a search for new sources. For example, the timber requirements of the British navy for ships' masts led to the building of ships specially designed to carry them across the North Atlantic. One particular tree, the Weymouth pine, is named after a Captain John Weymouth who was one of the early timber transporters. In 1608 Captain John Smith carried eight Poles and a Dutchman for the purpose of erecting sawmills in Jamestown, Virginia, and not long after that America began to export many goods, including timber.

Increasingly after this, the demand for timber rose, and by 1900 supplies were falling. This does not mean that people had not thought about forest conservation—as early as 1682 William Penn decreed that, in Pennsylvania, for every five acres of forest cut one acre must be replanted. However, these were small contributions, and adequate reafforestation schemes did not begin until much later.

Many suggestions have been made about how to maintain a credit balance in forests. The problem is basically very simple to diagnose. All we have to do is speed up forest growth and reduce timber consumption, cut out forest fires and cure tree diseases, as well as removing the insects which live on wood and woodpeckers which also damage it. But it is quite another thing to do something about reafforestation. Trees do not breed like rabbits. Time is essential. Tree farming means very much more than just planting the trees and then standing back to await the final crop. It involves raising seedlings, planting them out, and spraying them against tree illnesses and insects, and then thinning at the appropriate time. While all this is going on, steps must be taken to prevent damage by fire. Nowadays the paper industry replaces the trees that it uses at a rate which is adequate for our needs at present.

In addition to wood pulp, other fibrous materials such as cotton, flax and esparto, bamboo and straw are also used for making paper. Finally, waste paper is a secondary source of material and is used for certain grades of paper and much board. Trees are, however, the most important primary source for cellulose fibres.

It must also be borne in mind that several non-fibrous materials are used in making paper—minerals for loading and coating—dyes for colouring, and materials for sizing and water proofing.

Non-fibrous additives

Table 1.1. Non-fibrous additives to paper

Agents	Effects
Fillers or loading	improve opacity and brightness
Binders	increase strength
Sizes	reduce penetration of water and writing inks
for specific properties	whiteness colour formation
for manufacturing aids	filler retention anti-foaming

There are four main classes of non-fibrous additives.

1. *Fillers or loading agents* are white mineral pigments such as china clay. They increase the opacity and brightness of the paper, and improve the surface smoothness and printability, but reduce the strength properties if present in large amounts.

2. *Binders* may be starches, vegetable gums, synthetic resins and rubber

PAPER AND BOARD-MAKING

latices. Such additions improve the strength properties—tensile, tear and burst.

3. *Sizing agents* provide resistance to the penetration of aqueous liquids, such as water and writing ink. Blotting papers are unsized; writing papers are usually sized, and there are various degrees of sizing from "soft" to "hard". Rosin and wax emulsions are the main agents, but some special synthetic resins are also used.

4. *Miscellaneous additives*. A number of other materials are incorporated either to impart specific properties to the paper or to assist in manufacture, e.g.

Optical brightening agents for increased whiteness
Coloured pigments and dyes for coloured papers
Various gums to assist in formation on the paper machine
Anti-foaming agents
Filler-retaining agents
Wet-strengthening agents

Additives are inserted at several stages before the fibre suspension reaches the sheet-forming stage, i.e. in the pulping, beating or refining operations —where exactly depends on the additive.

What is wood?

Underneath the outer and inner layers of bark in a tree we have a layer which contains the plant foodstuffs, and it is through this layer that the tree itself grows. Then we have the main woody part which consists of bundles of cellulose fibres running vertically up the trunk, held together by a material which is called "lignin" (figure 1.1). Essentially

Figure 1.1 The structure of a tree

PAPER-BASED PACKAGING

the wood part of the tree consists of about 50% cellulose fibres, 30% lignin, 16% carbohydrates, and some 4% of other materials such as proteins, resins and fats. It is principally the cellulose which eventually becomes paper. It is composed of individual fibres which are finer than human hairs and are a few millimetres in length at the most. These fibres are about 100 times as long as they are thick. Lignin is a chemically complex substance which holds the fibres together. It is best, perhaps, to think of it as the glue holding the tree in one piece.

Pulping processes

To be useful to the papermaker, the raw material must be reduced to a fibrous state. The operation of doing this is called *pulping* and there are two basic methods:

(*i*) mechanical pulping
(*ii*) chemical pulping

For both processes, first of all the bark is stripped away from logs which have been cut to a suitable length at the appropriate stage in their growth. In the mechanical pulping process, the logs may be used directly in 1·2 m (4 ft) lengths or, alternatively, they may be chipped, i.e. converted into pieces of uniform size about 15–20 mm long. Two methods of mechanical pulping are employed.

In one the logs (figure 1.2) are pressed against the surface of a large

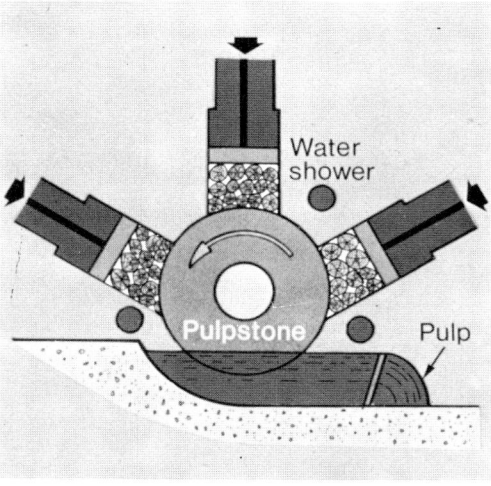

Figure 1.2 Mechanical pulping

PAPER AND BOARD-MAKING

revolving grindstone, kept wet by a stream of water which additionally removes the fibres. In the other system, the wood chips are passed between the two plates of a disc refiner with specially treated surfaces which are very close together and which are rotating at high speed. Thus the wood chips are reduced to individual fibres. In mechanical pulping the water-soluble impurities only are removed, and most of the lignin still remains. Additionally, by mechanical pulping, many fibre bundles and some damaged fibres are left in the pulp. Much of the grinder and disc-refined wood pulp is used for newsprint, although substantial quantities are employed mixed with chemical pulp for making certain kinds of board. Mechanical pulp is normally made from softwood, typically spruce.

Chemical pulping starts from chips, but removes all materials other than the cellulose fibres by chemical action and solution, the chemicals converting the lignin to a soluble form that can be removed by washing (figure 1.3). This type of pulping produces cellulose fibres of a higher purity than those produced by the mechanical processes. They are generally much less damaged and, in addition, the fibre bundles are fewer. Several different chemical pulping processes exist, and the quality of the pulp depends upon the process, as well as the kind of wood fibre used. For packaging purposes, three chemical processes are of major importance. These are the "kraft process" which retains most strength in the fibres, the so-called "sulphite" process which is less strong, and the "semi-chemical" process.

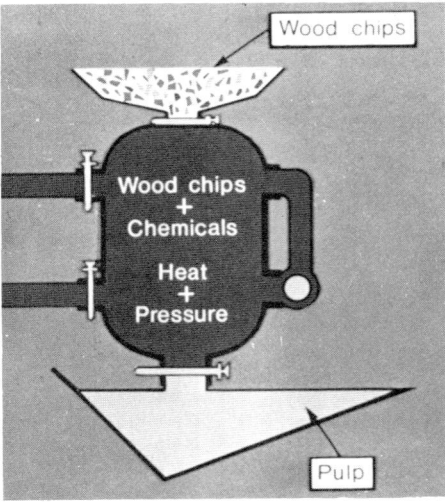

Figure 1.3 Chemical pulping

1.7

In the kraft or sulphate process, the wood chips are digested in a solution of caustic soda and sodium sulphate for some hours, but originally only caustic soda was used. Most of this, which dissolves out the lignin, can be recovered and re-used. In the recovery process the presence of sodium sulphate leads to the production of sodium sulphide as an end product. When put back into the digester and heated again under pressure, the sodium sulphide produces the necessary caustic soda for dissolving out the lignin, while avoiding excess. This method is known as the *kraft process*, the name being derived from the Swedish word *kraft* which means strong. In early days, kraft paper was always associated first with a brown colour and, second, with long fibres and what was called a "wild look through". This means that, because the fibres were long, they did not form such a uniform sheet and, when held up to the light, it looked a little "blotchy".

The sulphite process uses sulphur dioxide and calcium bisulphite. This is mixed with the chips in aqueous solution and heated to about 140 °C. Once again the lignin is dissolved out, leaving the fibres, and after digestion the mass is washed with water and then bleached with another chemical, such as calcium hypochlorite, before pressing into pulp sheets. This process gives a very pure cellulose fibre, although the resulting pulp is not as strong as that from the kraft process.

In the semi-chemical process the wood chips, which are usually from beech or birch trees, are partly treated by chemicals and partly mechanically to reduce them to fibres, hence the name *semi-chemical*. This semi-chemical pulp is often used for the manufacture of the fluting medium for corrugated board.

Beating

Once pulp has been produced, it may have to be bleached to make it white, or coloured, or treated in other ways. One of the most important processes in the pre-preparation of fibres for paper making is the so-called *beating process*. The object of most beating processes is to rub and brush the individual fibres, and cause them to split down their length in such a way as to produce a mass of thin fibrils which will enable them to hold together in the matted paper far more strongly. This process is called *fibrillation*. The greater the degree of fibrillation we can induce, the higher the strength of the paper can be. Different pulps respond differently to this treatment. Softwood fibres will fibrillate to a greater extent than hardwood or straw fibres. Hence softwoods are potentially able to produce better and stronger papers. Beating also cuts the fibres by the action of the bars in the beater against them. This is undesirable to some extent, although some cutting may be necessary for good running on the

PAPER AND BOARD-MAKING

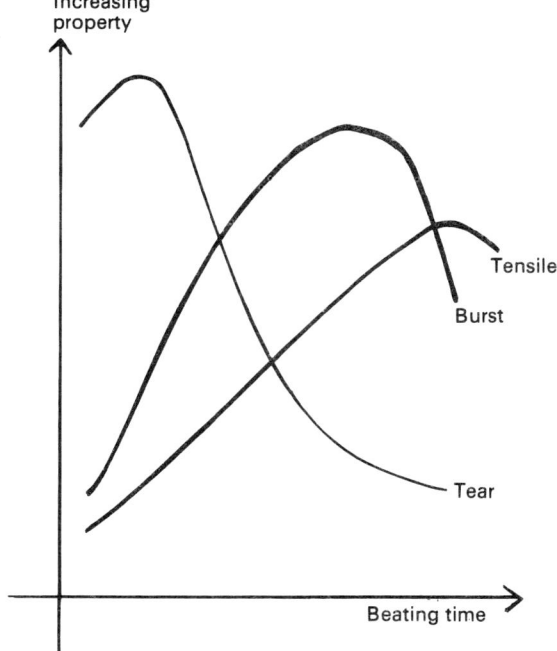

Figure 1.4 Effect of beating on strength properties

machine. Furthermore, the process breaks up any fibre clumps and refines them into individual cellulose fibres. The art of beating is to maintain a high proportion of fibrillation and a low proportion of cutting in the pulp, to give the desired properties in finished paper. Figure 1.4 gives an idea of the effect of beating on paper strength characteristics in respect of burst, tensile and tear strength.

Paper and board-making machines

Paper was first made in the United Kingdom by John Tait in 1496. He used a hand-making process producing a slurry of pulp fibres in water, inserting a wire "mould", lifting it and "couching" the matted fibre sheet onto a sheet of felt. The sheets were then stacked in a press where the fibre mat compacted and water was removed, and finally the paper sheets were hung over ropes to dry. All this was an extremely slow process. The first paper machine was invented in 1799 in France by Nicholas Robert but, because of the unsettled conditions in France, the invention was soon moved to England where, in 1807, it was taken up by

PAPER-BASED PACKAGING

two London stationers named Fourdrinier. They were relatively unsuccessful commercially and finally went bankrupt, but the type of machine has been given their name. The modern Fourdrinier machine has eight basic parts, four comprising what is called the wet end, and four the dry end. They are:

(a) stuff box or chest
(b) head box
(c) slice
(d) Fourdrinier wire
} the wet end

(e) presses
(f) dryers
(g) MG dryer
(h) calender stacks
} the dry end

Let us briefly describe the purposes of each in turn.

(a) Stuff chest

The stuff chest holds a slurry of pulp containing about 97 parts of water to every three parts of fibre. This is much too heavy to be formed into a paper sheet, and is diluted in a mixing chamber with more water to a consistency of about $\frac{1}{2}\%$. That is $99\frac{1}{2}\%$ of the "stuff" is water.

(b) Head box

The diluted pulp passes from the stuff chest to the head box, which is both a means of agitating the slurry and a turbulence reducer. Essentially it is designed to produce a uniform fibre suspension (the stock) and to feed it to the slice.

(c) Slice

The slice is part of the head box and consists of a narrow slot in its front face through which the stock flows onto the wire. Adjustments can be made in the opening to the slice, and it can be raised and lowered to adjust the flow.

(d) Fourdrinier wire

The Fourdrinier wire carries the stock from the slice up to the place where the sheet is removed at the so-called "couch" roll. The function of the wire is to allow the initial drainage of water to take place, helped by rollers underneath it and by suction boxes to remove the water, to get the mat of paper fibres into a sufficiently strong sheet to transfer to a felt which will move into the drying section. At this stage the mat of

PAPER AND BOARD-MAKING

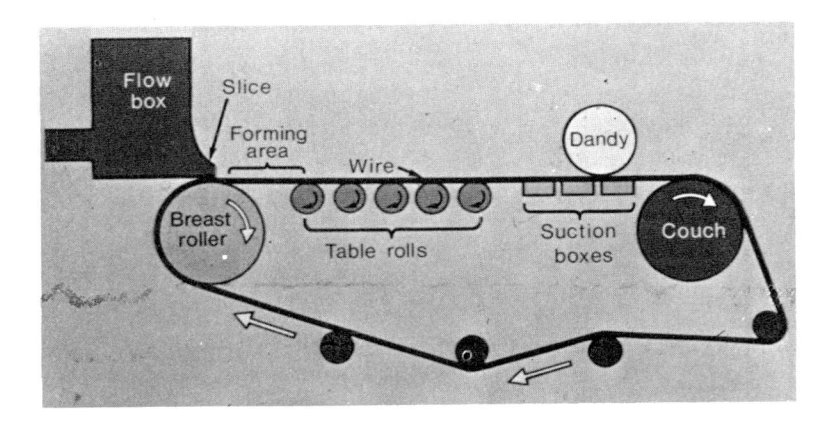

Figure 1.5 The wet end

fibre consists of about 75–80% water by weight, and the first operation is to pass it through presses to remove more water. The felts on which the sheet travels are made of wool; essentially they help to blot water from the sheet (figure 1.5).

(e) Presses

The presses (figure 1.6) may be either plain, which just catch the water in trays, or fitted with suction to help to remove the water by pulling it through thousands of tiny holes in a suction roll. After the presses the sheet still contains some 60% water as it passes into the dryers.

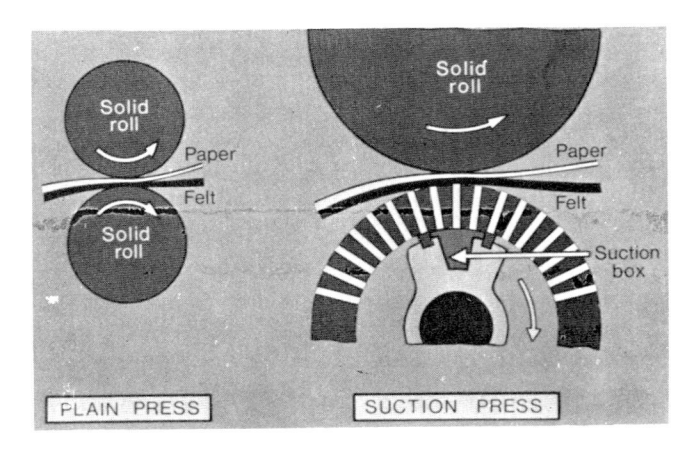

Figure 1.6 Plain and suction presses

1.11

PAPER-BASED PACKAGING

(f) *Dryers*

The dryers consist of a long train of heated cylinders which come into intimate contact with the paper and dry it by heat.

(g) *MG dryer*

In certain instances the paper may require to be *machine glazed*, and MG papers are given their gloss on massive cylinders which have a highly polished surface. This is a similar sort of process to the glazing process used with glossy photographic prints.

(h) *Calender stacks*

Machine finished (MF) papers are smoothed in what is called a *calender stack*. This consists of a series of rolls which iron the sheet by slippage. This occurs because the bottom roll in the stack is the only one which is driven, the others are turned by friction with the paper. The degree of machine finish is controlled by the number of rolls through which the paper passes, and it may be increased by applying a certain amount of surface water (from water boxes) to increase it. Finally, the paper passes to the reeler and winder, and is wound up into a giant roll.

The furnish, the degree of beating, the amount of filler, binders, sizes, etc., together with the operating variables of the particular paper-making machine and specific finishing processes, can be varied to produce many types of paper. The main packaging papers, their origin and uses are outlined in Table 1.2 (page 1.14).

Board making

As will already have been seen by the description of the Fourdrinier process, paper is made by pouring a slurry of fibres onto a wire sieve, when the water drains through leaving behind a mat which is ultimately consolidated into a sheet of paper. It can be easily realized that attempts to produce thick materials will cause difficulty in draining once the thickness exceeds a certain level. It is not possible therefore to produce thick sheets by the Fourdrinier process and other systems must be used. Improvements in Fourdrinier operation have been made by new flowbox designs, by using nozzles instead of a slice, and by improvements in the methods of drainage under the wire. Recently, however, a number of new methods of forming a web from the fibres have been evolved. The basic idea behind these is to form the paper web more rapidly, and usually

PAPER AND BOARD-MAKING

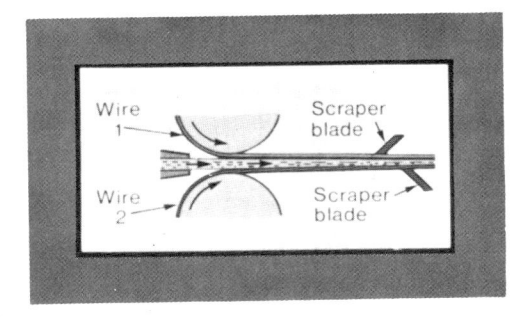

Figure 1.7
The twin-wire former

between two wires so that the drainage can take place from both sides. The twin wire former (figure 1.7) is in principle a method of letting the fibre suspension come out of the slice into a converging gap between two wires. Water is expressed from both sides of the sheet by the tension in the wire, and scraper blades pressing on the underside of the wires assist.

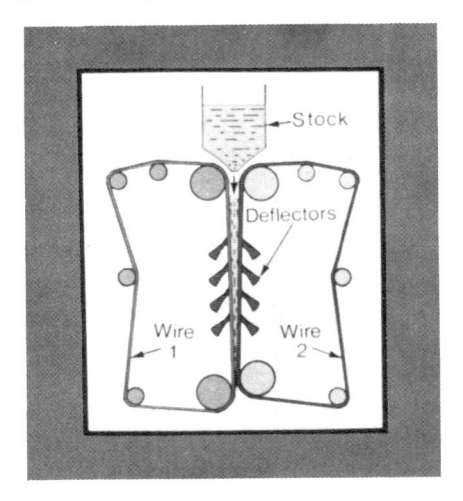

Figure 1.8
The Vertiforma

The Vertiforma (figure 1.8) is a similar system in which the two wires are arranged vertically. As well as allowing better drainage, the fact that there is a wire on both sides makes two-sidedness (differences in properties between the wire side and the felt side) impossible, and both sides of the paper are, therefore, similar.

1.13

PAPER-BASED PACKAGING

Table 1.2. Main packaging papers

Basic material	How made?	Weight range lb/1000 ft²	Weight range kg/1000 m²	Tensile strength lb/in width	Tensile strength kN/m	Properties and uses
Kraft papers	From sulphate pulp on softwoods (e.g. spruce)	14–60	70–300	MD 14–65 CD 7–30	2.4–11.3 1.2–5.2	Heavy-duty paper, bleached, natural or coloured; may be wet-strengthened or made water-repellent. Used for bags, multiwall sacks and liners for corrugated board. Bleached varieties for food packaging where strength required.
Sulphite papers	Usually bleached and generally made from mixture of softwood and hardwood	7–60	35–300	Very variable		Clean bright paper of excellent printing nature used for smaller bags, pouches, envelopes, waxed papers, labels and for foil laminating, etc.
Grease-proof papers	From heavily beaten pulp	14–30	70–150	MD 10–25 CD 5–12	1.7–4.4 0.85–2.1	Grease-resistant for baked goods, industrial parts protected by greases, and fatty foods
Glassine	Similar to greaseproof but super-calendered	8–30	40–150	MD 8–30 CD 5–16	1.4–5.2 0.85–2.8	Oil and grease-resistant, odour barrier for lining bags, boxes, etc., for soaps, bandages and greasy goods
Vegetable parchment	Treatment of unsized paper with concentrated sulphuric acid	12–75	60–370	12–80	2.1–14.0	Non-toxic, high wet strength, grease and oil-resistant for wet and greasy food, e.g. butter, fats, fish, meat, etc.
Tissue	Lightweight paper from most pulps	4–10	20–50	low strength		Lightweight, soft wrapping for silverware, jewellery, flowers, hosiery, etc.

MD = Machine Direction CD = Cross Direction

1.14

PAPER AND BOARD-MAKING

Table 1.3. Main packaging boards

Type of board	Construction	Chief characteristics	Principal uses
Plain chipboard	100% low-grade waste pulps (newsprint to corrugated fibreboard)	Low cost, not printable, light grey to tan in colour, poor folder	Rigid boxes and packing pieces (dividers, etc.) and fitments
Cream lined chipboard	Cream liner may be almost any quality of pulp from mechanized and chemical mixtures. Back as chipboard	Lowest-cost board capable of printing and folding	Lowest-quality folding cartons
White lined chipboards	White liner is 100% new pulp mainly chemical. Back as chipboard	Smooth-surfaced white board of good printability with good strength for folding	Quality folding cartons for shirts, detergents, etc.
Duplex board	Both liner and backing made from new pulps; backing usually a mixture of mechanical and chemical	Smooth-surfaced white board of good printability with good strength for folding	Quality folding cartons for shirts, detergents, etc.
Clay-coated boards	As white lined with clay-coated surface	Excellent printability and folding	Top-quality display cartons
Solid bleached sulphate or sulphite boards	100% sulphate or sulphite pulps	Strong boards of high whiteness, good folding qualities, odourless, suitable for food contact	Frozen foods, ice cream, baked goods

1.15

PAPER-BASED PACKAGING

Cylinder or vat machines

Such methods of improving drainage can only go so far and, where we want to make thicker material, say 4 or 5 mm thick in one pass, a vat or cylinder machine is used. Here the wire mesh takes the form of a cylindrical screen revolving in a vat filled with the pulp and water mixture. As the roller comes out of the mixture (figure 1.9) it brings with it a layer of fibres, and the water drains away through the wire. These fibres form a web which is picked up by an endless belt travelling over the top of the cylinder. Some of the water is squeezed out from the web by rollers, and then it passes on to another vat to join a web emerging from that one, and so on, it may be over four or five vats to build up the total thickness required (figure 1.10). In this process the vats are maintained at a constant level from a reservoir called a stuff chest, and the pulp/water mixture is maintained at about 99% water, 1% fibre. Once the sheet has been formed, similar drying systems to those employed on Fourdrinier machines are used.

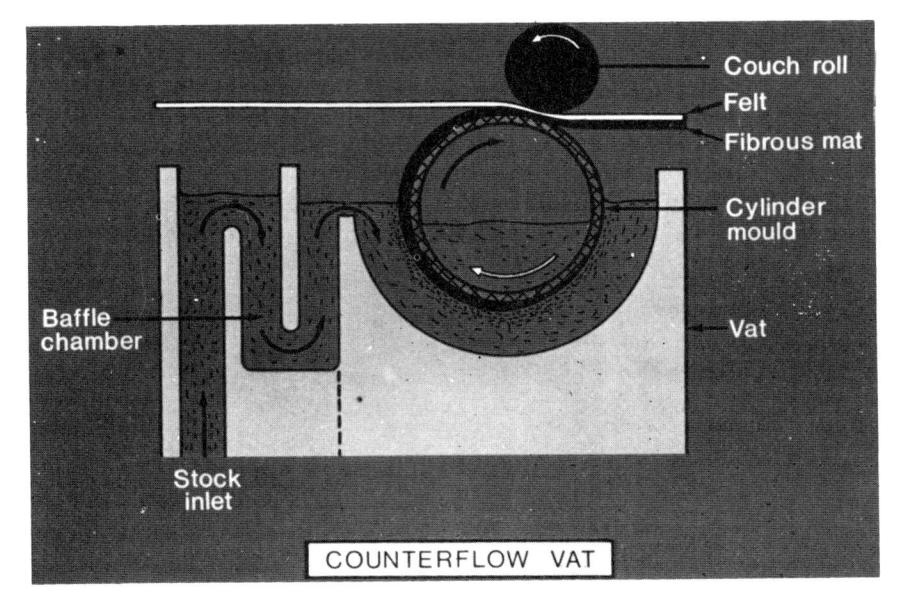

Figure 1.9 Counterflow vat

1.16

PAPER AND BOARD-MAKING

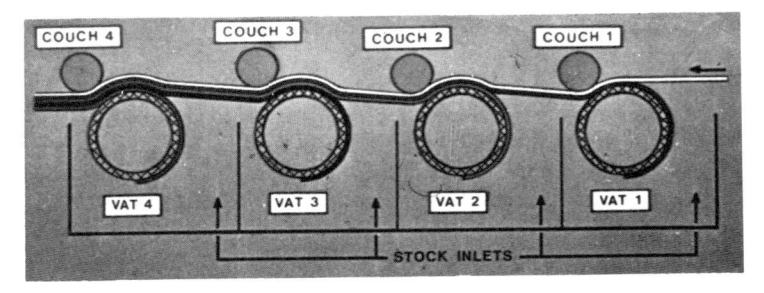

Figure 1.10 4-vat board machine

On a Fourdrinier machine, board is not built up in plies but formed as one thick layer. Drainage through this layer is very slow, and different layers of several types of fibre cannot be formed. Sheet formation is, however, more uniform.

There are some Fourdrinier installations that use several flow boxes to produce a multi-ply board. A base ply is formed on the first part of the wire. Some drainage occurs before a second flow box adds another ply. Water now has to drain through both plies. A third ply may be added in the same way.

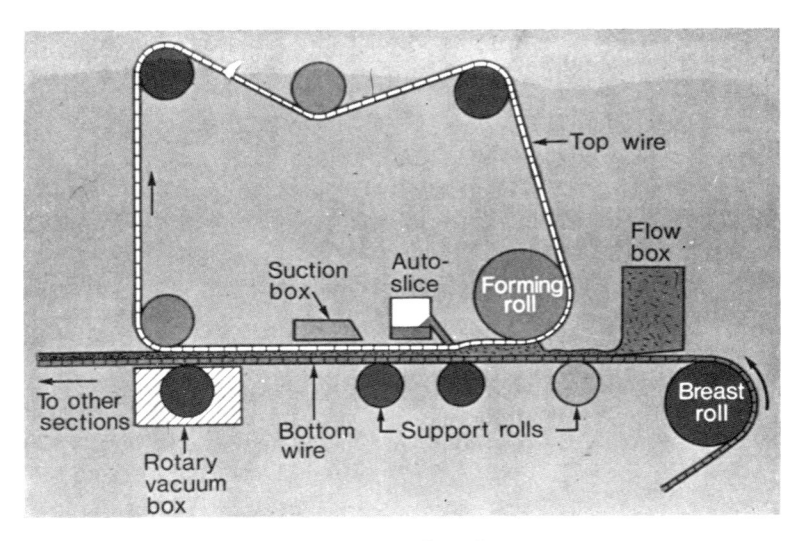

Figure 1.11 The Inverform system

1.17

PAPER-BASED PACKAGING

Combination machines form a base ply on a Fourdrinier wire and then combine this with several plies made on a cylinder mould machine.

A newer development in board making is the Inverform machine, in which the board is formed between two wires, water removal thus being possible from both surfaces (see figure 1.11).

Finishing processes

We have already mentioned the processes of calendering and machine glazing as operations in producing a paper sheet. Other types of finish can also be applied. For example, surface sizing can be carried out, usually with a solution of gelatin in water containing small quantities of other chemicals to make the surface more water-resistant and to improve its printing properties. Any sizing process can be applied on the paper machine or as a separate process afterwards. Paper and board are often coated with a layer of mineral pigment such as china clay bonded with a resin. Many boards these days are clay-coated in order to improve their appearance and their printing properties. The coating suspensions are either sprayed or fed onto the web from a rotating brush soaked in the material. The coating is then distributed evenly by other brushes, and the board conditioned in a steam chamber before calendering. Air knife techniques have also been used. Here an excess of coating material is applied to the paper and then metred off by a "doctor", which consists of a very fine jet of air blowing closely onto the paper surface. The main packaging boards are outlined in Table 1.3 (page 1.15).

Other surface treatments

Paper and board both have the disadvantage, from a packaging point of view, of being susceptible to changes due to moisture. There are many processes of impregnating, coating and laminating paper with other materials in order to improve the resistance to water. Probably the simplest and best known is waxing, and there are two main methods of carrying this out. In the first, *dry waxing*, the paper sheet passes through a bath containing the wax, and then through a heated section or hot nip rollers, which assist the wax to penetrate right through the paper. By this process the wax is impregnated uniformly and does not form a surface film. When it is desired to have a film of wax on the top of the sheet, the process employed is called *wet waxing*. The reason for this is that immediately after the wax film has been applied to the surface of the sheet, it passes into a bath of cold (usually refrigerated) water, which immediately sets the wax before it has time to penetrate into the sheet. The film of wax so applied provides much greater water-vapour resistance than the dry waxed sheet could ever do, but being a film on the surface it is much more easily damaged. The paper or board web may also be coated with many kinds of emulsions, varnishes, lacquers and the like, and can be laminated to other materials such as plastics, either using adhesives or by direct extrusion. The objective in all these instances is usually to improve the resistance of the basic sheet to water, water vapour, gases, greases or oils.

CHAPTER 2

Wrapping and Packaging Papers

R. R. Goddard

Introduction

This chapter will discuss those applications where paper-based materials are used in an unconverted form, i.e. as flat material. The largest use of paper in this way is for wrapping, in circumstances which range from counter reels in small shops, to high-speed automatic parcelling machines in factories. Other applications of packaging papers include interleaving and separation of materials, stiffening and supporting products, as well as in space-filling and shock absorption. Packaging papers are also used for the production of envelopes, sachets, bags and sacks, etc.

Wrapping papers

The wrapping operation may be described as the covering of an object or a collation of objects of a similar or different nature into a handleable unit. Paper in sheet or reel form may be used for this purpose. As in every area of packaging, the wrapping must perform certain clearly identifiable functions. The basic functions are to contain and protect, with subsidiary functions of communicating and providing convenience. Wrappers may be required to perform any or even all of these functions.

Containment

The wrapping of parcels of all types can involve contents which may be a single item, an assortment of different items, or a collation of identical items. The function of the wrapper for a single item may be to hold it compressed, e.g. bulky clothing; to present a smooth exterior surface, e.g. an engineering component; to obscure the markings on a transport pack to prevent confusion or to reduce the possibilities of pilferage.

In parcels for an assortment of articles, the function of the wrapper is primarily to secure these together in a single handleable unit. The more similar the articles are to one another in shape, the more easily this will be accomplished. For example, a collection of books of the same basic size but different thicknesses are relatively easy to parcel. If the various items differ widely in size, shape or density, then a simple wrap may not be suitable without first forming them into some other unit, e.g. by tying with string, or overwrapping them with special materials to bring them to a more acceptable shape. Almost inevitably the strength of the wrapper required in the second instance will be greater than in the first.

Protection

Protection is needed when the product, or its immediate packaging, is sensitive to some external condition or circumstance, or where the product is noxious or otherwise unacceptable when handled in an unwrapped state. The external factors likely to affect the condition of the product are: dust and dirt; moisture; light; heat; contact with other surfaces likely to cause punctures, scuffing or abrasion; and possibly oil or greases picked up during handling in works conditions. There may also be a need to prevent a component of a product from escaping out of the wrapper, causing either loss of product itself, or damage to both the pack and adjacent packs during transit. Examples of this include engineering items, which are frequently covered with oils or grease as protectives, blood from meat, and fatty materials from cooked foods.

Communication

Wrapping materials can be printed to identify the contents and the manufacturer, and also to provide a smooth opaque uniform background on which other information or instructions can be applied. Methods vary from handwriting through simple forms of printing to the attachment of labels and the provision of printed wrappers. Gift wrappers at Christmas time are an example of the last-named.

Convenience

Many of the less obvious aspects of packaging are covered in this area. Wrappers can conceal the nature of the contents, either to avoid embarrassment, or to deter thieves by preventing recognition of valuable contents. They may also prove that the item has been legitimately purchased in large department stores where theft can be simple without some security measures.

Performance requirements

Depending upon the functions which the wrapper is required to perform, a wide range of properties may need to be specified. These can be classified conveniently into physical and chemical properties, and they must always be within a specific price range.

Physical properties

A   *Description*

The material must be described in a recognizable and measurable manner, and frequently such a description is the only specification. This will cover the type of paper, and its thickness or grammage. These criteria will frequently give some performance indication within a particular range of papers. They are not critical criteria, except in respect of yield, but because the strength of a particular type of paper tends to increase as its weight and thickness increases, they do give a measure of strength. Weighing, therefore, is one of the simplest methods available of checking that the material supplied is of the nature required and this, combined with a count of the number of sheets, is frequently the only check of quantity and quality that is carried out. Even for this simple test there is a correct procedure, detailed in BS 3432. This is listed along with many other useful test procedures on page 1.36.

B   *Optical properties*

Colour, opacity, gloss and brightness are most important, and standard methods of testing are available (see p. 1.38). Colour in particular is a difficult attribute to specify, since it almost certainly introduces an element of subjective judgment. It would be preferable to specify colour by means of the "tri-stimulus value" but this is seldom done. In any event, reference to the three primary colours in this way would require sophisticated optical equipment. Such a numerical description could also be used, however, as a reference to assess the fastness of colours.

C   *Strength properties*

These are clearly related directly to the functions of containment and protection. The most commonly quoted properties are tensile strength (the force required to break a specified strip of material of known dimensions), stretch at the breaking load (a measure of the ability of the wrapper to absorb energy), initial tearing resistance (the force required to tear from an uncut edge), internal tearing resistance (the force required to continue a tear made from a clean cut in the edge of a specimen), puncture resistance (more commonly applied to the heavier materials) and bursting strength, which is the property most frequently quoted, the test being carried out either in a dry or wet condition according as to whether a wet-strength paper has been specified. Other strength properties which may be specified for particular circumstances, are folding endurance, abrasion resistance, and ply-bond strength, especially where laminated or multi-layer materials are concerned. (See pp. 1.36, 1.37 for specific test references.)

D *Barrier properties*

Where the product to be packed is sensitive to moisture or oxygen, a suitable barrier performance may be needed. Wrapping materials may also be required to provide barriers to liquid water, oils and greases, and these must also be specified and tested. (See p. 1.37.)

E *Miscellaneous*

Stiffness and surface friction relate to the ease of running of wrappers on automatic machines. Smoothness and air permeability, water absorptivity and dimensional stability may also be important if the wrapper is converted into a bag, envelope or pack by mechanical means, and when printing or securing with adhesives may be involved. Packaging materials which are intended to come into direct contact with sensitive products must also not contain any trace materials likely to affect the product, e.g. wrappers for highly polished metal items must be neither acid nor alkaline, nor contain unacceptable levels of sulphate or other corrosive chemicals. Wrappers for brass or silverware must be free from reducible sulphur, and in all these instances specified limits for acceptance of papers are involved. Levels and testing procedures are quoted in Proposed Procedures 42 and 49 of the BPBMA Technical Section.

Food wrappers must not contain more than minute traces of heavy metals to prevent the possibility of contamination. Vegetable parchment, which is frequently used for wrapping such materials as butter, must be virtually free from copper or iron, which catalyses oxidation of the fat and could lead to rancidity. There is a special British Standard Specification for this (BS 1820).

Interleaving papers

In this application the paper functions only as a separator to protect a product in flat sheet form. Typical products are sheets of window glass and metal, art prints, and similar materials where the surfaces of the sheets are sensitive either to physical damage or to chemical attack. Sometimes, however, the prime function of an interleaving paper will be purely to facilitate the separation. An example of this is that of sliced cooked meat separated by thin greaseproof papers.

Stiffening and supporting materials

Generally only the thicker grades of packaging material are used for these purposes, but even thinner materials wrapped around a fragile item, or part of an item, can provide sufficient support on occasions to protect it.

Shock absorption and space filling

Sheet packaging materials are used in some circumstances simply to fill irregular spaces in packs where voids occur because the product does not completely fill the box. Where this is so, it is preferable to use special sheet materials which will not readily compact, and which will produce the maximum degree of bulk. Examples of such materials are single-faced corrugated board, and "poppled" or foam-backed paper. Special laminated and structured papers are produced for furniture packaging, the various layers providing a smooth surface to prevent expansion, cushioning to reduce the effect of knocks, and continuity to prevent dust settling. Names such as Craterpak, Kushion Kraft, Custom wrap and Drakafoam are indicative of function.

Papers for conversion

Sachets, envelopes and bags consist simply of a prefabricated wrapper which only needs the product and the closure to complete the pack. Many packaging papers of various kinds, plain, printed and coated, as well as with multi-plies and laminates, are used for this purpose. Such prefabricated wrappers have always been useful where hand-filling operations are involved, and nowadays automatic bagging and enveloping machines have raised the potential considerably. Almost all types of paper can be used for these purposes, and the functions of such prefabricated wrappers are very similar to those already mentioned. When assessing the performance of such materials, the quality of any seals and joins must also be taken into account.

Types of wrapping material available

1. *Kraft paper*

Kraft paper is made by the Fourdrinier process from a pure chemical sulphate produced pulp. The term *kraft* is of Scandinavian origin and means literally *strong*. It is only applied to paper or board made entirely from paper-pulp produced by the sulphate process. It is available either *unbleached* (brown), which is used for general wrapping, or *bleached*, which is used for bag making and for wrapping clothing, food, medical goods and stationery. The bleached material is also used for dyeing to various colours and may have decorative applications in ribbed and striped form. Various surface finishes can be imparted during the final stages of the paper-making process according to requirements. Unglazed paper is smooth on both sides and does not have a gloss finish. Machine-finished (MF) papers are treated on the machine to produce a smooth, but again, not a glazed surface. Water-finished (WF) papers are produced by damping one or both sides on the machine, and this gives rise to a higher gloss and a smoother finish than MF. Machine-glazed papers (MG) have a high gloss finish on one side only, and super-calendered (SC) papers possess a high finish imparted by damping on the machine and subsequent treatment in a super-calender. This last variety is used extensively where attractive wrappers are required. All of these finishes may be supplied either plain or with a ribbed structure.

2. *Stretchable papers*

Stretchable papers come in two varieties, creped or uncreped. Creped paper is produced either on a paper machine or by a secondary operation. Extensible papers which are not creped are produced by a special process to give particular properties.

A stretchable paper will stretch under sudden strain and then exhibit a rapid rate of stress relaxation. This property of high energy absorption is responsible for a great many of its uses. Creped paper is produced by the removal of the moist sheet from a roll by the action of a doctor blade which piles up the paper in ripples parallel to the cross machine direction. Other methods of compacting the paper use other devices, but the result is similar. The use of elastomers in a paper can also produce stretch, but here we will consider only those produced by mechanical means. The characteristics of crepe papers are expressed by stretch, ratio, or texture. *Stretch* is the percentage increase in length; *ratio* is the ratio of the extended length to the average length of the creped test strip. Stretch is used in machine creping, ratio in secondary creping. The term *texture* refers to the surface of the creped paper, which shows a varying degree of fineness or coarseness. Measuring the number of hills per linear inch gives a numerical value. The texture and the ratio can be varied independently of each other within general limits, but in secondary creping a wider range of combinations is possible. The following generalization about the properties of crepe paper should be made:

1   The tensile strength of the uncreped base paper is normally higher than that of the creped paper.
2   The tear resistance is higher in a creped sheet than in the base paper due to the greater amount of material present.
3   The softness of the paper is increased, especially in dry creping, due to the increase in apparent density, in sheet compressibility, and a decrease in fibre bonding.

For many years the wet creping operation was the main on-machine method available for producing stretchable papers for packaging. Although such papers were satisfactory, the texture of the surface was rough and gave problems in handling. Such a texture also was coarse, and problems arose in printing. Consequently, attention was directed towards production of expansible papers which were not creped. The best known of these is probably that produced by the Clupak process. Here the sheet is similar in appearance to ordinary wrapping kraft paper but has the high energy-absorption properties of creped paper. This process employs a thick rubber belt instead of the creping blade on the machine. Figure 2.1 presents the process which is located in the dryer section at the early stages on a paper machine. The sheet meets the rubber blanket when it contains about 35% moisture, and it is then pressed between the rubber

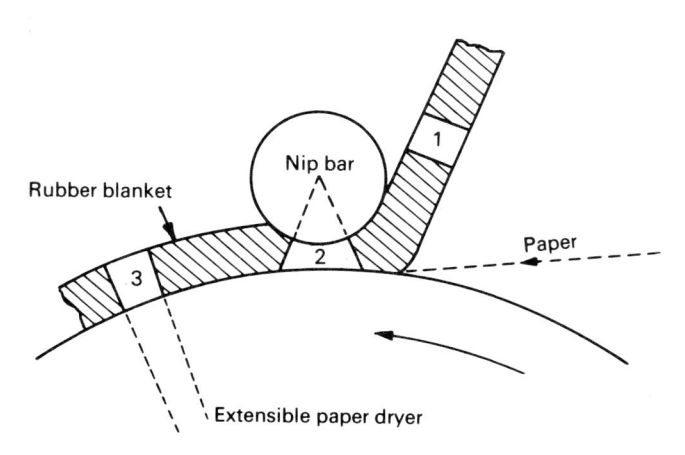

Figure 2.1   Essential part of the "Clupak" process. As the rubber blanket passes under the nip bar, the volume shown at 1 expands lengthwise to dimensions 2 and, when the pressure is removed, reverts to its normal dimensions as at 3.

blanket and the surface of the dryer roll. Under these conditions, the rubber blanket in the nip is expanded. As it leaves the nip it shrinks, compacting the sheet as it does so. The important variables in the process include the nip pressure, the dryer roll temperature, and the moisture content. No additives are used to assist creping, and the action is capable of giving a machine-direction stretch of 20%, although this is not frequently employed. In the cross-direction, the stretch is increased somewhat above the usual level, but by no means to the same extent. Such papers are frequently used in producing sacks, particularly where extra strength is required. Several other types of paper of a similar nature are produced, among them being Duo-stress.

3. *Wet-strength papers*

A wet-strength paper is a paper which remains strong when fully saturated with water. It is easy to define this in principle, but considerably more difficult in practice. We must first of all distinguish between wet-strength papers, waterproof papers, and moisture or water-vapour-resistant papers. The last-named must contain a barrier, usually a plastics or metal film, to resist the penetration of water vapour. A waterproof paper is one that sheds liquid water easily. Wet-strength papers, on the other hand, will often be saturated with water, and under these conditions they must still retain strength. By definition, wet-strength paper should retain at least 30% of its dry strength when these two strengths are measured by means of the bursting strength test. Wet-strength wrapping papers are made using kraft pulp to which water-soluble urea formaldehyde or melamine formaldehyde resins have been added. The condensation of these chemicals takes place in the drying section on the paper machine, and the reaction product (being highly water-insoluble) produces an effect on the paper sheet to increase its wet strength. Wet-strength papers are particularly useful for outside packaging as they are stable in adverse atmospheric conditions. They can retain sufficient wet strength for a long time under wet conditions and are also suitable for the packaging of goods which themselves are wet. They are again frequently used for paper sacks.

4. *Imitation krafts*

This type of paper covers a wide range of quality on which there is no general agreement. The description is applied to papers which are dyed to resemble kraft, to those made with a mixture of sulphate kraft pulp and waste pulp, and also to papers which are frequently used for the same purposes as kraft, usually in higher weights. "Rope browns" and "browns" are terms applied to wrappings composed mainly of waste fibres in both instances, and both of them have much less strength than kraft papers. Both are cheaper than the kraft paper which they are sometimes used to replace.

5. *Sulphite papers*

The principal sulphite papers used for wrapping purposes are usually machine-glazed and are often used for general-purpose bag making, while bleached kraft paper bags are used where greater strength is essential. Sulphite paper wrappers are also used for wrapping food and small items of textiles.

6. *Greaseproof paper*

Greaseproof paper, or imitation parchment, is made in the same way as ordinary paper, but the pulp is subjected to an intensive beating action for several hours before going to the paper machine. It is produced frequently as a substitute for vegetable parchment and is much inferior in "whiteness", bulk, strength and greaseproof qualities. Vegetable parchment may be distinguished from greaseproof paper by boiling in water. In a relatively short period, greaseproof paper and imitation parchment will be reduced to a soft mass which is easily defibred by pulling, while vegetable parchment will stand the action of boiling water for several hours with no sign of disintegration and when it is torn there is very little evidence of a fibrous structure.

7. *Glassine*

By passing a greaseproof paper over a super-calender, the paper surface can be given a very high finish, and the density of the sheet will be greatly increased. The resulting product is called a *glassine*. The close-knit structure and hydrated condition of the cellulose in glassines enables such papers to be used for wrapping purposes, not only for greasy products, but also to maintain flavours and aromas. It also, of course, will exclude any odours which might be contaminating. Untreated, however, glassines are not by any means moisture-resistant, and to achieve this quality, surface coating is applied. A smooth non-porous non-absorbent surface, such as is provided by glassine, is ideal for wax coating and for other types of coating which provide barrier properties to gases, water vapour and odours. The material therefore has considerable uses in the packaging field.

8. *Vegetable parchment paper*

Pure vegetable parchment is usually produced in two stages. First a pure chemical pulp stock is made into a paper on a Fourdrinier machine; this is then passed through a bath of sulphuric acid, the strength and temperature of which is carefully controlled, as is the time of immersion. The parchmentizing process is stopped gradually by washing in diluted sulphuric acid and water and, finally, the material is dried. Parchment is virtually a pure cellulose paper without additions. It is water-resistant, grease-resistant, resistant to fats and oils, and is odourless and tasteless. The usual range of weights produced is 40–75 grams per square metre, but lighter and heavier weights are possible. Most material is softened by the addition of glycerol, but unsoftened papers are sometimes employed. In addition to being used for wrapping, it is also made into bags and corrugated papers, and is often printed. It is extensively used to package fatty biscuits and other baked goods, butter, margarine and cooking fats, cheese, ice-cream, fish (wet, cured and fried with chips), poultry and meat. Packs for tea and coffee also often contain vegetable parchment as the immediate wrapper.

9. *Tissue papers*

These specially-soft thin papers, usually available in a low weight of between 17 and 30 g/m^2, are used for wrapping articles where the surface is susceptible to abrasion. All of them are generally white, but can be made available in a range of colours. There are three basic qualities available: acid-free, machine-glazed, and machine-glazed mechanical tissues. Acid-free papers are invariably unglazed and bleached, and they contain no chemical impurities of any great degree in the paper. They are particularly concerned with protecting goods that are susceptible to the action of acids, e.g. silverware, and also for wrapping expensive goods such as jewellery, to enhance their appearance. The machine-glazed papers are less costly than the acid-free and, although they have a very smooth texture, do not have quite the same degree of purity. They are extensively used for interleaving metals where soft paper is required. The machine-glazed mechanical tissue is a coarser tissue than the machine-glazed, and includes some mechanical wood pulp in its furnish. This can result in damage to goods with high finishes, but it is particularly suitable for counter sales of bread, and for protecting glassware where a cheap-quality tissue is suitable.

10. *Coated papers*

The production of synthetic polymeric materials capable of forming self-supporting films failed to supplant the use of coated materials, even

PAPER-BASED PACKAGING

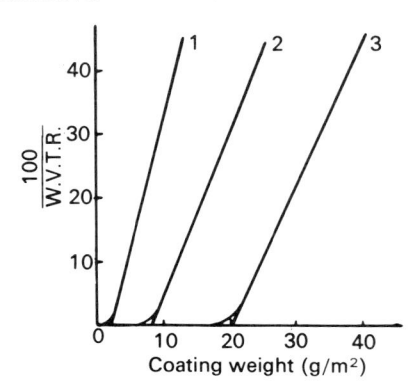

Figure 2.2 Behaviour of some substrates of different surface characteristics. Water vapour transmission rate (W.V.T.R.) in $g/m^2/24\,h$ at $38°C$ and 90% relative humidity.
(1) cellulose film   (2) glassine paper   (3) glazed imitation parchment paper

after the initial difficulties in converting had been overcome. In many instances the thickness of the film material required to give the requisite barrier properties is well below that at which it can be handled on filling and forming machines, and the cost becomes prohibitive. Coating a more expensive film material to a thicker and less expensive substrate to achieve the desired strength and handling properties is one way to overcome this difficulty. The commonly used substrate for this purpose is paper in one of its many forms. It should be stressed that the properties of the substrate, particularly in respect of its surface smoothness and absorption characteristics, are of great importance in achieving a continuous film of the more expensive barrier material at a low rate of application. The behaviour of some substrates of different surface characteristics coated with the same barrier material is shown in figure 2.2. It will be noted that the smoothest, and least absorbent, cellulose film gives the same resistance at the lowest rate of application.

The various coatings used in packaging are best considered according to the manner in which they are applied. These are:

(a) from aqueous solutions;
(b) from solvent solutions (lacquers);
(c) from aqueous dispersions;
(d) as hot melts;
(e) as extrusion coatings.

Schematic representations of the various methods used are shown in figure 2.3.

The main use of coatings applied from aqueous solution, apart from gummed tapes, is to make papers greaseproof and oil-resistant. For this

WRAPPING AND PACKAGING PAPERS

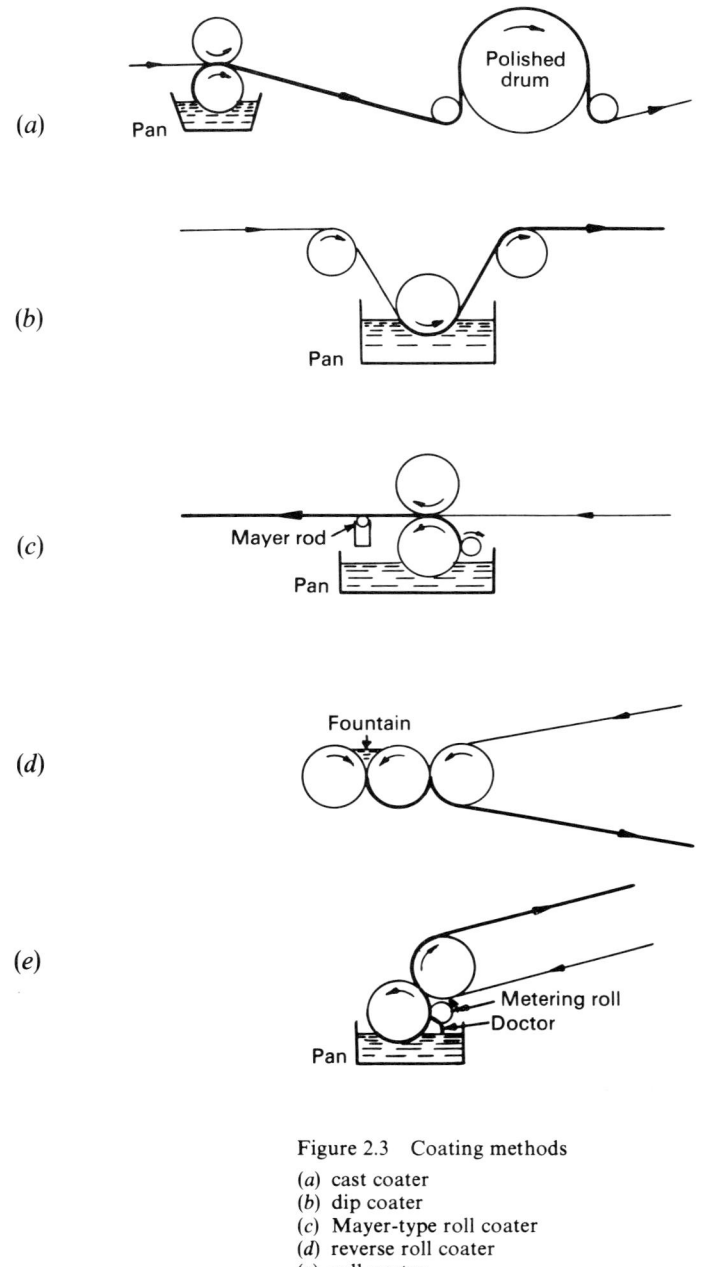

Figure 2.3 Coating methods
(a) cast coater
(b) dip coater
(c) Mayer-type roll coater
(d) reverse roll coater
(e) roll coater

1.31

PAPER-BASED PACKAGING

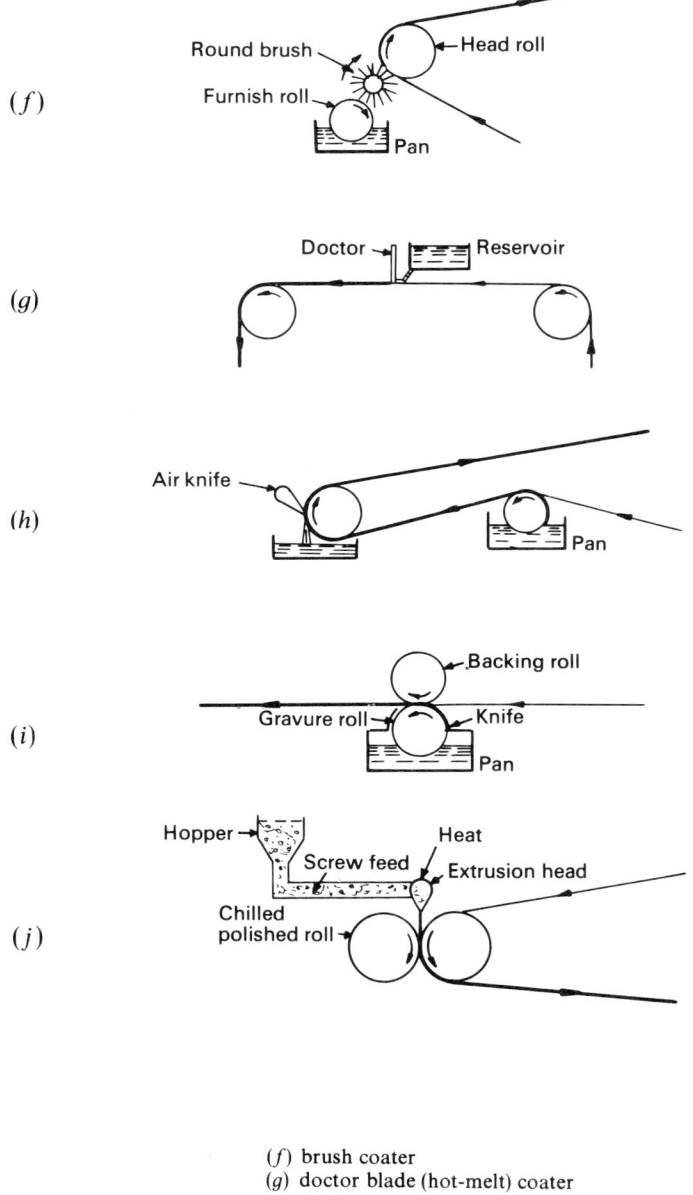

(f) brush coater
(g) doctor blade (hot-melt) coater
(h) air brush coater
(i) gravure coater
(j) extension coater

1.32

purpose the water-soluble cellulose ethers are still in use, although polyvinyl alcohol is tending to replace them in certain instances.

In the field of lacquer coatings, the older materials are still being used, although modifications both of the resins and plasticizers have taken place. This has assisted them to withstand the competition from recent additions to the resin range, and the improved methods of application developed to overcome the problems which faced converters wishing to use these newer materials.

Aqueous dispersions are making slow progress in packaging. Their main attraction lies in the fact that they contain little or no solvent, water being the carrier. Thus, when applied to a porous substrate such as paper, the evaporation is less than that experienced with lacquers. The use of air knife doctors has removed the need for improving such dispersions in order to apply them by conventional methods. To achieve the full performance from the resin used in such dispersions it is generally necessary to apply a number of coats to paper substrates, and to subject the coated web to a high temperature, in order to fuse the material into a continuous film. Dispersions, based on polyvinylidene chloride, find outlets in the treatment of papers, where a single application of low weight produces a low-cost heat-sealable paper with a fair degree of resistance to water vapour and fat-containing materials.

Hot-melt coatings comprise all the materials which can be applied in a molten state. This is probably the most attractive method of producing coated papers because no drying system is required. Providing the coating can be satisfactorily chilled after it has been applied, high production rates are possible. The coating material forms a film on the surface of the web which has properties very little different from those of the free film. Wax coatings are the oldest group of hot-melt materials in use today (figure 2.4) and, although paraffin wax is still used alone, the disadvantages which characterized such films (lack of durability, little resistance to creasing, scuffing and so on) have been considerably overcome by modification with various microcrystalline waxes, butyl rubber, polyisobutylene, polyethylene and copolymers of ethylene and vinyl acetate. Generally speaking, modified paraffin wax coatings contain up to about 15% by weight of the modifier, and these are used either singly or in combination to improve durability, crease resistance, scuff resistance, heat-seal strength and gloss.

The introduction of polyethylene, and the design of suitable extrusion equipment for coating polyethylene films on to other materials in reel form, is one of the major developments in packaging in the post-war period. Considerable tonnages of kraft paper, extrusion-coated in this matter, are used for bags and for one or more of the plies in multi-wall paper sacks, particularly for hygroscopic products and certain chemicals.

PAPER-BASED PACKAGING

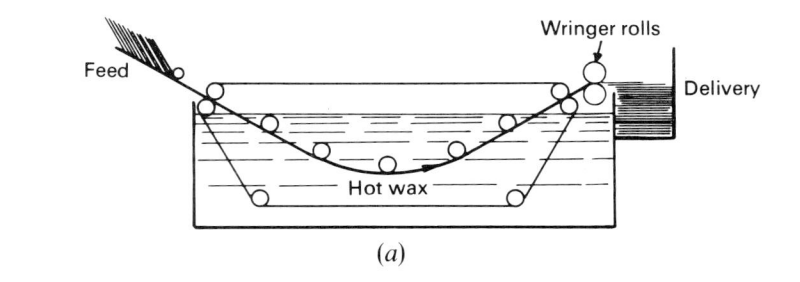

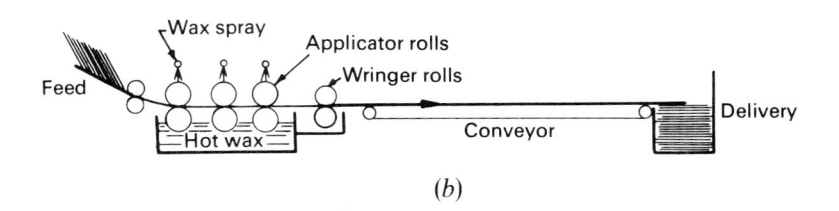

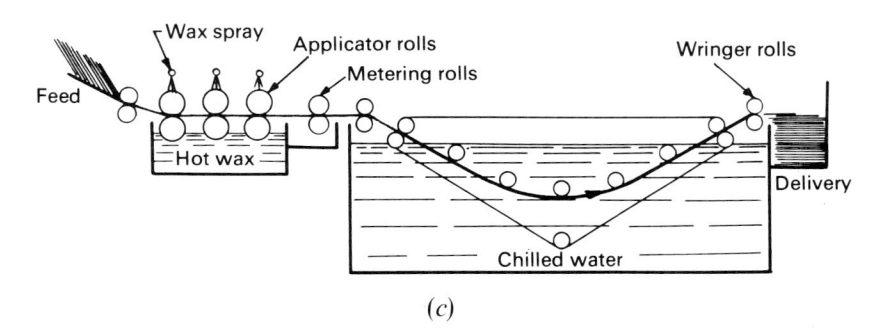

Figure 2.4 Waxing
(a) wax saturator
(b) hot waxer
(c) cold-finish waxer (wet waxing)

1.34

Polyethylene-coated paper-board is used for the manufacture of heat-sealed milk cartons, and as containers for fruit juices and frozen foods.

Storage of materials

All paper-based materials are sensitive to climatic changes, especially variations in humidity, although temperature does not affect them very much.

Plastics-coated papers may lose or gain moisture from one face, and as a result are particularly prone to curl.

Any smooth materials are prone to block if sheets are pressed together in stacks. Cast-coated materials may block, causing surface damage if stacked flat in conditions of high humidity.

Cellulose-films materials can embrittle at freezing temperatures; and uncoated grades are very moisture-sensitive.

Adhesive-coated materials may be unusable if activated by moisture and heat.

Reeled materials should be stored either suspended on bars pushed through their centres, or if overwrapped, by standing the reels on end. They should never be laid on their curved sides.

Some materials are sensitive to heat, or light, e.g. latex adhesives deteriorate with age. Adequate protection should be given, and stock should be rotated to minimize the storage time.

Many papers, particularly treated papers (e.g. waxed), are attractive to rodents and cockroaches.

The optimum conditions for storing packaging materials are constant temperature of about 20°C and 50% relative humidity. If the materials storage area is uncontrolled, and differs widely from the packaging area, then materials should be brought into the packaging area at least 24 hours prior to use and not unwrapped until equilibration has occurred as far as temperature is concerned.

1.35

PAPER-BASED PACKAGING

TABLE OF SELECTED TEST REFERENCES

The references given in the Table opposite are taken from

ISO/R	International Standards Recommendations
BS	British Standards
	In some instances the number of the standard is followed by a further reference, e.g. BS2782.509 is Method 509 in BS2782. An asterisk * indicates that a British Standard is in course of preparation.
ASTM	American Society of Testing and Materials Standards.
TAPPI	Technical Association for the Pulp and Paper Industry (USA) Standard. Where TAPPI Standards have been developed by ASTM, the latter reference has been used.
FEFCO	Fédération Européenne des Fabricants de Carton Ondule Test Methods.
BPBIF	Technical Section of the British Paper and Board Industries Federation Test Methods.
PIFA	Packaging and Industrial Films Association Standards.
BPBMA	British Paper and Board Manufacturers Association.

WRAPPING AND PACKAGING PAPERS

	Test, etc.	*Material*	*Reference*
Pre-test procedures	Sampling	Paper Board	BS 3430
	Conditioning	Paper Board	BS 3431 ISO/R 187

	See also:
	Laboratory Humidity Ovens
	    injection type
	    non-injection type
	Laboratories, controlled atmosphere
			BS 3718
			BS 3898
			BS 4194

Tests for mass and density	Grammage	Paper Board	BS 3432 FEFCO 2 and 10
	Mass	Wax on paper	BS 4685
Tests for strength properties	Adhesive strength	Adhesive and adherends	BS 847
	Burst strength	Paper and board Board Paper (wet)	BS 3137 FEFCO 4 BS 2922
	Folding endurance	Paper	BS 4419
	Heat-seal strength	Plastics/laminates	PIFA 2/74
	Ply-bond strength	Paper	TAPPI RC 364
	Puncture strength	Paper, board	BS 4816
	Stiffness Ring stiffness	Paper Paper	BS 3748 ASTM D-1164

	Tear strength
	  Initial
	  Internal
		Paper
		Paper
			ASTM D 827
			BS 4468

	Tensile strength and stretch	Paper	BS 4415

PAPER-BASED PACKAGING

	Test, etc.	Material	Reference
Tests for surface absorption and permeability properties	Abrasion (see rub-proofness)		
	Absorption of water by	Paper, board	BS 2644
	Absorption of wax by	Paper	ASTM D 688
	Friction coefficient $= \dfrac{1}{\text{slip}}$	Paper	TAPPI T 503
	Moisture content	Paper	BS 3433
	Permeability (see also water vapour transmission)	Paper to air BS 2925 Sheet materials to gas	BPBIF p. 13 ISO 2556 BS 2782-514A
	Resistance to grease	Paper	ASTM D 722
	Resistance to oil	Paper	BPBIF RMT 1
	Roughness	Paper	BS 4420
	Rubproofness	Paper	BS 3110
	Water vapour transmission	Sheet materials	BS 3177
Analytical tests	Acidity/alkalinity (pH)	Paper	BS 2924
	Ash	Paper	BS 3641
	Chlorides	Paper	BS 2924
	Contraries Copper, iron	Paper	BS 1820
	Sulphates	Paper	BS 2924
	Sulphur, reducible	Paper	BS 1820

1.38

	Test, etc.	*Material*	*Reference*
Dimensional tests	Curl	Paper	BPBIF RM 4.2
	Dimensional stability	Paper	BPBIF RM 53
	Thickness	Paper Board	BS 3983 BS 4817
Optical tests	Brightness Whiteness Reflectance Opacity Gloss	Paper	BS 4432
	Light fastness	Paper	BS 4321
Miscellaneous materials tests	Crease quality	Board	BS 4818
	Odour and taint	Packing materials	BS 3755
	Resistance to blocking	Paper	ASTM D-918

CHAPTER 3

Multiwall Paper Sacks

F. A. Paine

Prior to the 1939–1945 war the use of multi-wall paper sacks was confined to powdered materials such as lime, cement and basic slag. Today they are used for many different commodities.

There are few statistics published on the end-uses of paper sacks in the UK, but Table 3.1 gives an analysis of USA practice.

Table 3.1 End-uses of paper sacks (USA)

	%
Food and agriculture	37
Building materials (cement, etc.)	15
Chemicals (including drugs)	32
Minerals	10
Miscellaneous	6
	100

Paper sacks are made from several thicknesses or plies of paper nested inside one another, so that the total load imposed on the sack during handling and when subjected to strain is distributed over all the plies. The number of plies in any particular sack may vary from 2 to 6 according to its use; and it is this use of several thicknesses of comparatively thin paper, rather than one or two thicker ones, that gives the multi-wall sack its strength and flexibility.

There are two principal types of multi-wall paper sack: open-mouth sacks and valve sacks. Either type may be made with or without gussets, and have either a sewn or pasted closure. However, open-mouth sacks are commonly sewn, and valve sacks are normally pasted.

The various styles of sack in standard usage and the corresponding terminology are illustrated in figure 3.1.

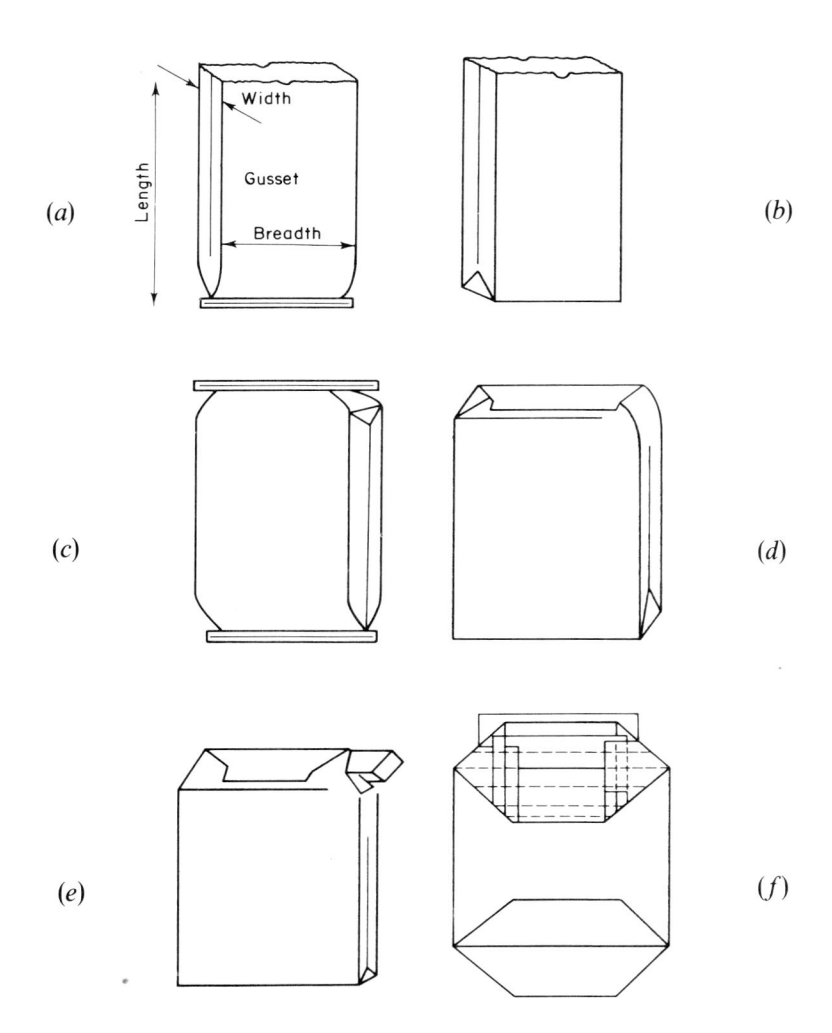

Figure 3.1
(a) sewn open-mouth sack
(b) pasted open-mouth sack
(c) sewn valve sack
(d) pasted valve sack: internal sleeve
(e) pasted valve sack: external sleeve
(f) stepped-end valve sack

PAPER-BASED PACKAGING

Materials used in multiwall paper sacks (see Chapters 1, 2)

Kraft paper usually denoted in specifications by the letter (K).

Low-stretch crepe kraft (LSCK) is kraft paper to which crepe has been applied on the paper-making machine before it is fully dried.

Extensible kraft (EK) is kraft paper treated by a special process to give a controlled degree of stretch. The best-known process is the "Clupak" process.

All these papers can be made "wet strength" by resins added to the paper stock after heating. They are then designated WSK, WSLSCK and WSEK, respectively.

Union kraft (UK) consists of two sheets, combined together with bitumen, wax or polyethylene. The results are known as bitumen union kraft, wax kraft union, and polyethylene union kraft respectively.

Manufacture

Paper sacks are manufactured on a machine called a *tuber*. The reels of paper, one for each ply in the sack, are arranged on a reel stand at one end of the tuber. Each reel is offset laterally relative to the next so that, when the several webs of paper are bent over to form a tube and stuck together, the resulting seams will be staggered (figure 3.2*a*). This results in a distribution of the "side line seams" over the back face of the sack, and gives better strength. When one or more plies in the sack require special protective or barrier properties, the reels from which such plies result are simply placed in the proper position in the reel stand at the end of the tuber (see figure 3.2*b* and *c*).

As has already been mentioned, it is very important that the plies be properly nested within each other, and the tuber crew keep careful watch on this during production. Quality inspection also checks this point by examination and tests.

The reels of paper are thus converted into a tube, and this is cut into pre-determined lengths as it comes off the tuber. For open-mouth sewn sacks, it is simply a matter of cutting the tube at right angles to its length; but for valve sacks a somewhat different "chop" is made, which allows the tube to be suitably folded at the ends and so formed into the finished sack.

Filling and closing sacks

Open-mouth sacks

Open-mouth sacks may be closed quite simply by "bunch tying" with wire using a hand-tool; potato sacks are so closed in the field.

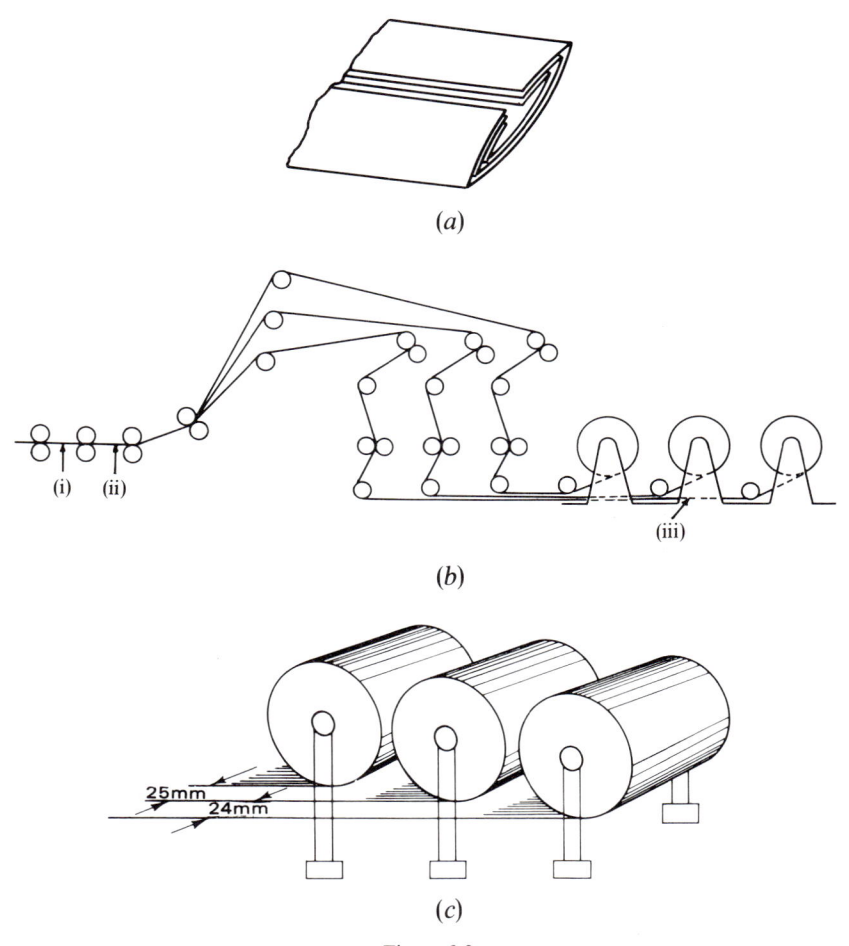

MULTIWALL PAPER SACKS

Figure 3.2
(a) staggering of back seams
(b) schematic drawing of a tuber: (i)—chop; (ii)—forming plates; (iii)—reel stand
(c) reel stack at end of tuber

Free-flowing products are usually closed with a simple manually-operated sack filler. A clamp holds the mouth of the sack open beneath a hopper with a feed gate or flow regulator.

After filling, the usual method of closing open-mouth sacks is by over-sewing through a creped paper tape. If the sacks are to be used for fine powders, and it is required that they are sift-proof, the closed end may be dipped in wax. Alternatively, the sack is closed by sewing only, and then a creped paper tape is stuck over the sewing line with a latex

1.43

adhesive. Open-mouth sacks can also be closed by pasting in the same way as valve sacks.

Valve sacks

Valve sacks are normally closed by folding and bending the ends over, and pasting them. Sometimes the ends are strengthened by cutting the plies in a series of steps so that, when the ends are bent over, the joins will be staggered, somewhat after the same fashion as the side seams formed on the tuber.

Both ends of a valve sack are, of course, closed by the sack manufacturer, and a valve inserted in one end.

To fill a valve sack, it is first placed on the filling spout and the filling operation takes place by one of several methods, depending on the flow characteristics of the product which passes through the filling spout into the sack.

Materials with good flow properties are filled by gravity alone, from an overhead hopper from which the product passes into a chute with a curved end and a filling spout. The sack is clamped on to the spout and supported while a feed gate in the chute is operated. An improved type of gravity packer (figure 3.3) can, with an easy-flowing product, give high filling rates (8–10 sacks per minute).

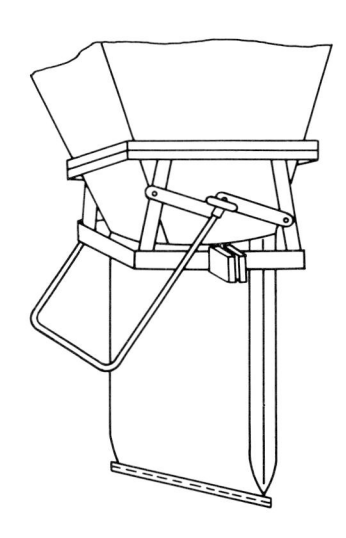

Figure 3.3 Vertical-fall gravity packer

MULTIWALL PAPER SACKS

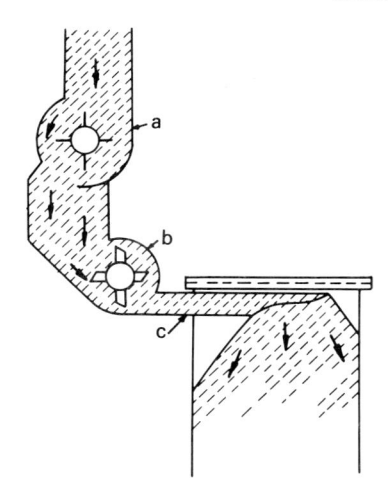

Figure 3.4 Rotary impeller (valve filler) for use with semi-free-flowing materials
(a) flow regulator   (b) impeller blades   (c) filling tube

Rotary impellers (figure 3.4) or screws (figure 3.5) are used for products which do not flow quite so easily. The rotary impeller, for example, handles such products as cement and limestone on multi-head fillers at 20 sacks a minute with one operator. Weighing is done on the spout with an automatic cut-off when the required gross weight is reached.

The screw filler can be fitted with an agitator above the screw for sticky products.

Granular or crystalline products which could be broken up by impellers

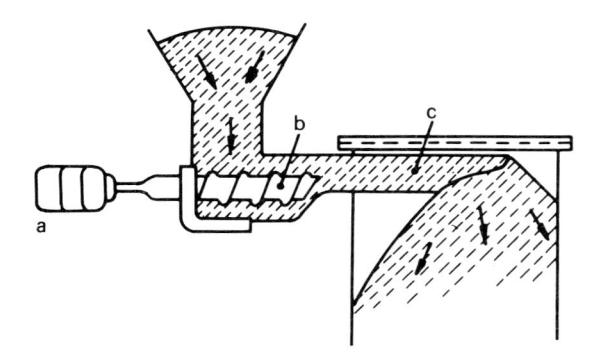

Figure 3.5 Horizontal screw packer (valve filler) for use with semi- and non-free-flowing materials
(a) motor   (b) filling screw   (c) filling tube

1.45

PAPER-BASED PACKAGING

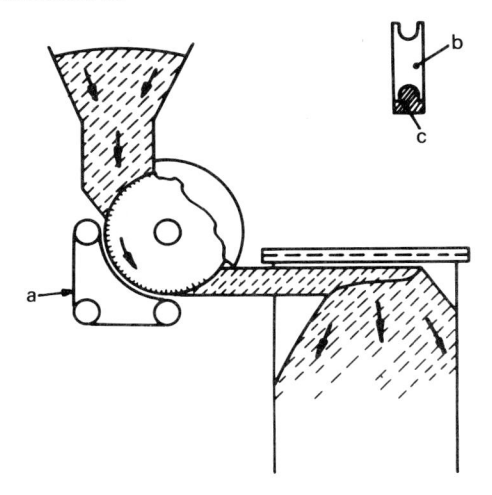

Figure 3.6 Centrifugal belt packer (valve filler) for use with free-flowing and semi-free-flowing granular materials
(a) continuous belt (b) side view of rotating pulley (c) belt and pulley groove

or screws may be filled by a belt packer (figure 3.6). The charge is pre-weighed and falls onto a fast-moving endless belt which delivers the product into a horizontal spout and thus into the sack.

Difficult powdered or granular products may require fluidizing to make them flow, and here a pressure packer is used. The product is fluidized by introducing air through the porous base of a storage/pressure chamber, and then pushed into the filling tube, either by the weight of the product in the storage chamber, or by a secondary air supply above the porous base. Weighing is done on the spout with an automatic cut-off when the required weight is reached (figure 3.7).

Storage of sacks

The strength of the paper, and hence the performance of the sacks made from it, varies with the moisture content of the paper. The strength increases with moisture content up to a certain level, and dry conditions tend to lower the strength of the sack.

Now the temperature at certain places on the tuber (because of frictional effects) during manufacture of the sacks tends to dry out the paper and thus reduces its strength. If such sacks were used immediately after manufacture, the breakage rate during handling, which is normally negligible, could increase beyond acceptable limits. Accordingly, it is usual to store the sacks until they come to equilibrium with the surrounding

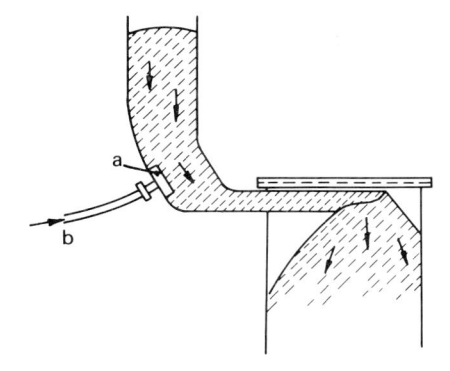

Figure 3.7 Air float packer (valve filler) for use with free-flowing materials
(a) air pad   (b) air

atmosphere, and the moisture content of the paper is restored to its previous value. The period of storage required is usually about two weeks, though this can be accelerated by installing a moisture spray in the store or keeping it at a suitable controlled humidity.

Handling of sacks

The filling operation and subsequent handling procedures vary considerably from factory to factory. However, sacks must be able to withstand the more severe rather than the milder conditions, as it is not always known how they may be handled after leaving the factory. Sacks are subjected to a considerable number and variety of hazards. Firstly they are filled from a hopper. After filling, they fall off the filler spout, vertically if they are open-mouthed sacks, and almost horizontally if they are valve sacks. They then proceed through the packing shed to the loading bay or warehouse on a series of conveyers and chutes. Finally, they are either stacked on a pallet or loaded on a lorry or rail wagon, and during such operations they may be dropped more than once. Also, if they are stored in a warehouse, they will probably be stacked to a considerable height.

At a later date the sacks will travel to their final destination by road, rail or sea. During this journey they will be shaken about; they will be handled again during unloading, and then probably re-stored until ready for use. At each loading and unloading they may be dropped intentionally or otherwise, and each and every hazard imposes a strain on the paper from which it never fully recovers. The first drop when falling off the filling spout of the hopper, during which the contents may consolidate, often produces the greatest strain, all subsequent strains being less but

PAPER-BASED PACKAGING

Table 3.2. A comparison of sack breakages on a plant and industrial basis*

Type of failure	Animal feed industry		Fertilizer industry			Sugar industry			
	Factory (1) (II)	Factory (2) (II)	Factory (1) (I)	Factory (2) (I)	(III)	Factory (1) (I)	(III)	Factory (2) (I)	(III)
Body burst	14	24	7	14	33	30	21	15	18
Gusset burst	19	21	86	17	33	17	18	17	20
Top sewing	17	6	1	6	13	5	14	21	27
Bottom sewing	4	6	2	29	11	4	10	13	11
Snagging	33	29	1	34	9	22	27	20	20
Torn corner	8	9	3	—	1	18	8	13	4
Unidentified	5	5	—	—	—	4	2	1	—
Total	100	100	100	100	100	100	100	100	100

(I) Sacks dispatched direct from production line, or equivalent to direct dispatch.
(II) Sacks stored temporarily in piles of 8 to 10 high prior to dispatch.
(III) Sacks stacked in the full sense of the word prior to dispatch.
* These tables are taken from a lecture given by P. N. Harvey to the EUROSAC Conference in Lucerne, 1963 (see Ref. 4).

of an additive nature. Thus, the total strain imposed on a sack during its lifetime may be considerable.

It is to be expected that paper sacks will fail occasionally, but the percentage of failures must be kept to an absolute minimum, and reduced to zero if possible.

Tables 3.2 and 3.3 are an analysis of breakages observed during a survey of damage in a number of factories packing three different products and totalling 330,000 sacks.

Thus, it is very important to check that a particular style of sack will be strong enough to withstand the handling it is likely to receive. For this purpose it is essential that a laboratory assessment be carried out both of the package and the paper from which it was made.

The assessment of paper and sacks

At some stage prior to the manufacture of the sack, the paper from which it is to be made is examined for a number of physical properties. This examination may be at the paper mill laboratory, or in the quality control laboratory of the sack factory, or both.

It is obviously important that the paper has strength when under load, and also that it should stretch under these conditions. Thus, the tensile strength and stretch characteristics of the paper are important, and these qualities must not fall below a certain level. Other properties such as the tearing resistance, the porosity of the paper, and its degree of sizing are also measured.

Table 3.3. Mean breakage distribution patterns (at manufacturing plants)*

	Methods of dispatch			Grand average
	(I)	(II)	(III)	
Body and gusset bursts	51	39	48	46
Sewing line failures	20	17	29	22
Snagging failures	19	31	19	23
Others	10	13	4	9
	100	100	100	100

* These tables are taken from a lecture given by P. N. Harvey to the EUROSAC Conference in Lucerne, 1963 (see Ref. 4).

The sacks themselves may be assessed by drop testing. There are a number of ways in which this can be done, but the one most commonly used and which best correlates with field performance involves the repeated dropping of each sack from a fixed height until failure occurs. The sack is placed on a trap-door type platform (either flat or standing on one end) and raised to a pre-set height, when the trap opens and the sack falls onto a selected type of floor. The sack is then replaced on the platform, in the same position as before, and the drop is repeated. The procedure is continued until the sack finally fails. The number of drops before failure occurs, and the nature of the failure, are recorded. At least ten and preferably more sacks of the same type must be tested before the average number of drops causing failure is calculated. This is known as the *drop number* for the particular style of sack under investigation.

By testing different types of sack, using the same contents and dropping from the same height, a comparison of their strength can be made. It is important, however, that sufficient sacks of each type are tested to ensure that the results are representative. Also, testing should be carried out in a conditioned atmosphere, so that variability due to different moisture contents of the papers involved is kept to the minimum.

The sacks should be left in the conditioned atmosphere in which they are to be tested for sufficient time for them to come to equilibrium with it. This usually means overnight before the day on which the test is made. If a conditioned laboratory is not available, the sacks should be tested alternately, i.e. one from the first set, then one from the second set and so on, until one sack from each set has been examined. The second one from each set is then tested, followed by the third one from each set, until all have been examined. In this way, variations in atmospheric conditions will be spread throughout the various sets as evenly as possible.

The drop number of a sack depends not only on the paper from which it is made, but also on the nature of the contents. It is, therefore, very important that the same contents are used for each type of sack to be compared. Moreover, it is advisable to use the product for which the sacks are intended as the contents during testing.

Laboratory testing, if carried out carefully by an experienced operator, can (within limits) determine the suitability of a particular style of sack for packaging any product. However, handling procedures in practice can vary, and a field trial is very useful before full commitment to a particular type is made.

So far only the mechanical strength of the sacks and the corresponding mechanical hazards have been discussed. Sacks, however, may have to withstand other hazards. The most important of these is moisture, either as liquid or in the vapour state. In some instances it does not matter if the contents of the sack get wet, provided the package still holds together when wet. Here it is necessary for the outer ply at least to be made from "wet strength" paper. This is a paper so treated that it does not disintegrate when wet or lose all its strength under such conditions. Certain products must be protected against both liquid water and water vapour. To achieve this, one or more plies of a bitumen-laminated kraft paper (union kraft) or a polyethylene-coated kraft paper are incorporated into the make-up of the sacks. The latter is a better barrier, and can also be used as a means of preventing the contents becoming contaminated with paper fibres if it is inserted as the innermost ply, with the coating facing inwards. Polyethylene-coated kraft also behaves better over a wider temperature range, being able to withstand a higher temperature without the melting experienced by bitumen, and not suffering from the latter's fault of cracking at low temperatures. However, polyethylene-coated paper is more expensive than bitumen-laminated paper, and so the latter is often used in preference where packaging costs have to be kept as low as possible.

If a water-vapour barrier is incorporated in the plies of a sack, it is important that moisture is prevented from entering through the ends of the sack. This is particularly important with sewn sacks, as moisture can enter through the sewing line. The moisture enters down the space formed by the two sheets of barrier material in the gaps between the stitch holes, rather than through the holes themselves. However, an improved result can be effected by the technique of sewing and overcapping as previously mentioned.

Other speciality or barrier papers such as wax-laminated paper or silicone-impregnated paper may be used as plies in a sack. Paper may also be sprayed with flame-proofing agents or rodent-repellents. These papers are, of course, more expensive than conventional kraft, and sometimes the special chemical treatment can weaken the paper.

Present and future use of paper sacks

The total number of paper sacks used in Britain has not increased greatly over the last 10–12 years, and there have been changes in the pattern of usage. A greater use of bulk handling has captured some markets, e.g. cement previously packed in paper sacks, and is also making inroads into the market for flour. The introduction of plastics sacks has made large inroads into the packaging of fertilizers in paper sacks and other products requiring high weather resistance.

Against these areas of reduction in the use of paper sacks, new markets have opened up. One such market is the use of paper sacks for refuse disposal. Here a two-ply sack, at least one ply being of wet-strength paper, hanging from a holder, is used in place of a conventional metal or plastic dustbin. Such containers are very light and clean to handle and, being completely disposable, save the dustman a return journey to each house. As the refuse is completely enclosed in the sack, it may be collected on an open lorry instead of the more specialized and costly vehicle normally used.

The physical requirements for a refuse sack are somewhat different from those for sacks used in packing powdered or granular materials. These sacks are not sewn up at the top, nor do they have to be handled on packing lines and appear undamaged, after rough handling, at the other end. Minor damage, though undesirable, is of secondary importance provided there is no spillage, and the sack can be easily handled on to the lorry or other vehicle. However, during the period in which the sacks are hanging from holders outside the house they may be subject to the weather, including rain, and good wet strength is imperative. Plastics sacks, particularly of black polyethylene, are also used for this purpose.

Baler bags

Baler bags (figure 3.8) may be defined as flexible containers for a number of smaller units or consumer packages. As their name implies, they were first used in connection with the packaging of bales of wool, where a suitably styled large bag is required, though not necessarily of great strength. Bales of wool do not have jagged edges or corners which may poke their way through the sides of the bag, and the bulk density is not high. Also, little protection against deformation is required, and thus a lightweight bag forms an excellent container.

However, baler bags are also used for packing a variety of small units such as bottles and cartons. In other words, they not only serve instead of a piece of wrapping paper to provide a stronger package, but they also replace corrugated cases and sometimes wooden boxes.

PAPER-BASED PACKAGING

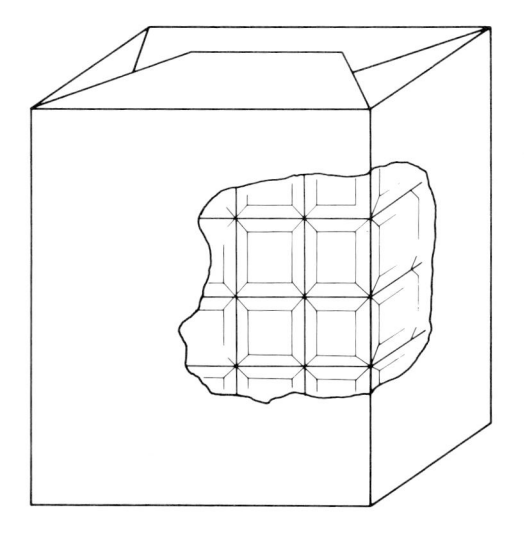

Figure 3.8 Baler bag

As with any outer containers to carry smaller containers, the style of a baler bag and, in particular, its dimensions must be decided with greater care than if the bag were to contain powders or granules. It must be tailor-made to fit the contents, so that they are held securely. The method of closure will also depend on the contents. Thus the style, construction and method of closing a baler bag will be determined by the size, weight, nature and number of the smaller units contained in it.

BIBLIOGRAPHY

1. Howarth, P. and Rothman, M. (Dec. 1959, March 1960 and Oct. 1960), "Drop Testing of Multiwall Paper Sacks", *Packaging*.
2. Rothman, M. (1963), "Multiwall Paper Sacks—a General Assessment", *Paper and Board in Packaging*, 179–96, Pergamon Press.
3. Rothman, M. (1963), "Basic Research into the Behaviour of Paper Sacks", *Paper Packs and Packaging Review*, October.
4. Harvey, P. N. (1963), *Sack Hazards and Breakage Rates in User Plants*, Eurosac Conference, Lucerne, June.
5. *Evaluation of Package Performance*, Proceedings of PATRAPAK Conference, Oxford, 1963.

CHAPTER 4

Folding Box-board Cartons

D. J. Hine

The folding boxboard carton is familar as a retail pack in the distribution of food stuffs, confectionery, toiletry and cosmetics, tobacco products, light engineering goods, and toys and games. Usage in these ways accounts in the United Kingdom alone for an annual conversion of over 600,000 tonnes of board into cartons.

Prior to 1879, paperboard boxes were made by cutting, folding and gluing with much hand work but, in that year, Robert Gair in America took out a patent for the mechanized process of cutting and creasing board which is essentially the same as that used today. Initially, converted printing machines were used; now, cutting and creasing presses are purpose-designed with outputs to match the present-day demands for precision and productivity.

Definitions

The process of carton making is so versatile that the styles offered are legion.[1,2]

Any definition of a folding carton should set it apart from rigid paperboard boxes. Folding cartons are:

(a) made from paperboard of thickness between 300 μm and 1100 μm and
(b) delivered in a flat collapsed state for erection at the packaging point.

The thickness limitations set cartons apart from regular slotted fibreboard cases, which are sometimes also called "cartons" because they are delivered flat.

Materials

The boards used for cartons are mainly made on cylinder mould or Inverform machines. Only the thinnest boards are made on single and twin-wire machines.

The two main methods give many possibilities, as all these boards have a ply structure. All plies may be repulped waste as in a chipboard. Replacement of the outer ply by better-quality pulps gives in succession cream-lined chipboards, and second and first-quality white-lined chipboards. Duplex boards without any waste pulp at all are also available. Board made entirely from chemical pulp (solid white boards) are produced for frozen food packaging and other applications where the board must be waxed. The cost of these boards obviously rises with the quality of pulp used.

Where special properties such as moisture resistance or grease resistance are required, coated and laminated boards are employed. These may be wax-laminated where the wax is the moisture barrier, glassine-lined for grease resistance, or plastics-coated for special properties, including heat-sealing. Paperboard cartons, even if made from plastics-coated boards, unless of special construction, cannot be expected to give much moisture protection to the contents, and are therefore used with inner bags.

The external appearance and the printing quality of carton board is greatly enhanced by the use of coated board. Clay and other minerals are coated on to the lined outer surface of the board. The coating is applied either during the board-making operation or subsequently. A limitation of the boards which are coated on the board-making machine is that the coating must generally be of a somewhat thinner nature and have a less glossy surface than coatings applied off the machine as a separate operation. The latter include cast coated boards which have a very high gloss and brightness. In many instances, the extra cost of these boards limits their use to the packaging of comparatively expensive goods, such as cosmetics.

Foil-lined boards are also used for various types of carton, not only where protection is of importance but also to give particular display effects.

Chipboard manufactured from repulped waste materials (newsprint, corrugated and solid fibreboard cases, etc.) accounts in its various forms for more than half the current cartonboard consumption in the United Kingdom today. *Plain (unlined) chipboard* is only utilized for the very simple type of carton for holding stocks of materials which do not require any great display. It is almost unprintable in terms of quality, and its strength is somewhat lower than that of any other variety. Some chipboard cartons are produced, however, and are lined with printed paper wrappers.

White lined chipboard, with its improved appearance, printability and folding qualities, is used in greater quantities than any other grade. It is used for many cartons where the internal appearance is unimportant, and where no contact with foodstuff occurs. Taint is possible when foods containing flour or fats contact chipboard directly. Obviously, if a separate bag is used inside a carton, the contact between its inner surface and the food is prevented and, under these conditions, white lined chipboard may be usable.

With cartons which are opened repeatedly, the better-looking inner face of a *white lined manilla* or *duplex board* may be required, although white lined chipboard is often used for these types as well. With certain exceptions, most foods must avoid direct contact with chipboard, so that either pure pulp or duplex boards are used; or alternatively, glassine, sulphite or other types of paper may be laminated to the chipboard side of a white lined chipboard where foods are concerned.

Pure pulp boards are particularly popular for materials such as flour confectionery and chocolate, where staining of the carton by fats, etc., is possible. In this context also, boards lined with greaseproof paper, vegetable parchment or glassine are often used.

For a considerable number of products, protection against moisture and moisture vapour, together with grease resistance, is required, and here paraffin wax used in a number of ways can provide a low-priced solution. A great amount of work has gone into producing grades and blends of paraffin wax with various other components for different applications.

Wax-laminated white-lined chipboard is widely used for detergent or soap powder cartons if moisture loss or entry is to be avoided. Such waxes are also used as the laminating adhesives for glassine and other greaseproof liners referred to above.

Wax coating on one or both sides of the board can also be carried out at any stage after printing, and waxed cartons are particularly used in the frozen-food industry, although many are employed in an unprinted state, with a printed paper or film overwrap.

More recently, polyethylene and other extruded polyolefines have been employed, both for coating boards and for laminating various types of liner to chipboards and other boards.

Developments in polymeric materials are adding continually to the range of material available for combining with paper and paperboard in all forms. Note the increased use of materials such as polyethylene, polypropylene, polyvinylidene chloride and nylon, either in direct combination with board or as ancillary materials. By and large, rigidity, which is one of the prime factors required by all unit retail packages, can most economically be supplied by paper and paperboard, whereas barrier properties, such as moisture resistance, oxygen resistance, and grease resistance, are readily supplied by comparatively thin films of plastics. The combination of the two (rigidity and resistance) in coated paperboard can result in highly protective packages for many goods.

The use of barrier materials in cartonboard is principally restricted by the inability of the normal types of carton closure to prevent the ingress of moisture directly without passage through the barrier. Because of this, a considerable amount of work has been directed towards the development of one-piece functional cartons, aimed initially at the frozen-food industry where the protective requirement was originally demanded. Further designs have led to the use of cartons for liquids of various kinds.

Manufacture

Preliminary stage

As a general rule, the manufacture of folding cartons is a "bespoke" business. Cartons are not manufactured and sold from stock, but only for a specific purpose. All operations from the design to the delivery, including the size of sheet on which the cartons are printed and from which they are cut, are specific solely to the production of the one type of carton which has been ordered by the user. Consequently, there is always the necessity to allow sufficient time between the requirement for cartons and the date by which the order must be placed.

The user prepares a design brief, setting out details of the product and any special requirements on protection and method of use, the available machinery for packaging and the part the carton is expected to play in the distribution and sale of the contents. This brief guides design agencies, or the design departments of carton makers, in the submission of samples and artwork. Once the sketch and sample have been accepted, and agreement has been reached on the type of carton, the preparatory work can begin. Negotiations will have insured that an adequate specification for both the construction and the graphic design has been reached. In other words, a specific price has been accepted by the carton user. Such a proposal automatically implies that the method of printing the cartons has been considered, and selected in conformity with the requirements of the job, as well as the availability of the equipment. This is important, because any of the main reproduction processes (letterpress, offset lithography, gravure and flexography) can be used to print folding boxboard cartons. Each process, however, requires its own special properties in the board and processing, and the method selected can make considerable differences to the result.

Depending on the printing process, work starts on the preparation of the finished artwork. This covers complete working drawings and prints of any pictures that are needed, and the text matter set in type and so on. Full-scale drawings giving the exact dimensions, with the provisions for the creasing and cutting, register marks and colour separation details, must all be produced for subsequent handling by the appropriate department of the carton manufacturer. At the same time, the board must be obtained in the correct sheet or reel size and grade for the particular purpose, printing plates or cylinders must be produced, and a cutting and creasing forme prepared. All this preparatory work usually precedes the actual commencement of the job of producing the carton.

Production stage

The conventional methods of carton making involve first printing the cartonboard and then cutting and creasing it to allow the subsequent folding to shape, the stripping of any waste material which is not required in the final construction, and finally the finishing operation of putting in the joins where necessary, either by gluing or by stitching.

Printing and cutting and creasing are, in the main, sheet-fed processes but, for certain long-run work, rotary web-fed presses are used. There are in addition web-fed reciprocating-platen cutting and creasing presses which can be combined with printing either before or after cutting.

Printing

All the major printing processes have been used to print cartons. The method preferred is by no means the same in the different countries of the world, but in the United Kingdom probably the majority of medium-sized runs are now printed by offset lithography. As can be seen in figure 4.1, the lithographic process involves the offsetting of the ink film from the plate onto a rubber blanket, and the transfer of this ink film from the rubber blanket to the surface to be printed.

The letterpress process, on the other hand, is generally a direct printing process, the raised surfaces which carry the ink being directly pressed against the surface to be printed (figure 4.2). During recent years, a process which is referred to variously as dry offset, letterset or indirect letterpress has been used for certain types of carton. In this process, the ink from a relief surface is offset on to a rubber blanket, and the transfer of the ink to the board is achieved by pressing it against the rubber blanket. No water is involved, however, hence the use of the term "dry".

FOLDING BOX-BOARD CARTONS

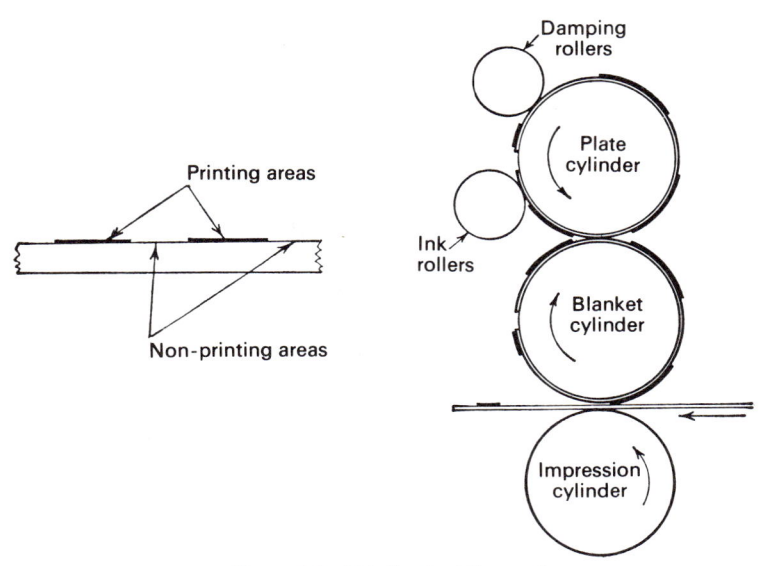

Figure 4.1 Printing by lithography

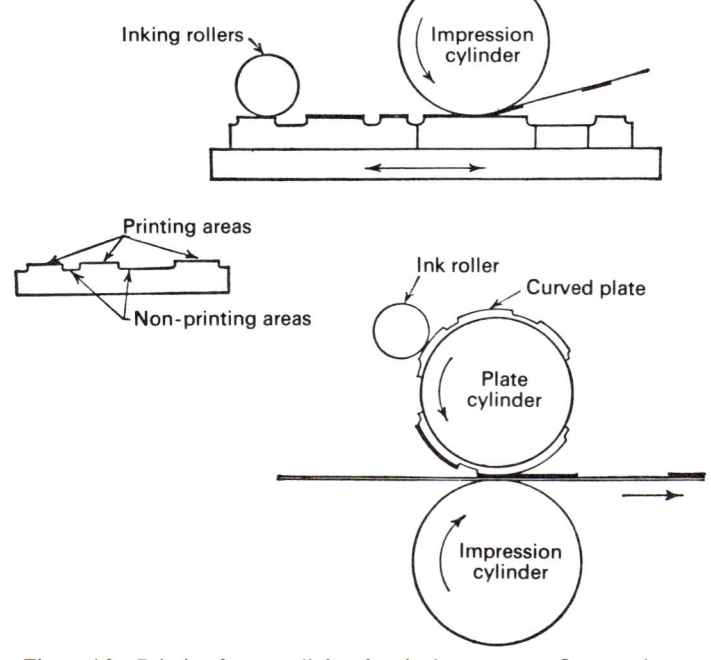

Figure 4.2 Printing from a relief surface by letterpress or flexography

1.59

Flexographic printing, which is essentially a relief printing process, is largely confined to those cartons where the demands of print quality are less stringent than normal.

In the gravure process shown in figure 4.3, the ink is carried in recesses in the printing surface and transferred to the board when it is pressed in contact with the surface. This process can be either web-fed or sheet-fed, and is used for the longer runs, due to the cost of making the etched recessed printing surface on gravure cylinders. Over recent years, however, newer methods of producing gravure cylinders have been developed, and the cost of gravure-printed cartons is now by no means prohibitive in what would in the 1960s have been regarded as short runs.

The cartonboard must be chosen to make a pack suitable for the particular application, but the printing processes also impose restrictions on the choice of material.

For example, in lithography, the high tack of offset inks makes demands on the surface strength (pick resistance) of the board used. The surface reaction of the board (pH) must be controlled to avoid ink-drying difficulties, and the (oil) absorbency of the board surface must be suitable for the inks used, so that the penetration of the ink vehicle into the board takes place at the right speed. Too rapid absorption, or too slow penetration, of the ink vehicle can lead to difficulties, such as powdering of the ink surface or set off on to the back of the board. Letterpress printing involves similar but generally somewhat less stringent requirements. In addition, the lithographic process uses water to prevent ink reaching the non-printing areas on the printing plate; this will result in the further requirement that the surface to be printed must be able to withstand the slight wetting which will occur as it passes through the press. This is particularly important where clay-coated boards are used. Although gross deterioration of the board surface is now a very rare occurrence, continual slight leaching of soluble coating material can cause

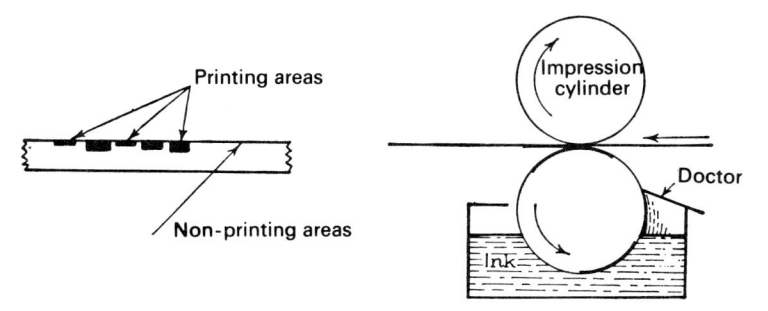

Figure 4.3 Printing from a gravure cylinder

considerable difficulties during the printing process, particularly with multi-colour machines.

The above considerations do not apply to gravure printing, because this uses solvent-based inks which possess negligible tack. The drying of the ink also is largely by evaporation, rather than the partial oxidation of the ink which occurs with most letterpress and litho inks. However, surface irregularities in the board can cause trouble, and result in individual cells in the intaglio surface of the gravure printing cylinder or plate failing to transfer their ink (see figure 4.3). This causes a deterioration in the print quality (called "speckle"), and generally means that the property of surface smoothness measured at printing nip pressures is most important.

Whichever printing process is used, the resulting print must have the required resistance to rubbing during the handling that it will subsequently receive in the filling and closing operations on a packaging line. It must also be resistant to the rubbing encountered during transport and distribution. Much can also be done by careful design of the print to keep those areas of the carton surface subjected to rub on machines, etc., as free as possible from ink.

Inks which are non-toxic and give prints of low residual odour must be used on food cartons.

The cutting and creasing operation

The purpose of the cutting and creasing operation is to enable the carton blank to be detached from the sheet or web with creases correctly located, so that the carton may be readily glued or stitched as required, and subsequently erected to the correct shape.

In the early days of carton making, the press was almost exactly the same as the printing press which produced the printed design. However, instead of a printing plate (which in those days was usually a letterpress printing plate) a cutting and creasing die, consisting of creasing rules and knives to impress the design of cuts and creases upon the board, was attached to the bed of the machine. Nowadays, reciprocating platen presses are used by most carton makers in the United Kingdom, with rotary presses reserved for long runs of carton designs for cereals, cigarette packing and detergents.

The formes for platen and flatbed cylinder presses are flat and usually made from special plywood. In a one-piece forme, the sharp-edged cutting rules and rounded-edged creasing rules are inserted in slots cut in the plywood using fretsaws or more recently laser beams. The alternative, a block die, is made from pieces of plywood, metal or

PAPER-BASED PACKAGING

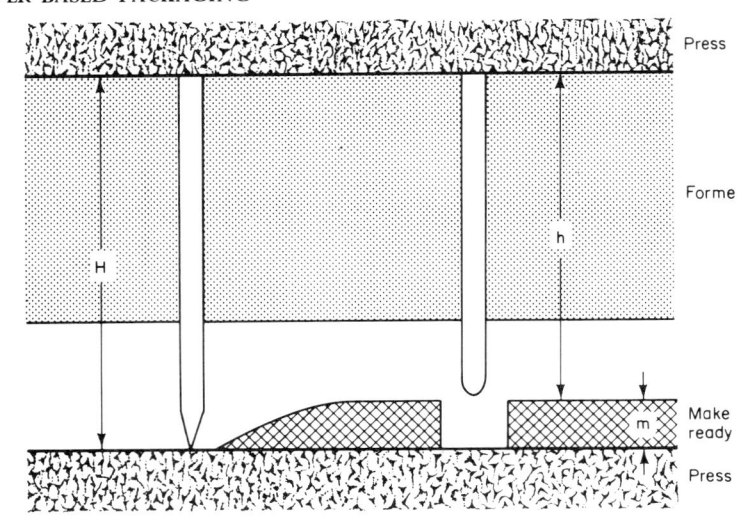

Figure 4.4 The cutting and creasing forme

plastics cut to the sizes of the carton panels, with the rules clamped between the blocks.

A forme has rows of cutters arranged to correspond to the printed images on the sheet. Accuracy in the panel sizes of the individual cartons and in the spatial arrangement of cutters in the forme to $\pm 250\,\mu m$ is considered necessary for machine-packed cartons.

Figure 4.4 shows diagrammatically a forme mounted on one platen of the press, and a cutting rule contacting the steel of the other platen to sever the board. The crease is made by the rounded-edged rule pressing the board into a groove in the makeready or counter. These creases have two purposes: they fix the lines at which the board will fold easily, and prevent the board cracking at the bends.

It is necessary for the board to be partly delaminated by the crease in order that a clean roll is formed on the inside when the back of the board is displaced for folding (see figure 4.5). The outside of the fold is, naturally, of very much greater importance, since here any defects would be easily seen. Both the depth and the width of the crease have to be suited to the board to get this desirable result.

The performance of a carton on a packing line can be influenced by the quality of the creasing. This may show itself as a high resistance to opening of the glued carton, which will slow down hand erection or seriously disrupt automatic filling. Poor crease quality can also produce flaps which spring open and panels which bow. The key factor is the

FOLDING BOX-BOARD CARTONS

Figure 4.5 Stages in the folding of a good crease

resistance to folding of the creases, and there are test methods by which this can be measured and limits set.

Stripping

The result of the cutting and creasing operation on a sheet of board is to produce a number of cartons held together by an intact front edge and small "bridges" at various points on the sheet, so that it is still handleable as a sheet and does not fall completely into its component cartons. The next operation is to remove the material which holds the sheet together, a process termed "stripping". A certain amount of automatic stripping is done as the sheet passes through the cutting and creasing press; this, in general, removes unwanted central portions of the cut sheet which cannot easily be removed by subsequent operations. Complete stripping is possible on some modern presses, but generally this operation is still performed by hand with the aid of some machine tools such as mechanical hammers. Essentially, the operator uses a

PAPER-BASED PACKAGING

rubber-headed hammer and knocks the waste material away from the cartons which are "torn" out of the sheet in blocks. After brushing the edges to remove fluff and loose material, the blocks are passed on to the next operation.

Finishing

The carton blank may be treated no further by the carton maker if it is destined for certain packing operations. Cartons which are erected and secured by interlocking flaps, or cartons which are erected and the seams glued on the packing machine, are examples. Many cartons are, however, glued by the carton maker along the side seam to form a tube. Prior to gluing, the carton may have passed through various processes, e.g. windowing or waxing.

Alternative means of erecting cartons can be provided by the carton maker. He can print on to areas of the blank, heat-sealing or pressure-sensitive adhesive patches. Cartons using these methods of securing flaps are usually covered by patents, either of their design or of the equipment used in erection.

When a collapsed tube is formed on the side seam gluer, two diagonally opposed creases are folded 180°. The majority of gluers have a pre-folding operation which momentarily folds the other two main creases to between 90° and 180°, and then unfolds them. By this means, the subsequent erection of the carton on the packing line is made easier. The operation is referred to as *prebreaking the creases*.

It is important to remember that prolonged storage of cartons, particularly under load when stacked in cases, can reduce the effect of the pre-folding operation, and almost return the carton to its original condition. Complaints sometimes occur which can be traced to the fact that cartons have been kept for too long under pressure in stacks, so that the subsequent erecting operation on the forming and filling machine has been unsatisfactory. Modern carton gluers can also make other styles of cartons, such as spot-glued trays and collapsible styles, as well as straight-side-seam glued tubes. Once these operations have been completed, the cartons are packed ready for dispatch to the user. For cartons to work well on the packing line they must not be distorted in store, nor exposed to extremes of temperature and humidity. Therefore, both the bulk containers and the storage conditions need proper selection. In the relevant clauses of section 7, part 12 of BS 1133 it states that cartons are best bundled, and paper banding is preferred as it causes less damage.

For transport, bundled cartons arranged on edge in outer containers with separators between the rows and layers will prevent distortion and scuffing damage. A lower-cost alternative is overwrapping of the bundles, using paper tape to secure the wrapper.

Advice on storage conditions generally for packaging materials is given in BS 1133, section 2. In particular, for cartons, excessively damp or dry storage should be avoided, or the performance of cartons at all stages will be impaired. The section indicates maximum storage periods of 2 to 3 months for certain cartons, the critical time depending on the type of board and the sensitivity of the cartoning machine.

Board requirements for converting and finishing

During creasing and folding, cartonboard is subjected to complex stresses, and the ability of a board to make a good carton depends on its rigidity, its ease of ply delamination, and the stretch properties of the printed liner. If the stress (which is essentially a tensile stress) impressed on the crease by folding is greater than the top layer of the board can stand, cracking will occur, producing an unsightly result, particularly if the crease has been completely printed over with a dark ink. It is an important requirement, therefore, that the surface layer on the top of the board is of an elastic nature and relatively high strength compared with the properties of the underlying layers which are in compression.

These stresses become very much more critical where parallel creases are placed close together on the surface of the board, because during the actual creasing operation, higher tensile stresses can be created between the two creases. The demands on the board become even more stringent when creases meeting at an angle are placed in close juxtaposition one to the other on the surface of the blank.

The cutting operation is performed with a cutting knife instead of a creasing rule. Here the requirements are largely that the board shall cut cleanly and not be friable after the cutting operation has taken place.

A practical creasing and folding test is described in BS 4818 which enables the suitability of a board for carton making to be assessed in numerical terms. This standard details the use of the Pira Carton Board Creaser, a bench-mounted hand-operated platen press fitted with three parallel creasing rules and a corresponding set of grooves. Once the width of the creasing rules and their depth of penetration into the grooves has been set according to the thickness of the test board, the groove width is varied. When the resultant creases are hand-folded 180°, the minimum and maximum width of groove giving acceptable folds can be determined by reference to a set of photographs of defective folds. This test is made both along and across the machine direction of the board, and the range obtained for each direction. The range of groove width becomes greater the more suitable is the board for carton making.

The principal finishing operation of gluing (usually at high speed) the hem of the carton to the flap on the other side, to form a flattened tube uses adhesives of very high tack, and this requires even greater board resistance to ply separation than the printing inks referred to earlier. Uniform and correct absorption of the adhesive (which is usually an aqueous adhesive) is necessary if gluing is to be satisfactory. Depending on the design of the carton, board of a low rigidity can considerably hamper the gluing operation, because the less rigid the board the greater the effect the crease must have in order to produce a difference when the fold is made in the gluing machine.

Carton specifications and quality control

The specification agreed between carton user and supplier can be as simple or complex as the application demands. The carton maker will always seek to deliver a commercially acceptable product, but the critical requirements will obviously differ between a small order for single-colour printed cake cartons, hand-erected at a local bakery, and a multi-million carton order for running on high-speed packaging machinery.

Besides production checks to see that the carton meets the customer's requirements, the carton maker has to make quality acceptance tests on his raw materials to check their suitability for the converting processes and ability to give the desired carton quality.

Printing quality may be specified as colour matches to specimen trios of prints, the target and the acceptable variants either side and, for example, a minimum level of rub resistance.

Creasing quality starts with the forme and the checking of the dimensions of every die on it. It must be appreciated that the final carton size will depend on other factors, including the sensitivity of board dimensions to change with variations in the moisture content of the board, which may be affected by the relative humidity of the surrounding atmosphere. When carton production starts, cartons are examined for clean cutting, crack-free folding, and correct assembly of the carton. If necessary, crease stiffness can be measured. These checks continue throughout the run, to see that wear of the makeready does not cause a deterioration in crease quality.

Where the carton incorporates an opening device relying on perforations, as do many soap powder packs, a maximum value for the burst strength of these perforations may be included in the specification, to reduce complaints from housewives that they break their thumb-nails before the carton opens.

During side-seam gluing, samples will be examined, and checks made for extraneous glue spots inside the tube, and for good adhesion and

FOLDING BOX-BOARD CARTONS

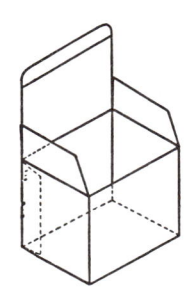

Tuck-end construction with one end closed.

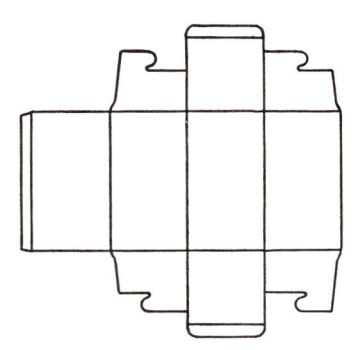

Combination tuck-end and lock-end closure (flat blank).

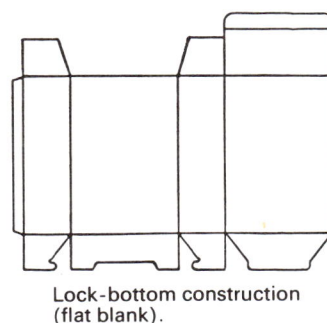

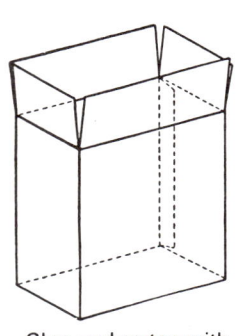

Lock-bottom construction (flat blank).

Glue-end carton with one end closed.

Figure 4.6 Typical tube styles

skewing of the side seam. Gluer adjustments may be made so that cartons meet special critical requirements, such as even stacking when loaded into a packing-machine hopper.

Many of the tests that may be made on board, inks, print and cartons are covered by industry and national standards. Early discussion of the carton application between supplier and user is desirable, so that the necessary quality can be established and appropriate tests selected.

Common carton styles and their uses

The purchasing specification agreed for a carton should include a dimensioned drawing of the selected style. Where the style is simple, it may be sufficient to refer to the number in the European Carton

PAPER-BASED PACKAGING

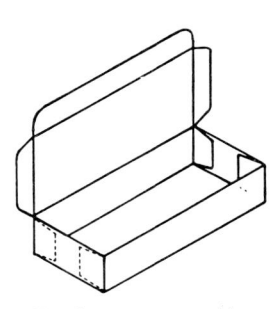

Glued-corner tray with tuck cover.

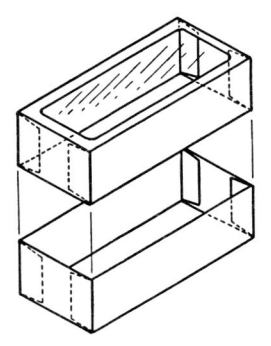

Two-piece tray with transparent window.

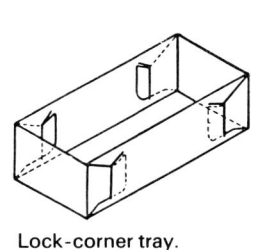

Lock-corner tray.

Glued-corner tapered tray with locking top flaps.

Figure 4.7   Typical tray styles

Makers Association code and give the major dimensions in the approved sequence. An alternative is to use the styles named in the Packaging Code BS 1133.

Cartons are made in a great variety of styles and are generally named from the method of closure, their shape, their end use or other special features. Nevertheless, while there are literally hundreds of different types of folding carton, the entire range can be considered as having been evolved from the two primary concepts of a tube (figure 4.6) and a tray (figure 4.7). We may explain these two terms, *tube* and *tray*, in the following way:

> A tube consists of a sheet of paperboard folded over and glued at the edges to form a (rectangular) tube, the ends of which can be sealed or locked in a variety of ways.
>
> A tray consists of a sheet of board with all four sides folded at right angles to the main sheet, and locked or adhered together at the corners. One panel of this tray can be extended to form a cover, if desired, which will act as a lid.

1.68

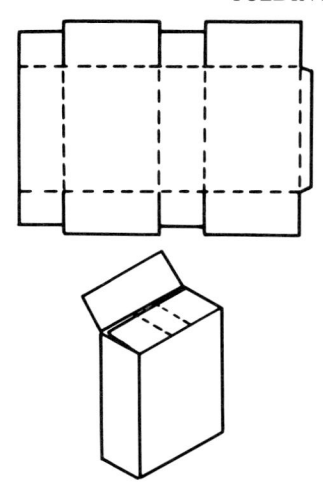

Figure 4.8 Glue-end carton, flat blank, and partly closed box

Bearing in mind that virtually all cartons are produced on one or other of these principles, we can consider some typical styles.

Undoubtedly the style accounting for the largest carton tonnage is the familiar glue-end carton (figure 4.8). This is produced from a single blank, and is delivered by the manufacturer in the form of a collapsed rectangular tube with four flaps at each end for sealing. The style is particularly suitable for high-speed automatic packaging, and is used for

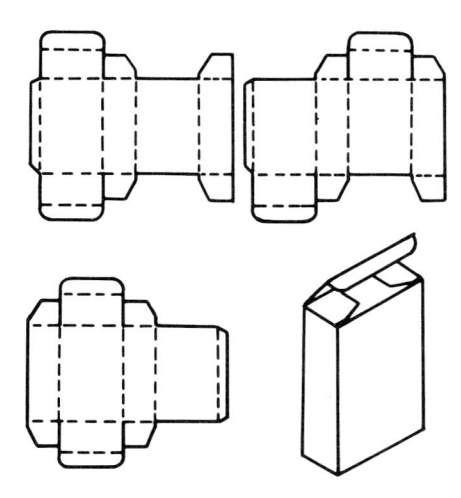

Figure 4.9 Three shapes of blank for a tuck-end carton

granular materials, including common foodstuffs, soap and detergent powders.

The tuck-end carton (figure 4.9) is widely used where a reclosure is required. This is often the case with many household goods, clothing items, toys, etc., where customer inspection may be desirable. With this style one of the four flaps of the glue-end style is omitted, and on the opposite flap is provided an extension which can be tucked into the body of the carton once the tube has been produced. This reclosure facility also makes it popular where repeated dispensing of things such as tablets, drawing pins and other small articles may be required.

Heavier articles often require a safer reclosure device than is provided by the tuck-end style, and for these many forms of the lock-end carton (figure 4.10) have been designed in which a tongue or tongues on one of the main flaps engages with corresponding slits in the opposite flap. All the above styles of carton, as will be readily realized, are based on the tube.

The shell-and-slide carton (figure 4.11), which consists of an outer tube (shell or hull) and a tray-like slide with tuck-in ends, is familiar to all of us in the cigarette pack. The recent tendency to replace this shell and slide with a so-called "flip top" carton in the cigarette industry is an example of the modern tendency to reduce all types of carton to a one-piece blank. Shell-and-slide cartons are particularly suited to the frequent dispensing of solid objects which substantially fill the carton at the beginning. The flip-top carton is obviously also suited to this purpose.

The largest single remaining class of cartons is based on the various

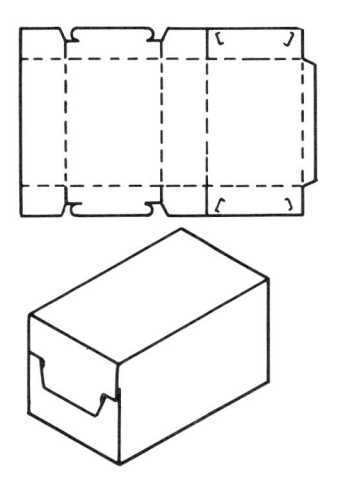

Figure 4.10 Lock-end carton and blank

1.70

FOLDING BOX-BOARD CARTONS

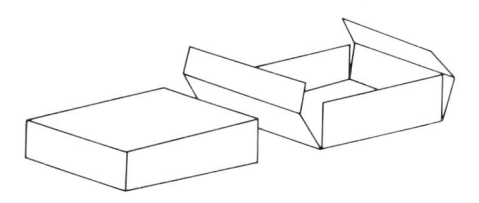

Figure 4.11   Shell and slide

types of tray (figure 4.7). These consist of an unbroken bottom with the sides folded at right angles and secured either by spot-gluing, stitching, locking or folding over in some manner to complete the tray. Their principal advantage is that they have a larger area for initial loading, and they also provide the same larger area for visibility of the contents where this is required. They are thus particularly useful where it is difficult to insert the contents through the ends. They find wide uses for fruit, biscuits and cakes, and are often over-wrapped with a transparent film. The addition of a hinged lid to the basic tray forms a fully-enclosed one-piece carton which is again used for similar types of products.

A number of the tied carton systems, in which the user purchases cartons to be filled on equipment leased from the carton maker, give packs of the tray type.

The lid-tray type consists essentially of two trays, one of which is made slightly bigger than the other, to act as a lid. This is sometimes referred to as a *semi-rigid box*. This particular type of carton is often used for packing shirts (figure 4.12) and other clothing which folds to a rather small thickness with a large area and needs display of some description. These trays can be provided with considerable rigidity where necessary by utilizing fold-over ends and by gluing them. This is particularly useful where heavy goods are concerned, or where the goods are easily crushable, and where rigidity in the walls of the carton is essential. The retailing of boots and shoes is a good example of the use of this type of carton.

Window cartons have become part of present-day marketing and are produced by having some area in a main panel cut away and covered (usually with a transparent film) to provide greater visibility of the contents. The window style can be incorporated into virtually any other style of carton. The limitations are generally on the actual type of window that can be used.

All of the carton styles mentioned above are commonly produced with folds and creases which are based on straight lines. During the recent past the ability to handle curved creases has grown, and a considerable

PAPER-BASED PACKAGING

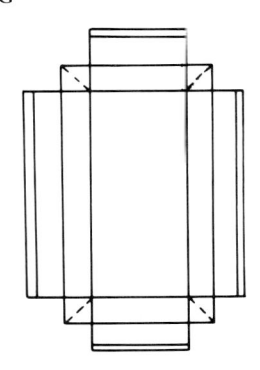

Figure 4.12 Blank for one half (lid) of shirt box. The other half is similar in construction but just slightly smaller when made up.

amount of experimentation on shapes based on the use of curved creasing rules is in progress. This can given considerable scope for display purposes, but usually entails complications in terms of the packing of the retail units into their outer cases.

Where enhanced barrier properties are required, a number of patented styles are available, incorporating a loose liner which forms a bag or pouch inside the carton on the filling machine. These pouches can be liquid-tight. With appropriate machinery, liquid-tight cartons can be obtained in single-wall construction.

Cartoning systems

A "cartoning system" is in every sense a package deal. The carton user negotiates with one source for the cartons, the machinery on which to erect, fill and close them, and the servicing of that machinery. Many of these cartoning systems are obtained through carton suppliers rather than machinery manufacturers.

The pack styles available as systems cartons extend from simple creased blanks erected into lock-corner trays, through side-seam glue-end loading styles, to cartons with an inner liner that can contain liquids or function as vacuum packs. A range of single-wall cartons often made from the web are also available for liquid packing.

The machinery available for handling cartons can vary from simple hand-fed erecting machines (which are then filled by hand on a conveyor and closed by hand by a simple tucking operation) to a sophisticated two or three-headed erecting machine, one or more automatic filling stations and closing lines, coupled with film overwrapping equipment and means for packing the cartons directly into cases for dispatch. More recently, developments in shrink wrapping have led to collating a number of cartons into units for dispatch by these methods.

Advantages of cartoning systems

From the point of view of the package user, there are several advantages in utilizing a cartoning system. Firstly, there is a simpler administration operation involved. One company is responsible for the supply and servicing of both machinery and cartons. Since the particular supplier of cartons will normally be producing them for the type of machine used in the system on a relatively large scale, he should be familiar with the detailed requirements of the carton blanks or flats, and the limitations of the carton handling machinery concerned. The package user need not, however, be completely tied to one carton supplier, as most of the systems available are licensed for more than one production unit. In some instances there may be two or three sources of supply within one group of companies, and with other systems competitive companies who are both licensees for a particular system can provide aternative sources of supply.

The tied cartoning system is principally applicable where versatility and flexibility rather than full automation is needed, and this is the requirement with many food lines.

A cartoning system using leased equipment has the further advantage that the capital outlay on machinery is avoided in situations where the packaging requirements of a company may change from year to year.

Developments in carton styles

Each of the cartoning systems has particular applications, and usually a feature which is patented. In the early systems this was often a locking tab and slit (Kliklok and Sprinter for example) which could be incorporated in various ways to give a lock-corner tray, with or without an integral lid also closed with a lock. Variants on this lidded tray style used adhesive sealed flaps to make the closure (e.g. Diotite).

The erection of these cartons was on relatively slow-speed intermittent-motion machines, often coupled to simple hand or machine filling lines.

When higher packing speeds (over one per second) are needed, a continuous-motion fully automatic machine is usually required, and vertical and horizontal cartoning systems using end-loading cartons were developed. These use side-seam glued blanks.

Although the resulting cartons can be sift-proof, they are neither liquid nor gas-tight.

The first approach to a liquid-tight carton erected carton blanks, and then waxed them by dipping. Later these could be pre-waxed and heat-sealed into a carton (e.g. Perga and Pure Pak). More recently, in the Tetrapak, Zupak and Brikpak systems, the cartons are made and filled continuously from a reel of preprinted plastics-coated board. Developments here have given the possibility of aseptic filling.

The incorporation of a liner which can be sealed into a bag can give a carton able to contain a liquid, or give gas and water vapour protection to the contents. Examples of developments of this style are the Cekatainer, Hermetet and Pemplex systems.

Further developments have incorporated into systems cartons with various tear-off, easy-open and dispensing devices.

Factors to be considered when selecting a cartoning system

Product factors

Obviously, as with all packs, the nature of the product and the way it is presented for sale will be prime considerations in the selection of a system from the available alternatives.

(a) What sort of protection does the product itself require from moisture, from oxygen, from outside odours, etc.? Is it greasy, or wet, or otherwise able to affect the board from which the carton is made?

(b) What sort of variations are likely to be expected in the product itself? How easy will it be to control the size of the product, and within what limits?

(c) How can the product best be filled into the carton? What experience is there of packing similar products?

(d) What further processes follow filling? Do the cartons have to be overwrapped for extra moisture protection or deep frozen for example?

Machinery considerations

(e) What is the required production rate of the machine, both immediately after installation and at the expected peak of production?

(f) What is the number of package sizes which may be involved?

(g) What is the frequency of size changes?

(h) Will the system have to cater for alterations of the product type or number within the size variation anticipated?

(i) What are the labour requirements, and how do they vary with production rate, and can they be met locally?

(j) What sort of space is required for putting in the machinery, bearing in mind the possible increase in production later on?

The choice of board to meet the product protection requirements can have an influence on the efficient functioning of the cartoning line, feeding, erecting and closing the packs.

Marketing considerations

(k) Can the graphic design and print requirements of marketing the product be met with the system?
(l) How will the carton affect distribution and consumer acceptance of the product?

General considerations

(m) Does the proposed new equipment have to link up with any other equipment already in existence, such as over-wrapping machines, or case packers or filling heads?
(n) What sort of advice can be obtained from the carton supplier on such other ancillary equipment with which he may have had experience on his cartons on other lines?

When we consider the possibilities of making all these many styles in a variety of board qualities and rigidities, and of their decoration by printing and lamination, the versatility of the carton as a retail pack can well be understood.

REFERENCES

1. European Carton Makers Association Code.
2. Section 7, BS 1133. Paper and board wrappers, bags and containers.
3. Section 2, BS 1133. Introduction to packaging.
4. BS 4818.

CHAPTER 5

Rigid Boxes

P. B. Robinson

This chapter outlines the main sequence of operations through which the components of a paper-covered rigid box must pass during the course of manufacture.

Until a few years ago the number of rigid boxes manufactured in the United Kingdom had declined considerably. However, there has recently been a considerable increase in demand for rigid boxes. The demand comes mainly from the public for a stronger and more permanent type of packaging for many items. It is difficult to define the principal market for rigid boxes, but they are used by the hosiery and footwear trades and for small hand tools, and also for the expensive perfume and cosmetic markets. It is impossible to lay down hard-and-fast rules in rigid box making. Conditions, layout, etc., vary so much from one firm to another, that methods tried and found successful by one company may be hopeless in another set of conditions.

Board

The choice of board used depends very largely on the size of the box, the type and weight of the contents, the intricacies of manufacture, and the degree of accuracy required.

During recent years, change has taken place within the rigid box industry, as to the type of board used. Originally two main types were used. Firstly, Dutch Strawboard was the traditional material used by the majority of box makers engaged in the mass production of machine-made boxes. It is a stiff material, and makes a good rigid box. However, because of various drawbacks (which include its propensity to absorb moisture,

PAPER-BASED PACKAGING

and to vary in size, and for it also to give off a "musty" smell when damp) box makers looked for alternative materials.

The original chipboard offered by British board mills, although not as rigid as the Dutch Strawboard, offered the advantage of not shrinking to quite the same extent as Strawboard. It therefore was more reliable for ensuring that the correct fit was obtained between boxes and lids, and for minimizing the acute warping of shallow lids. The main disadvantage was the lack of rigidity, and often a thicker board was required to obtain a similar rigidity to the Dutch Strawboard. This led the British board mills to produce a new grade of board known as *rigid board*, as a substitute for Dutch Strawboard. It was intended to combine the advantages of both Dutch Strawboard and the home-produced chipboards. This board is now widely used in the United Kingdom and is available in thicknesses ("calipers") ranging from 500 to 2900 microns (μm).

No strict rules can be drawn for the rigid box industry as to the type of board to be used. Generally speaking, the box maker will decide on the best board for the particular job in question, and this may vary from a plain unlined rigid board, through the complete range to a high-class food-quality duplex board.

The covering papers used are numerous, though generally in the rigid box industry there are a number of standard types used. These include enamels, flints, tints, leatherettes, and many others. The covers can, of course, be printed, embossed, or gold blocked and, as with board, the type of paper used depends on what the customer requires. The main consideration for the box maker is to know the behavioural characteristics of the various types of papers, and how to overcome them if they have adverse affects on production.

Having, therefore, selected the type of board, and decided the thickness and weight suitable for the job, the box-making operations can begin.

Cutting

First of all the board must be cut into the correct blank size. This is usually done for the straightforward type of rigid box on a rotary cutter and scorer.

There are several types of machine available with hand or automatic feed, straight-line or right-angle. Straight-line machines cut and score the board in one direction at a time, passing through the machine. Right-angle machines carry out the cut-and-score operation in both directions in one pass through the machine. The scoring operation is, however, usually carried out in the same way on all machines.

A scoring wheel (see figure 5.1), which has a cutting edge, is fixed at a pre-determined distance from a roller, and the board is passed between

RIGID BOXES

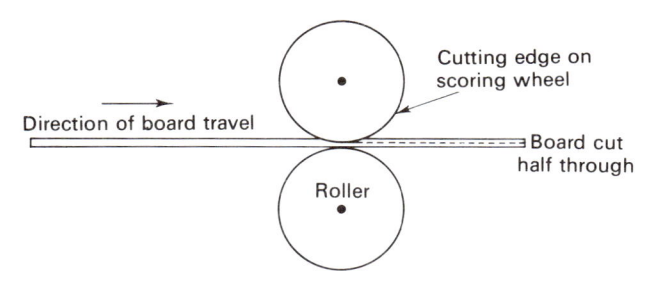

```
                    ___
                   /   \       Cutting edge on
                  |  •  |  ←   scoring wheel
    Direction of  \___/
    board travel   ___
    ────────→    ╞═════╡ Board cut
                  ___    half through
                 /   \
                |  •  |
                 \___/
                 Roller
```

Figure 5.1 A scoring wheel

the two. The distance between the roller and the cutting wheel is such that the cutting edge penetrates only halfway through the board. Cutting can be carried out either by setting a scoring wheel sufficiently low to cut right through the board or on some machines it is done by a shear cutting action, brought about by two wheels partially overlapping each other. This produces a piece of board, the overall size of which will correctly make up into the box, and on which there are two sets of scores, one set at right angles to the other (see figure 5.2). These scores divide the box into its length, width and depth dimensions. This operation, and that of corner cutting which follows it, may also be carried out on a cutting and creasing press using a metal die, in a similar manner to that employed for folding cartons. Using this method results in producing an improved quality of blank, as well as alleviating the need for the separate process of corner punching. The blanks processed by this method make

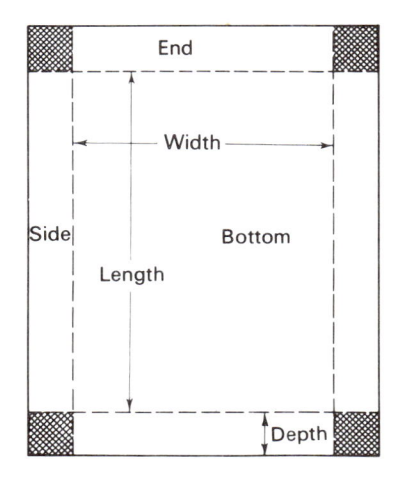

```
            End
    ┌──┬──────────┬──┐
    │▓▓│          │▓▓│
    ├──┼──────────┼──┤
    │  │          │  │
    │  │← Width →│  │
    │  │          │  │
    │Side│ Bottom │  │
    │  │ Length  │  │
    │  │          │  │
    │  │          │  │
    ├──┼──────────┼──┤
    │▓▓│    ↕Depth│▓▓│
    └──┴──────────┴──┘
```

Figure 5.2 Board showing two sets of scoring lines at right angles to one another

1.79

up into a better rigid box, as the improved accuracy allows for better production on the quad staying and automatic covering machines. Consequently there is a strong tendency for rigid box makers to move away from the standard rotary cutting and scoring machines to using hand-fed and semi-automatic cutting and creasing platens. However, on smallish quantities it is often not economic to pay for the cost of the metal die and the additional cost of makeready of the cutting and creasing press. In addition, especially on thin boards, the difficulty of pressing a length of metal rule into a piece of cardboard (which may vary in thickness) just sufficiently to give the effect of a score, can be problematic.

Corner cutting

The next operation is to cut out the corners of the blank (see figure 5.2) to allow the sides and ends to be folded at right angles to each other, and to the bottom of the box. It is essential that extreme accuracy be maintained in this cutting operation, otherwise the box will not make up square. It is done on a machine which can be designed either to cut one or two corners at the same time. Nowadays there are machines available which will cut out the corners at the same time as they fasten them together with stay paper.

Corner staying

The object here is to stick gummed paper around each corner, thus making the box rigid. There are various types of machines to do this job. The simplest machine, which will only stay one corner at a time, is hand-operated—the operator having to fold up the sides and ends of each box by hand. The stay paper used can either be a gummed paper tape, made sticky by the application of water, or a thermoplastics coated paper which is heat-activated. The choice of paper depends entirely on the type of machine. Thermoplastics stay papers have advantages, particularly when staying a deep box (say 100 mm) or more) and when the operator is learning the job. With gummed paper, the rate of consumption of stay paper can be so rapid that insufficient time is available for tape to develop the necessary tack to enable it to stick properly; whilst at the other end of the scale, when a trainee is operating the machine, she may be slow or fumble a box to such an extent that the paper dries out before it is applied. However, it is not always possible to obtain sufficiently thick "thermo" stay paper to meet all customer requirements.

The quad stayer is virtually 4 single stayers combined, with the addition of an automatic feed. Here also there is a choice of machine using either water or heat as the means of creating tack. One point which must be borne in mind when using a quad stayer which offsets to a certain extent some of its advantages, is that a separate forme or block is required for every size produced, although adjustable formes can be obtained. This block is used for forming the blank and is normally made of wood, metal or a plastics material. (There is also a device which may be built in to new quad stayers or added to existing machines in some instances, which will corner-cut all four corners of the blank before folding and staying them.) Finally, the boxes are nested one within another automatically, straight off the quad. The box is now ready to be covered with paper.

Alternatively, in instances where the box size is small and where strength is not required, the work can be delivered to the covering machine with the sides bent and not corner-stayed. The strength of the box is obtained by the covering paper holding the corners together, as will be explained in the box covering section. The blanks can either be bent by hand or by an automatic bending machine, which can be linked directly to the automatic covering machine.

Paper slotting

The covering paper must first be processed so that when wrapped around the box on the automatic covering machine it produces a neat and tidy turn-in. This process is called *mitring*, and a label punch which operates on a similar principle to a corner punch is used.

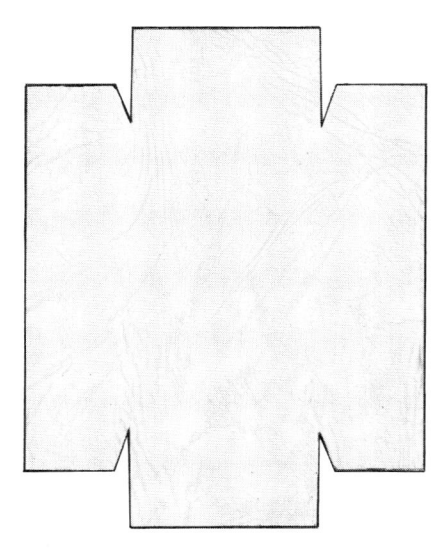

Figure 5.3 Shape of box cover paper, showing mitring

PAPER-BASED PACKAGING

Firstly the cover is guillotined to a size which allows for a turn-in. The cover is, therefore, normally about 22 mm larger than the dimensions of the blank. Secondly the cover is mitred (see figure 5.3).

This enables the cover to be turned in automatically on the machine and at the same time produces a suitable finish. This process is necessary for producing covered boxes on an automatic covering machine.

Box covering—simple lid/tray boxes

The board has been cut and corner-stayed, and the paper mitred. The next operation is to amalgamate the two, in other words to cover the

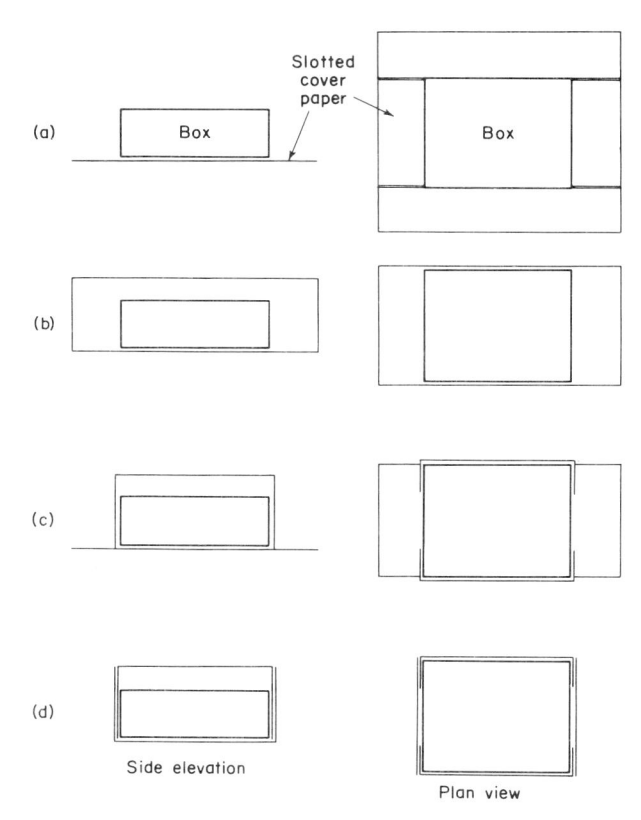

Figure 5.4 Box Covering
(a) box stuck to slotted cover paper
(b) long sides of cover paper folded up and stuck to long sides of the box
(c) four end-flaps bent round at right angles
(d) two short cover sides rolled up

1.82

box with the paper. Most machines for this job work on the same principles, the basic movements all being similar.

First of all the paper cover is glued on its under side. This can be done either by an operator manually passing the paper over a glue roller and placing it on a table, glue side up, or as it is more frequently done nowadays on an automatic gluer. In this a stack of covers is placed in the gluing machine, which picks them up one at a time and feeds them over a glue roller, from which they pass on to a long moving belt, which is perforated to allow suction to hold the paper in position and flat. It also helps to prevent the covers from curling due to the application of glue, which would make it more difficult for the operator and machine to handle them. The length of the belt, coupled with the timing of the automatic gluer, is designed to create a sufficient time-lag between the application of the glue and the use of the paper for the required degree of tack to develop.

When the cover reaches the operator, a corner-stayed box is placed on it. The accuracy of positioning the box is important, for on this largely depends the quality of the final product. The box with the cover stuck to the bottom is then placed by the operator on the forming block fitted to the machine. A separate block is required for each size run on the machine, and may again be made of wood, metal or plastics material.

The forming block, with the box on it (figure 5.4a), passes down into the machine, where the two long sides of the paper are folded up against the side of the box (figure 5.4b). The four end-flaps of the long paper sides are then bent round at right angles against the ends of the box (figure 5.4c). Next the two short sides of the cover are rolled up against the ends of the box (figure 5.4d). Then the 11 mm of projection paper all round (the paper is normally 22 mm larger in both directions than the board) is turned inwards on all the four sides, at right angles to the side of the box. This is possible because the top and middle blocks on the machine separate at this stage, and the paper is inserted between the two blocks. After this, the top block comes down to join the middle block again, thus turning all four edges of the cover inside the box. At the same time, equal pressure is applied externally to all four sides to ensure perfect adhesion. Finally, the block rises up out of the machine, bringing with it the covered box, which is ejected on to the packing table (figure 5.5), where it may be thumbholed or "lidded-up", or have any other ancillary operation performed on it that may be required.

Thumbholing can be carried out with a hand punch, a treadle-operated punch, or a fully automatic machine where the box is fed straight into the thumbholer off the covering machine without being handled by an operator.

There are even more advanced machines than this. On one, the boxes

PAPER-BASED PACKAGING

Figure 5.5 The box-covering process. The packing table, showing box on slotted cover paper (as in figure 5.4a), and machine with forming block.

are automatically fed after they have been placed on the cover by hand, but the basic movements are the same.

Another system allows for the linking up of a quad staying machine with an automatic covering machine, with an intermediate machine taking the corner-stayed boxes from the quad and spotting them automatically on the covering paper prior to the wrapping unit. This requires no direct operator, and the machine can run at well over 2000 boxes per hour.

Other than purely hand-made boxes, of which there are still many produced, there are many instances in a rigid box factory where hand work is necessary for the finishing off of automatically made boxes. This includes the production of hinged-lid boxes, where a calico or paper hinge is applied to the box and lid, or where racks or other platforms must be fitted to boxes.

There has been an increase in the demand for rigid boxes which contain vacuum-formed plastics fittings; these must also be applied by hand.

Flanged and envelope-wrapped work

Apart from performing the normal tight-wrapped work already described, some machines are capable of producing variations in the form of flanged and envelope (or loose) wrapped boxes.

For a flanged box, a flange card (with or without a pad to give a domed

1.84

effect) is cut and stuck to the scored box, either before or after corner staying. (The domed effect can also be obtained by moulding the board on special machinery which is steam or electrically heated.) The covers are slotted slightly differently. The operation of gluing the cover and placing the box on to it then follows the same pattern as already outlined, the only difference being that the rollers used for turning up the sides and ends of the covering paper have a longitudinal recess in them to accommodate the projection of the flange card; and instead of a wooden bottom block, a brass or similar material plate is used. This supports and protects the cardboard flange, and gives a sharp edge to the paper where it is wrapped around the projection.

For envelope-wrapped work, the covers are not slotted at all, and what is more, they are glued only along all four edges. The box, therefore, cannot be stuck to the cover initially, because there is no glue in contact with the bottom or sides of the box. The paper is centred on the machine against two lays (side and end) and the box is put on to the block. The block then descends, taking the paper with it into the machine. This movement, which in the tight-wrapped box turns in the ears of the cover, folds the envelope-wrapped cover into two "pleats" at each end. Finally the cover is stuck to the board all the way along the turn-in. The glue may be applied to the cover by a simple hand stencil or, alternatively, automatic gluers are available which are provided with an attachment which wipes the glue off the application roller mechanically everywhere, except where it is to be applied to the edges of the cover.

Another variation is to machine-cover a flanged box so that the cover ends about halfway up the outside of the box, in other words forms a capped flange cover, and then band a piece of contrasting paper around the sides, either by hand or on a banding machine.

Hand-made boxes

The hand-making of fancy boxes is a craft, and requires a great deal of experience and training before it can be satisfactorily performed, particularly the more intricate types. Round boxes are still sometimes made by hand, although mechanization is utilized for the production of large quantities.

Wire-stitched boxes

There is one other category of rigid box which should be mentioned, and that is the wire-stitched box. This is the crudest form of box, and its method of manufacture is very simple, the board being guillotined, bent instead of scored, slotted instead of corner-cut, and then stitched up to make it rigid. It is a purely utilitarian type of box, useful as a means of transport, and to a certain extent for protecting loose articles.

Problems of rigid boxmaking

1. *Variations in the dimensions of board due to shrinkage or expansion*

As far as possible all materials should be reasonably dry, and in equilibrium with the works atmosphere before use, to prevent shrinkage or expansion. The most satisfactory method is to expose the material to the workshop atmosphere for as long as possible before use, but away from any sources of direct heat. This process can be accelerated by hanging the board in specially constructed ranks, often suspended from the ceiling.

2. *Curl of paper due to application of aqueous adhesives*

Curl is often due to the disproportionate expansion of one side of the paper compared with the other, brought about by the application of moisture, with or without heat, when glue is applied. Varnished papers often curl worst of all, due to the coating of varnish being less affected than the plain paper. As yet there is no satisfactory method of controlling curl on all occasions. The automatic gluer with the suction belt helps considerably, but is not always enough. There are several other factors, all of which are relevant on occasions in dealing with curling troubles, for example:

The glue must not be too hot (approximately 60°C).
The water content of the glue must be correct.
The grain of the paper also affects the curl. On a long deep box it is better to have short-grain covers, i.e. covers where the shorter dimension is cut in the machine direction, so that the paper, which will tend to curl more with the grain than across it, will have the tendency minimized on the longest measurement.
The time-lag between applying the glue and handling the glued cover must be sufficient to allow adequate penetration of the glue's moisture. It is common knowledge that paper will often curl rapidly when one side of it is moistened, and then flatten out again after a further period.

3. *Bowing of boxes*

Bowing is often due to damp board being used or, alternatively, due to the cover paper shrinking to a greater extent than the board. Board as thick as possible should be used, to offer the maximum resistance to the tension so produced. Another cause can be the gluing of different-substance papers to either side of a piece of board. Thus, when lining the inside of a board with a fancy paper, and then covering the outside with another material, the substances of the two papers used should be as near to one another as possible, so that the pull exerted on one side is equalled by the resistance offered by the opposite side. Lidded boxes, particularly shallow ones, may have lengths of board folded up "zig-zag" fashion placed on edge inside the box before lidding up. This will help to prevent any inward bowing as the box dries out.

4. *Tearing of papers*

Customers who insist on having boxes covered with weak material because it is cheaper are often deluding themselves. The boxmaker may point this out when the order is placed, but knows that he will experience higher spoilage than normal, and makes an appropriate allowance for wastage.

The rigid box as a package

Disadvantages

1. Relatively high cost compared with folding cartons. This gap tends to widen as labour costs increase, due to the higher labour content of the rigid box compared with the folding carton.
2. High transport costs because of the high volume of air held within the box.
3. The same factor also applies to storage, rigid boxes requiring much space.
4. Rigid boxes cannot be printed pictorially to the same extent as cartons. For example, it is essential to avoid tight-register printing between the sides and the ends of a box since, as the box must be sited manually on the paper, variations are inevitable.
5. A relatively slow rate of production coupled with storage problems make it difficult for a manufacturer to stockpile big quantities in readiness for a large scale advertising campaign for instance.

Advantages

1. The possibilities with regard to the use of covering papers are infinite. By the judicious choice of papers, very pleasing and satisfying results can be obtained.
2. The same comment applies with regard to the variations possible in the design of a rigid box, compared with a folding carton.
3. The protection against crushing rigid boxes afford their contents is nearly always greater than that provided by folding cartons.
4. In conjunction with this last point, the greater rigidity allows the boxes to be stacked to a greater height in the warehouse.
5. For the smaller manufacturer, or the producer of the exclusive type of article, the rigid box is valuable because it can be obtained economically in much smaller quantities than printed cartons, since the initial costs are very low.
6. The most important of all considerations is that the rigid box can be made to sell and present its contents in a fashion which cannot be equalled. It can be displayed on the shop counter, with the lid open, increasing the sales appeal of the article inside. It can be so designed and covered as to be suggestive of a quality and standard of luxury which is difficult to achieve by other types of container.

CHAPTER 6

Solid and Corrugated Fibreboard Cases

A. R. Lott

The first use of corrugated material for packaging purposes was the subject of a patent (No. 122 023) granted to an American named Albert Jones in December 1871. It claimed "A new and improved corrugated packing paper" that could be used for wrapping fragile objects such as glass vials and bottles to "present an elastic surface...more effective to prevent breaking than many thicknesses of the same material in a smooth state".

In August 1874, Oliver Long patented an improvement on this, adding a lining to prevent stretching, etc. The Jones patent was acquired by Henry D. Norris and the Long patent by Robert Gair. At about the same time, Robert Thompson was also making a cork or shredded-paper lined packing material and, in 1875, Thompson and Norris combined forces to develop these materials.

Between 1878 and 1888, a legal battle developed between Gair and the Thompson and Norris combine, the latter charging the former with infringements. In 1888, a compromise settlement was reached, in which Gair recognized the validity of the Thompson and Norris patents in return for manufacturing rights on a royalty basis. The compromise may have been reached because the litigants had discovered an English patent of 1856 in the names of Healey and Allen concerned with the corrugating of paper and other materials for lining or cushioning the sweat bands of hats.

The best material found in these early days for making corrugated paper was butchers' strawpaper, and the development of machinery to add one or two liners to the corrugated sheet was the principal concern of the manufacturers up to 1890. About this time, the idea of using the material for making cases rather than limiting its use to inner packagings was conceived and, in 1894, the Thompson and Norris Company introduced its "cellular board boxes".

By 20th May, 1895, these boxes were becoming accepted by the carriers, as is evidenced by a letter from a Wells Fargo agent that states, "The new cellular board shipping box which you sent to our office has been tested and I am pleased to say has borne, without damage, such handling as it would probably be called upon to stand in ordinary transportation. I consider it superior to the wooden crates commonly used around pasteboard boxes. Our drivers will be authorized to receive goods packed in these boxes for transportation."

It is interesting to speculate on the tests that were made and to note that they would seem to be performance tests. The prominence gained by the successful use of these boxes for express shipment, together with the expiry of the original patents, left the way clear for rapid development.

There was considerable secrecy in the early days about the methods of making corrugated paper, but it seems likely that the idea was copied from the fluted rolls used in laundries for ruffled lace collars. The sheet of paper was first dipped in water and then passed through a pair of meshed rolls driven by a hand crank and heated, first by gas jets and later by steam. One of the first machines made produced 6 m of fluted paper per minute about 0·6 m wide.

Considerable advances in technique have been made, and today a modern corrugating board machine can produce double-lined board 2·5 m wide at speeds in excess of 300 m/min.

Solid fibreboard cases were first produced in the United States about 1902 and were introduced to the United Kingdom some seven years later.

The use of both types of case was influenced by what is now known as the Pridham decision in 1914. The Interstate Commerce Commission of the United States at this time made a ruling that the railways would no longer discriminate against fibreboard cases, and that they could be carried for the same charge as timber cases. From this time onwards the development of fibreboard was rapid and in the United Kingdom fibreboard cases account for 15% of the total spent on packaging. A ratio of about 2 : 1 existed for some time between the production of corrugated as against solid fibreboard for packaging purposes, but this has changed in favour of corrugated board.

Materials

(a) *Solid fibreboard*

Solid fibreboard is composed of paperboard (usually chipboard) lined

PAPER-BASED PACKAGING

on one or both faces with kraft or similar paper between 0·13 and 0·30 mm thick. The total caliper of the lined board ranges from 0·80 to 2·8 mm.

(b) *Corrugated fibreboard*

Corrugated fibreboard consists essentially of two flat parallel sheets (liners) of paperboard, with a central fluted or corrugated sheet between them. The combined board is held together with adhesive applied to the crests of the fluted sheet. The materials used for any of the three basic components can be in a variety of weights and type, and the flute configuration can be varied.

In the United Kingdom corrugated fibreboard is available commercially with liners of "grammage" between 125 and 410 g m^{-2} and fluting medium of nominal 113 or 127 g m^{-2} is used. The three-component material already described is known as single-wall board. Double-wall board consists of two fluted layers, separated by a flat sheet and faced on both sides with a liner (i.e. 5 components). For heavy-duty application triple-wall board is also made, consisting of 3 fluted layers and 4 flat sheets.

The corrugating medium—materials, flute height and configuration

As previously mentioned, butchers' strawpaper was the first material used for corrugating medium, the stiffness of the material being approximately what was desirable in those early days. A modern corrugating medium must have properties that can be classified under three heads:

Combining properties. The characteristics of the sheet must be such as to allow it to pass easily through the corrugator and to accept the flute configuration, then to be adhered to both liners at speeds in excess of 300 m/min.

Conversion properties. After combining into board with the liners, the corrugating medium must possess properties that will allow it to perform satisfactorily in respect of scoring, both at the take-off end of the corrugator and in the printer-slotter. It must also successfully resist the stresses induced during these conversion processes.

Case performance properties. The ability of a corrugated case to remain a rigid container when subjected to the normal hazards of transport is dependent, to a large extent, on the ability of the fluiting medium to keep the liners apart and thus retain the stiffness of the board.

The fluting structure used for commercial corrugated fibreboard consists of four size ranges. In each size range the material should have an approximately sinusoidal fluting form. The fluting ranges are given in Table 6.1.

A, B and C flute corrugated fibreboards are widely used industrially for cases for the transit of goods. E flute board is widely used in display

Table 6.1. Fluting ranges for corrugated fibreboard

Flute configuration	No. of flutes per metre	Flute height (mm)	Minimum flat crush (N m^{-2})
A (coarse)	104–125	4·5–4·7	140
B (fine)	150–184	2·1–2·9	180
C (medium)	120–145	3·5–3·7	165
E (very fine)	275–310	1·15–1·65	485

cases and similar applications, usually combined with high-quality printed liners. Several types of paper are used for manufacturing commercial corrugating medium.

(a) Semi-chemical papers made by treating wood chips with chemicals to achieve pulp of the desired properties.
(b) Straw papers made from furnishes of between 25% and 75% straw with various quantities and grades of waste pulp.
(c) Kraft paper used in the United Kingdom on some grades of weather-resistant board.
(d) Secondary fibre, chip or waste papers of various grades can be used in their own right after treatment or be added to any of the materials above to give a balance of properties at a moderate cost.

The liners

Three main types of liner are used: pure kraft, the so-called jute or test liner (which consists of a lower grade of pulp, frequently waste, backed with a pure kraft sheet) and chip, straw or bogus papers (which are principally used for making fittings). Chipboards are sometimes combined with a thin kraft sheet, or may be bleached or dyed to look like kraft.

The required liner properties for modern corrugated board can be divided into the same three groups as with the fluting medium, and similar considerations will apply.

Boards

Single-wall board is commercially available in A, B, C and E flute construction. Double-wall board is normally available commercially in AB, CB, AA and AC fluting combinations. Triple-wall board is usually produced commercially in AAB, CCB and BAE constructions.

Manufacture

Solid fibreboard is made by pasting plies of board (usually chipboard) together, and lining one or both faces with either a kraft or a test sheet. Corrugated board is made by running the fluted sheet through corrugated rollers and applying a facing sheet of either kraft or "test" liner to both sides of the fluted sheet.

Once the board has been produced, the processes of conversion into cases are very similar for both materials. Many variations of style, quality and properties are possible in both materials.

The manufacture of corrugated board and cases

The basic operations carried out in 1890 have not changed in principle.

The fluting medium is plasticized by steam showers (figure 6.1) and a preheating roll, passes into the corrugating rolls, and the tips of the flutes are then coated with adhesive. The first lining sheet, which may be passed over one or more preheating rolls, is stuck to the flute tips. The single-faced board so produced passes up into and over a bridge between the single facer and double-backing unit. This bridge holds a store of material festooned on a belt conveyer to allow time for adjusting the machine at points between the various units of the corrugator and for reloading fresh reels of the components.

The single-faced material is drawn from this "store" and passes, corrugations downwards, over an adhesive applicator that coats the flute tips with adhesive (figure 6.2). The second web of liner passes over preheated rolls, is then stuck to the flute tips between a belt conveyer and the top of a heated plate or "table". The sheet is now no longer flexible, and must be kept flat as it travels over the heated table beneath the "blanket" belt conveyer which holds the board under a slight pressure. After passing through this drying section, the board enters the second half of the maturing section, where it cools as it passes between "blanket" belt conveyers towards the slitting and scoring units and the cut-off.

Thus, the corrugator produces "blanks" cut to the outside dimensions

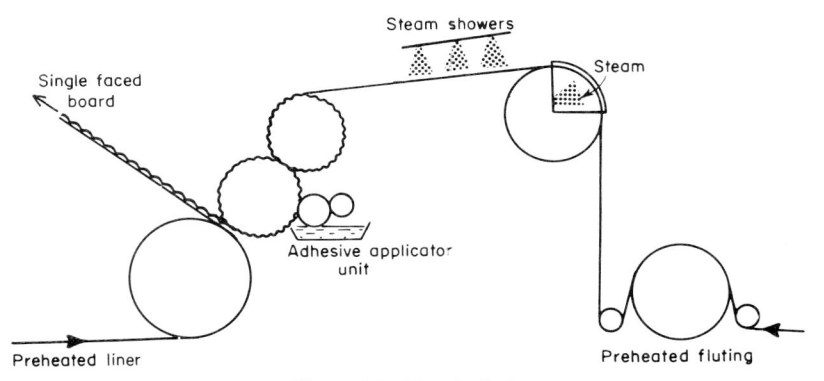

Figure 6.1   The single facer

SOLID AND CORRUGATED FIBREBOARD CASES

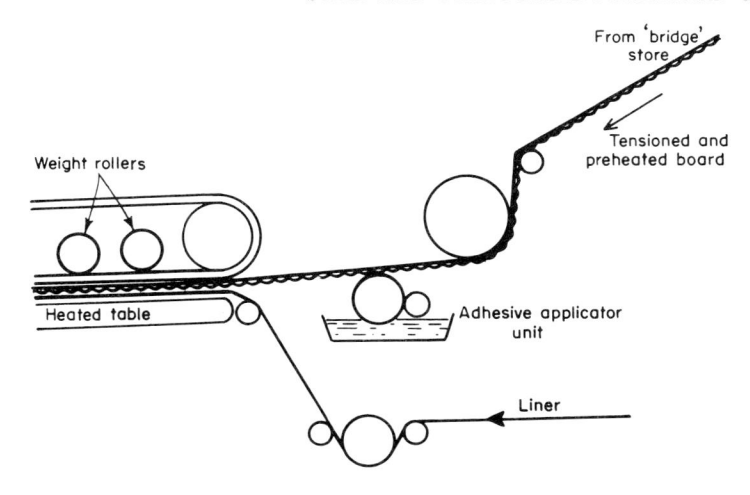

Figure 6.2 The double backer

required and, if the blanks are for cases, with the machine-direction scores already made.

These blanks pass to the printer-slotter (figure 6.3) where they are printed, the cross-direction scores added (the blanks pass through the machine in the cross-direction) and, if stitched or glued, the joining flap cut. The result is the complete printed blank for the familiar one-piece case (slotted container), which requires only the addition of the joint on the fourth corner.

The board enters the machine as a rectangular blank with the two scores produced on the corrugator, and emerges as a printed "cut-out" that merely requires folding and joining to turn it into a case. Figure 6.3 shows

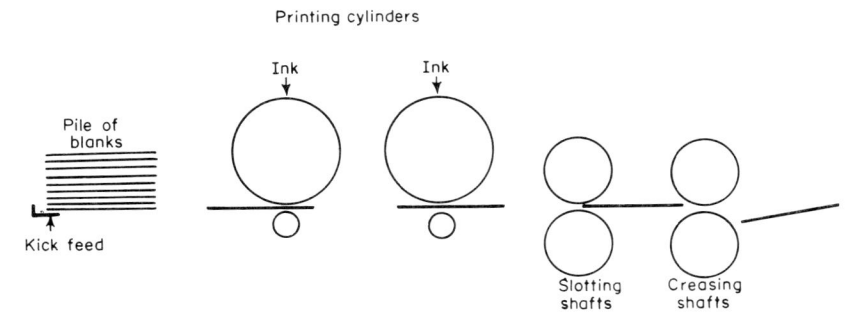

Figure 6.3 The printer-slotter

1.93

PAPER-BASED PACKAGING

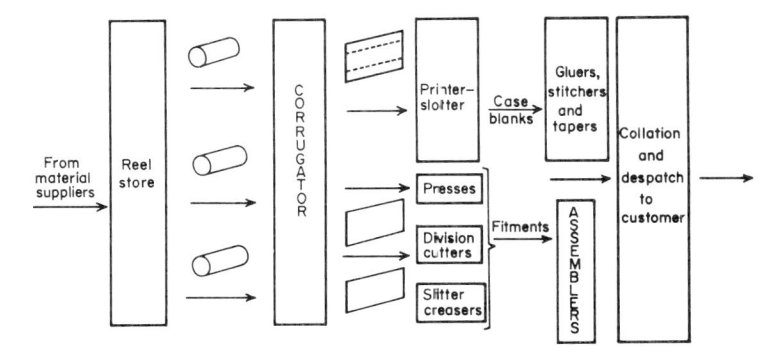

Figure 6.4 Flow chart for container production

a section through a two-colour printer-slotter, but three- or four-colour prints are possible. The blanks are fed one at a time into the machine by a kick or suction feed from the bottom of the pile, and pass between a series of rollers which control their progress. They are first passed beneath the printing cylinders of the rotary letterpress type, using rubber or synthetic dies, and then between two pairs of shafts, where they are scored and slotted.

Three types of joint are possible and, from this point, the blank passes either to the stitcher, the gluing machine or the taping machine.

Other styles of case are possible, and these are produced by platen presses that cut and score the board in a manner similar to that used in making folding boxboard cartons. A flat or circular "forme", holding cutting knives and creasing rules, is pressed on to the board and produces the blank.

Fitments are also produced from board for strengthening the case for supporting various articles packed inside. Figure 6.4 shows a typical flow chart from raw materials to finished products.

The manufacture of solid fibreboard cases

The manufacture of solid fibreboard containers is very similar to that of corrugated containers. The board itself is made on a machine called a *paster*. Most pasters are equipped to run as many as five plies. The board combinations are usually a kraft liner at the top (and sometimes the bottom) and one, two or three plies of chip paper for the centre fillers. The reels of board are placed on reel stands, one reel being pulled from the top, the next from the bottom, thus causing the edges of the web to push against each other to prevent curling. Modern machines have two adhesive stations. In making five-ply board, no adhesive is applied to the top liner; generally, the adhesive is applied to the top and bottom of the first ply of chip, and to the bottom of each of the second and third plies. The top and bottom liners are joined to the fillers just prior to a series of five or more squeeze rolls. Each set of rolls has a higher degree of pressure applied consecutively, in order that the adhesive is not squeezed out along the edges of the board. To keep the board even between each set of rollers, the rolls differ in diameter by approximately 0·15 mm with the larger roll first, making each set of rolls pull just a little faster than those preceding. The trimmed sheets come off the cut-off on a conveyer and are piled in loads ready for the next operation.

Fibre presses perform three operations: creasing, slotting and printing. Some machines are hand-fed, others have a slitting attachment so that there can be two or more blanks cut out at the same time. The original sized sheets are fed on to a conveyor that has spacer bars to guide the blank on to the rolls of the first section of the press. Thus the horizontal scores, the tapered slots and the stitch flaps are accomplished in one operation. Rolls then pull or carry the sheet under the printing cylinders, which have either metal or rubber plates, commonly known as *dies*, which have been previously mounted on a cylinder.

The printing process is exactly the same as that of a corrugated container. So far as this type of board goes, the manufacturer's join is usually stitched.

Closures

To provide a useful transit container, a fibreboard case must have a satisfactory and effective closure. There are three main ways of providing this. The flaps of the case can be secured in place by the application of adhesive tape, by stitching or stapling, or by gluing.

The adhesive tape can be of the self-adhesive type, with a plastics film base (usually unplasticized PVC) or based on kraft paper requiring water to activate the adhesive prior to sticking. For heavy case weights (25 kg or more) heavy-duty adhesive tapes are available based on polypropylene and polyester films or kraft paper, with reinforcing filaments. Adhesive tapes are usually applied along the line where the case flaps meet. For greater strength H-taping along the flap join and the case edge can be used. In most operations, case-sealing tapes are applied by machines which can seal cases with tape at a rate of tens of cases per minute, applying tape to top and bottom of the case in one pass.

Some closures are made to fibreboard cases by applying stitches or staples to the case flaps, thus securing the inner flaps to the outer ones. The stitches can be applied either by hand-operated guns (working on spring pressure or pneumatically) or by machinery similar to that used for stitching the manufacturer's joint. Before deciding on a case closure by stitches, it is necessary to consider the nature of the contained product, since the application of the stitches or vibration of the product against the stitches during transit can cause damage to certain products.

In some instances the closure is made by applying adhesive to the surface of the inner flaps of a fibreboard case, and pressing the outer flaps against them. This can be done either manually or by machine, and is usually the cheapest closure system. Generally, if more than about 2000 cases of a given type are dispatched weekly, it is economic to install a gluing unit.

Suppliers of fibreboard containers are always able and willing to provide information about the best methods of closure, according to the quality of case used and the contents to be packed.

Uses of solid and corrugated board

Solid and corrugated fibreboards are widely used for the packaging of all types of merchandise. Foodstuffs and household goods generally are probably the largest field of use, but considerable quantities are absorbed by such diverse goods as furniture, domestic appliances, chemicals, nails in bulk, light engineering components, etc. It is probably true to say that, apart from heavy machinery and the like, they are a suitable packaging medium for almost any product. The most common packages, however, lie within the weight range 2–50 kg. Special types of fitting and reinforcement extend this range to about 250 kg for normal goods; and for the bulk packaging of powdered commodities, special packages in corrugated board will carry up to 1 tonne of material. Limitations on the use of fibreboard may be imposed by its resistance to climatic conditions. Weatherproof boards of various types are made, but the degree of weather resistance varies from type to type, and also increases the cost.

Generally speaking, fitments to secure the goods within the outer case are constructed from corrugated board. In a number of instances other materials, such as timber, hardboard, insulating boards and various cushioning materials, are employed. As usual, the final cost of the package frequently determines whether or not the particular material used will be successful.

Ever since the early part of the century, when paperboard packing cases began to be used instead of the more traditional timber, wickerwork or steel containers, manufacturers have been comparing the advantages and disadvantages of solid board as against corrugated board.

It will generally be agreed that, from a technical point of view, either solid or corrugated fibreboard can be used to give a satisfactory package for almost any type of commodity, and normally the deciding factor as to which is preferable is the relative cost. In certain instances, however, there may be an advantage in using one material rather than another, and the following considerations which can apply in specific circumstances are worth noting.

1. Solid fibreboard is more resistant to puncturing, both from external hazards encountered in transport, and also when the article packed possesses sharp projections. Very irregularly shaped objects, such as engineering components, steel bars, nuts and bolts, are more likely to puncture a corrugated case than a solid one.
2. Solid fibreboard is more resistant to wet or damp conditions than corrugated fibreboard, although the water resistance of both materials can be improved by various technical means. In particular, wax impregnation and coating with polyethylene are used to improve the weather resistance of board. At least one UK company manufactures a resin-and-glass-fibre-reinforced corrugated board which can be used to produce cases shipped by deck stowage.
3. Although both types of case are principally used as single-journey containers, there are a number of instances where several journeys are required from one container. This type of usage occurs principally where the manufacturer of the product has a closed distribution system in which the chances of loss of an empty container are relatively small. In this type of distribution system the case is delivered with its contents to the retailer, and empty cases from the previous delivery collected at the same time. Solid fibreboard generally gives better results for two main reasons. Firstly, it lasts longer, since the board cannot become crushed and, secondly, slight damage to the container is more easily repaired with gummed paper tape.
4. When the density of the product packed is high, solid fibreboard frequently gives better results, because under these conditions corrugated board is liable to become crushed and to lose its strength.
5. Solid fibreboard in the United Kingdom is normally made from waste pulp lined on one side with a pure kraft sheet. The waste pulp, particularly when the manufacturing mill is located near tidal waters, is liable to contain excessive amounts of chloride, and this can cause trouble where metal articles are packed by promoting corrosion. This disadvantage of solid fibreboard is, to some extent, offset by the fact that corrugated fibreboard bonded with silicate of soda can be equally corrosive if in contact with certain metals, such as aluminium. In both instances, however, arrangements to prevent contact between the board and the metal are fairly easily made.
6. Where the contents of the package do not fill the case completely, corrugated fibreboard may have some advantage because of its greater rigidity. This greater rigidity is also of advantage in respect of the resistance of the container to the crushing forces produced by stacking, and therefore such contents as cartoned goods might be preferably packed in corrugated board.
7. For packing canned goods, opinion is divided; some users consider that solid board is better, since the rims of the cans bury themselves in corrugated board and cause looseness of fit within the container; others prefer corrugated board because of the better-quality interior surfaces. In all probability the answer here is dependent on the method of marketing.
8. When the product to be packed is relatively light in weight, e.g. such commodities as breakfast cereals or soap flakes in cartons, the corrugated fibreboard case is almost invariably cheaper than the lightest-weight solid fibreboard case producible. Efficient protection of the contents is achieved with a very light corrugated board because of the great rigidity of this material.
9. In certain instances, a combination of solid and corrugated is a better solution than either alone. Corrugated cases with solid board divisions, and solid cases with corrugated surrounds, are examples.

In modern industrial packaging applications, corrugated cases are often used with plastics liners for liquids. There is also a growing use of systems using fibreboard and shrink wraps, e.g. in the packaging of liquids in glass bottles it is possible to eliminate dividers between the bottles, holding them tightly together in a tray with a shrink wrap.

The hazards of transport and the assessment of fibreboard packages

The four main hazards of transport (drops and impacts, vibration, vibration under load and compressive forces in stacks) apply almost universally in all forms of transport. The hazards that cause most damage to corrugated cases are drops and compressive forces. It is not intended to deal with the subject of package performance, but it should be borne in mind that, for the best use of fibreboard cases, a careful study of the hazards occurring in the particular distribution system concerned is necessary, so that the laboratory transport test correlates fairly accurately with the actual conditions of transport.

Essentially, four correlations need to be known before a complete understanding of what is required can be obtained. The first correlation is between the actual distribution system and package evaluation tests made in the laboratory on a filled container. A second correlation between the laboratory transport test on the filled container and simpler tests on cases without the merchandise is then required for quality control purposes. The test on the empty containers (or tests, if more than one is required) must then be correlated with tests on the materials used for making the container.

Having ascertained those properties of board that contribute towards the strength of the case, a fourth and final correlation between these properties and the variables of board-making that control them would provide a complete understanding of what is required between the board-making operation and the performance of the package in the distribution system. These factors are not always appreciated.

The first correlation between actual journey hazards and laboratory transport trials is reasonably well understood, and the performance of a package which has been tested in the laboratory can be predicted fairly accurately. The second correlation between the laboratory journey and some simpler test or tests on an empty container is by no means so well understood. For fibreboard cases it would probably be generally agreed that some form of drop test with a compression test is necessary. There is no simple relation here, however, since the conditions that cases meet in transport vary considerably. Nevertheless, the importance of the compression test in evaluating corrugated fibreboard cases is universally recognized.

Workers in this field have largely concerned themselves with relating the compression strength of a sleeve of rectangular cross-section to:

(a) the compression strength of a case, and
(b) the "stiffness" of the board.

The difference between the sleeve and a case is provided by the flap-forming operation at the top and bottom, for the bending scores lower the

SOLID AND CORRUGATED FIBREBOARD CASES

stiffness of the board in this area, allowing it to roll and bend under compression. Thus, the load that the case can sustain is less than that of the corresponding sleeve by an amount dependent on the quality of the materials used in making it and on the efficiency of the score-forming operations.

Case styles

An international fibreboard case code has been agreed between the various fibreboard case-making organizations of many countries. The code is known as the FEFCO/ASSCO Fibreboard Case Code and is widely used as part of a case specification, and for description and ordering purposes.

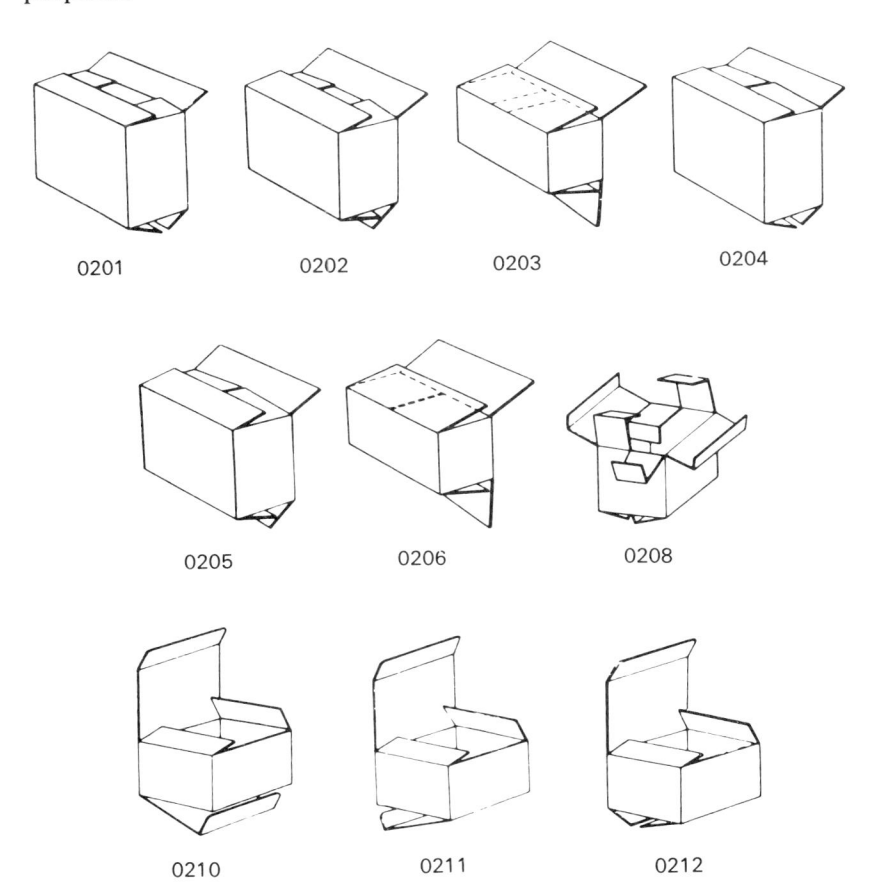

Figure 6.5 The international case code. Styles of fibreboard cases—0200 Code

PAPER-BASED PACKAGING

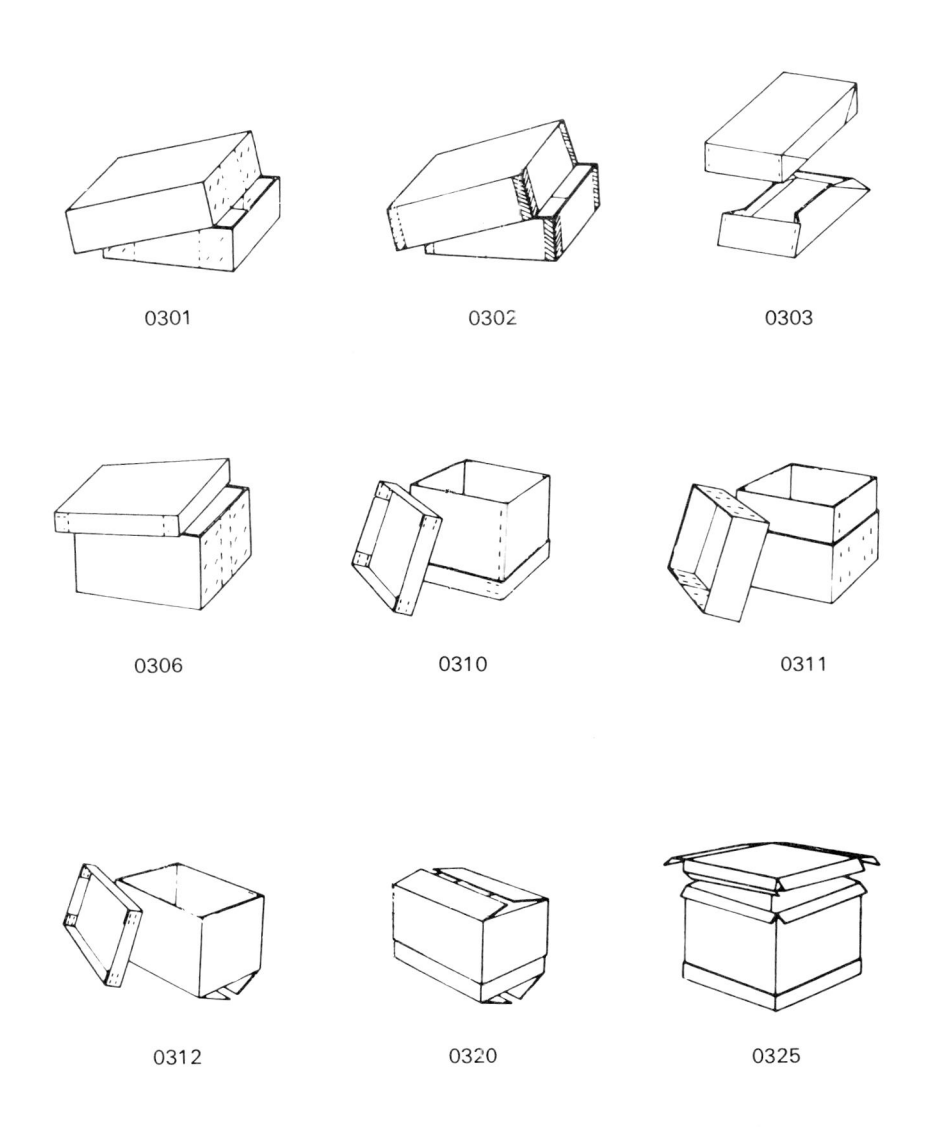

0301

0302

0303

0306

0310

0311

0312

0320

0325

Figure 6.6   Fibreboard case code—styles 0300 Code

SOLID AND CORRUGATED FIBREBOARD CASES

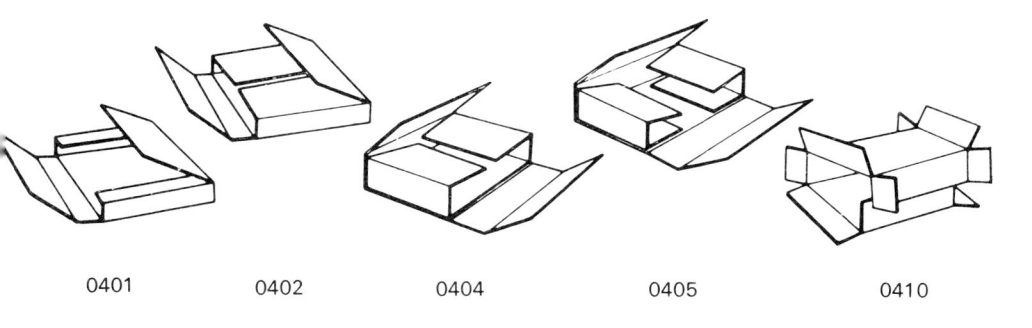

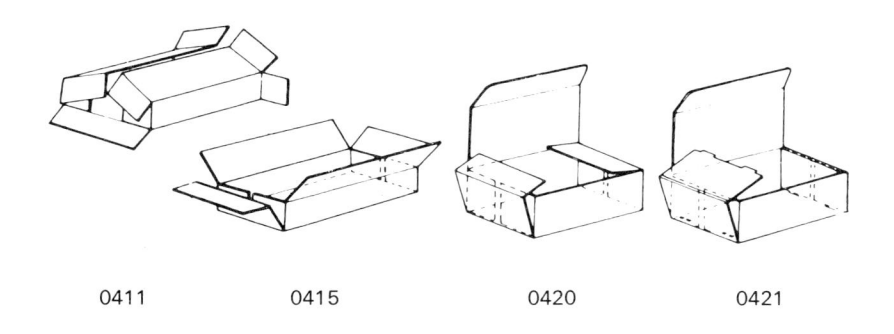

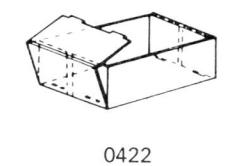

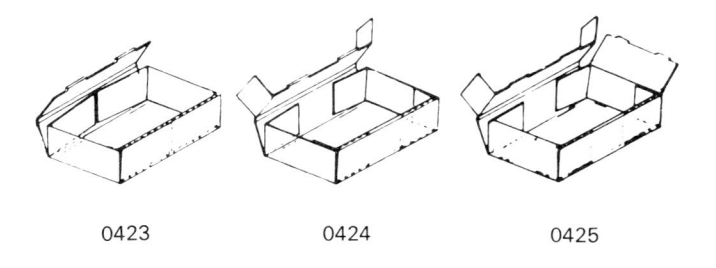

0401   0402   0404   0405   0410

0411   0415   0420   0421

0422

0423   0424   0425

Figure 6.7 Fibreboard case code—styles 0400 Code

PAPER-BASED PACKAGING

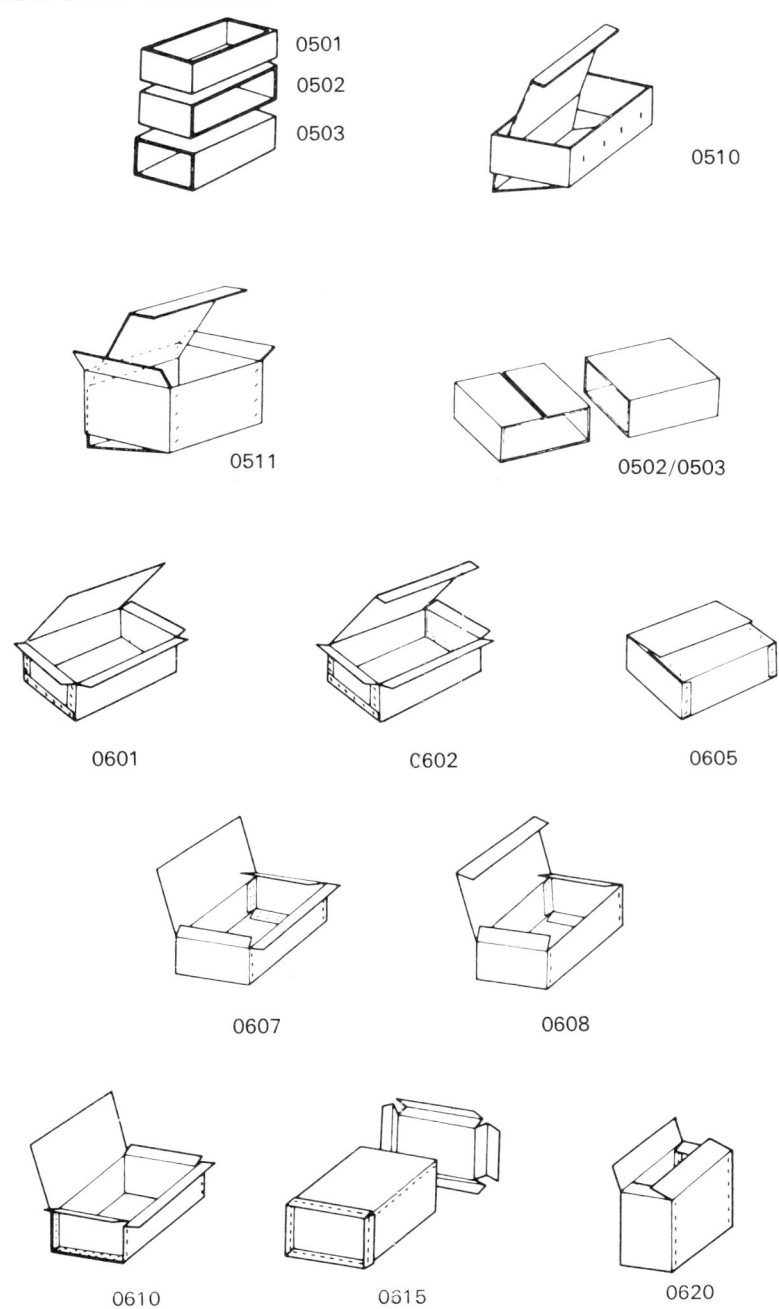

0501
0502
0503
0510
0511
0502/0503
0601
0602
0605
0607
0608
0610
0615
0620

Figure 6.8 Fibreboard case code—styles 0500 and 0600 Code

SOLID AND CORRUGATED FIBREBOARD CASES

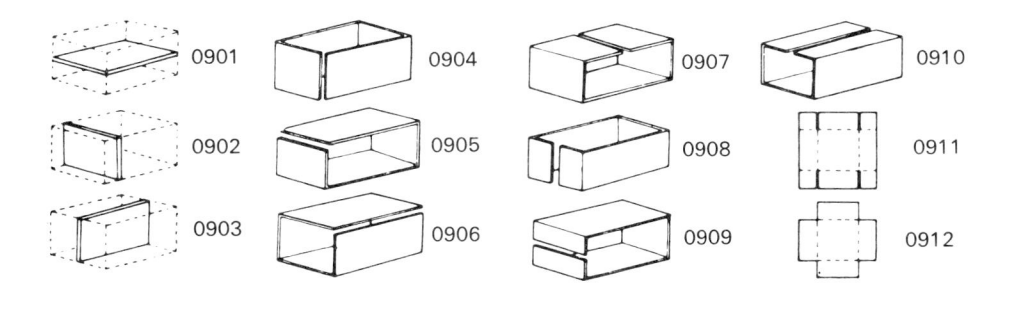

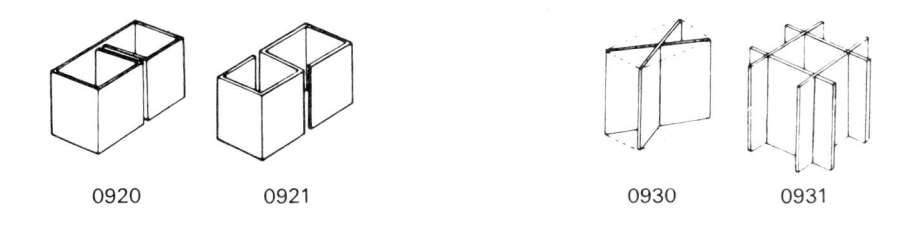

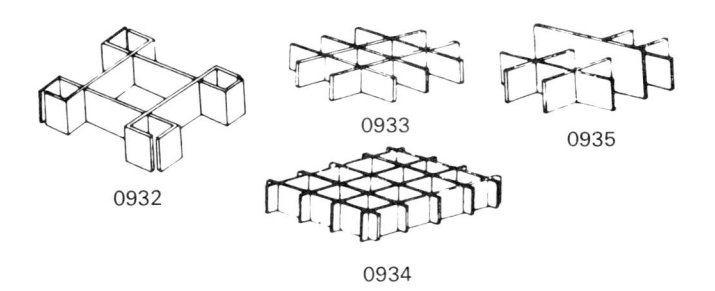

Figure 6.9 Fibreboard case code—fitments (pads, surrounds and dividers)

1.103

PAPER-BASED PACKAGING

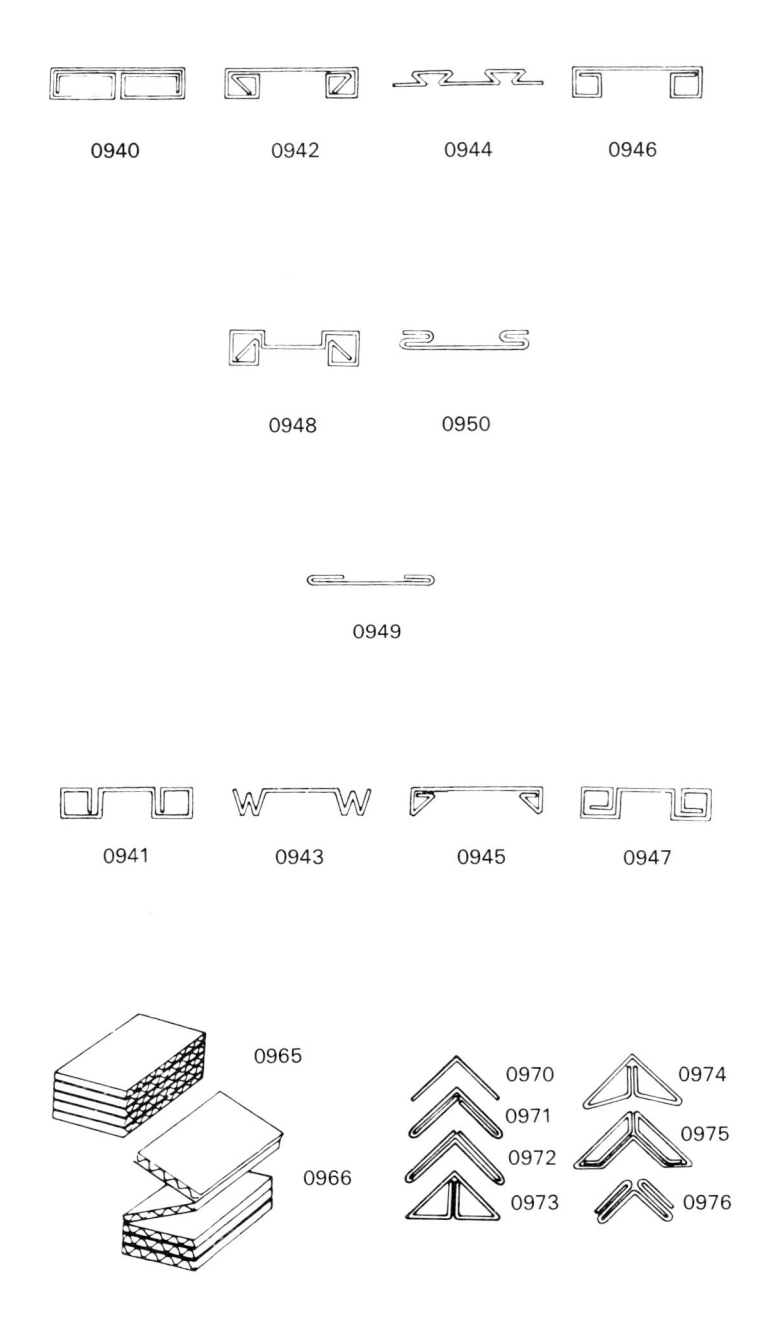

0940   0942   0944   0946

0948   0950

0949

0941   0943   0945   0947

0965
0966
0970
0971
0972
0973
0974
0975
0976

Figure 6.10   Fibreboard case code—fitments (platforms, pads and corner pieces)

Case quality and quality control

The performance of the fibreboard case pack is dependent on the quality of the materials and the efficiency of processing at three stages in production.

(a) Before processing—raw materials: liners, corrugating medium, and adhesives.
(b) After combining into board.
(c) After case-making is finished.

The sheet materials used for making both solid and corrugated board are usually bought on a description, basis weight, and caliper (thickness) specification. The tests normally carried out are (in addition to the basis weight and caliper) for bursting strength and ring stiffness. Some method of assessing behaviour in respect of absorption of water is also used. The Cobb test, the water drop test, and other sizing tests are common for this last property.

Fluting medium is also assessed by the Concora fluting medium test (CMT). It consists essentially of taking a piece of the fluting medium and passing it through heated corrugating rolls at a rate of about 3 metres/minute to produce a fluted sheet. This sheet is then affixed to a backing medium, usually adhesive tape, and is then subjected to a flat crush test.

The combined board is usually assessed in terms of its basis weight, caliper and Mullen burst strength, its materials of construction and the adhesive used to bond its components. Typical values are given in British Standard (BS) 1133, Section 7, and in the Home Trade Packaging Standards published by the Fibreboard Packing Case Manufacturers Association of the UK (FPCMA).

Corrugated cases are often specified in terms of the style, internal dimensions, and board grade and quality. However, any experienced fibreboard case user will inspect closely a sample of the cases coming in to his plant to ensure that the print quality is good (class, correct colour, in register), that the case is squarely and cleanly cut and assembled, that there are no splits in the inner liners, and that the manufacturer's joint has been properly made.

If the right materials have been specified and good-quality processing performed, both corrugated and solidboard fibreboard cases offer economic and effective transit containers for the storage and transport of a wide range of products.

CHAPTER 7

Fibre Drums

Bowater Packaging Ltd.

Fibre drums are non-returnable transit packages. Their size can vary very considerably; the diameter ranges from 180 mm to 600 mm, while the height can be anything from 75 mm to 1·80 m, or more in special instances. Although they are designated one-trip containers, many manufacturers using their own transport regularly between factories have found they can be employed safely on a returnable basis for several journeys.

As the name implies, they are made from fibrous material (paper or paperboard) and since this is of lower tensile strength and rigidity than the materials used for their predecessors (conventional wooden barrels or steel drums) it is essential that the maximum strength be derived from the shape of their final form. Fibre drums are, therefore, always cylindrical with parallel side-walls, and a cross-section taken across the drum at 90° to the vertical axis is a circle. Fibre drums always have the side walls made from fibrous material (paper or paperboard); the ends at the top and the bottom may be constructed of fibrous material, but can equally well be produced in other materials. Steel, wood, plywood and hardboard are the common alternatives, and almost any combination of these may be used; e.g. we can have a fibre base with a steel lid, or a wooden base with a fibre lid.

The side-wall construction is achieved by winding onto a mandrel one or more plies of fibreboard, the plies being firmly adhered to give a solid side-wall. All fibre drums, irrespective of their final form, start in this way from a cylinder or tube of laminated fibreboard. The side-wall material used can be in reel form, several individual sheets or one single sheet. When a reel is used, the side-wall of the drum may be either wound straight (often referred to as convolute winding) or spirally wound.

These two methods are defined as follows:

Spiral winding.—A process of producing a cylindrical tube by winding several plies of paper around a stationary mandrel, each ply being wound in a helix or spiral overlapping, superimposed upon, and bonded to the ply below.

Straight winding.—A process of producing a cylindrical tube by winding several individual rectangular sheets or a web of paper around a rotatable mandrel, so that the several sheets or the several turns of the web are directly superimposed upon and bonded to the sheet or ply below.

Whichever method of winding is used, it is essential that all plies shall be firmly glued together overall to form a solid side-wall. The adhesive used will depend upon the use for which the drum is being made. The normal adhesive used is solium silicate, but starches, dextrines or special waterproof adhesives are also employed.

Spiral winding is a more rapid operation than straight winding, and therefore is usually cheaper, but the method does not give such a strong cylinder as straight winding. Usually spiral cylinders require more layers or plies of paper to give the same strength as straight-wound cylinders. It is also important to note the regulations within overseas countries when exporting goods in fibre drums, e.g. in the United States, drums with spiral-wound side-walls are not acceptable. Adoption of the UN recommendations will extend the US regulations to include fibre drums with spirally-wound side-walls, and side-walls manufactured from a single ply of solid fibreboard.

Drum ends and lids

The materials used for the making of the end are often determined by the type of product to be packed, and also the relevant handling conditions for filling, storing and transporting. It is sufficient to say that, if fibreboard ends are used, and they are made from two or more discs, they must be secured together with a suitable adhesive, and fixed into the ends of the cylinder sufficiently firmly to give the finished drum a satisfactory performance. The method of fixing the ends varies between manufacturers, and basically the method used gives two distinct types of drum. The first is a flush-ended drum in which the flat part of the top and/or base is level with the edge of the drum side-wall (figure 7.1a). In the other construction the flat part of the top or bottom is recessed below the edge of the drum side-wall (figure c).

If steel ends are used, they must be suitably ribbed to give additional rigidity. They can be attached to the side-wall either with adhesive (figure 7.2a) or by crimping the metal (figure 7.2b), thus trapping the

PAPER-BASED PACKAGING

ends of the side-wall firmly by purely mechanical means. Laminated fibre bases can be secured in a similar manner by crimping a metal hoop onto the drum base (figure 7.2c).

Metal lids are secured by several means, e.g. if the lid edge is straight, it can be secured to the side-wall with self-adhesive tape or gummed tape, as in securing a fibre lid of the pillbox type. It is also possible to use lugs—strips of metal fitted at equal intervals near to the top edge of the side-wall, around the periphery, which are turned over to hold the lid in position. This is similar to a traditional closure used on some types of steel drum. The third method of securing a steel lid is by means of a locking band, sometimes known as a *closure ring*. In this method of closure it is necessary for the top edge of the drum side-wall to be

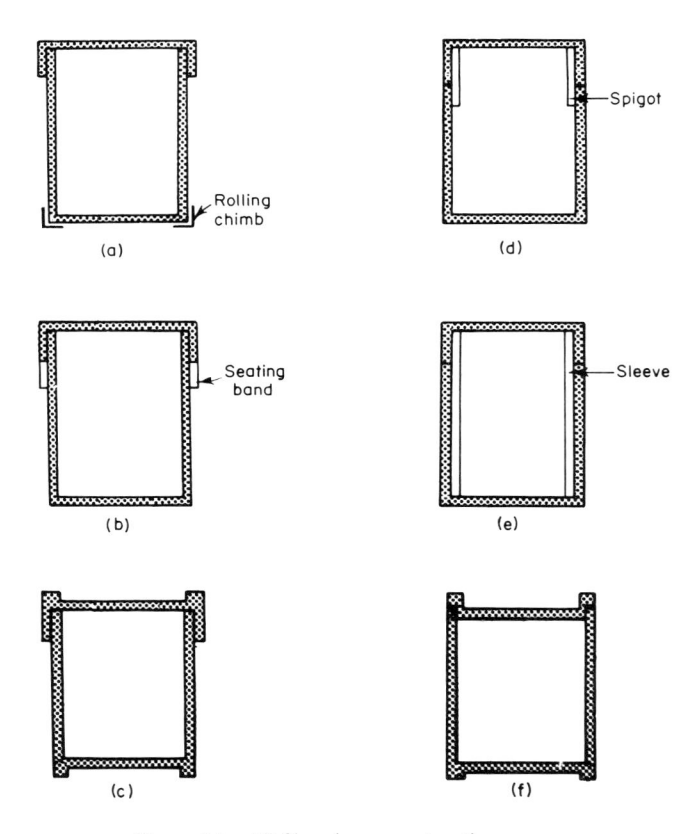

Figure 7.1 All-fibre drum construction
(a) flush-ended
(b) flush-ended with lid seating band
(c) recessed end, pill box lid
(d) flush-sided with spigot
(e) flush-sided with inner sleeve
(f) recessed end, plug lid

FIBRE DRUMS

reinforced with a metal ring, to give a lip under which the locking ring is clamped when closed. The closing of this ring can be effected either by a clip type of closure, or by a lever-action closure (figure 7.3). Both of these methods are also used on steel drums with full-open tops.

With all-fibre lids, various forms of tape can be used, the method selected depending upon the weight of the contents in the drum. For extra-heavy weights it may be necessary to use a gummed cambric tape rather than the self-adhesive plastics tape used for lighter loads.

When sealing drums with any form of tape, it is preferable to have a "flush" side-wall to avoid wrinkling the tape at the juncture between the lid and the drum side-wall. This flush side can be achieved by several methods. The first is by the use of a spigot (figure 7.1d) which consists

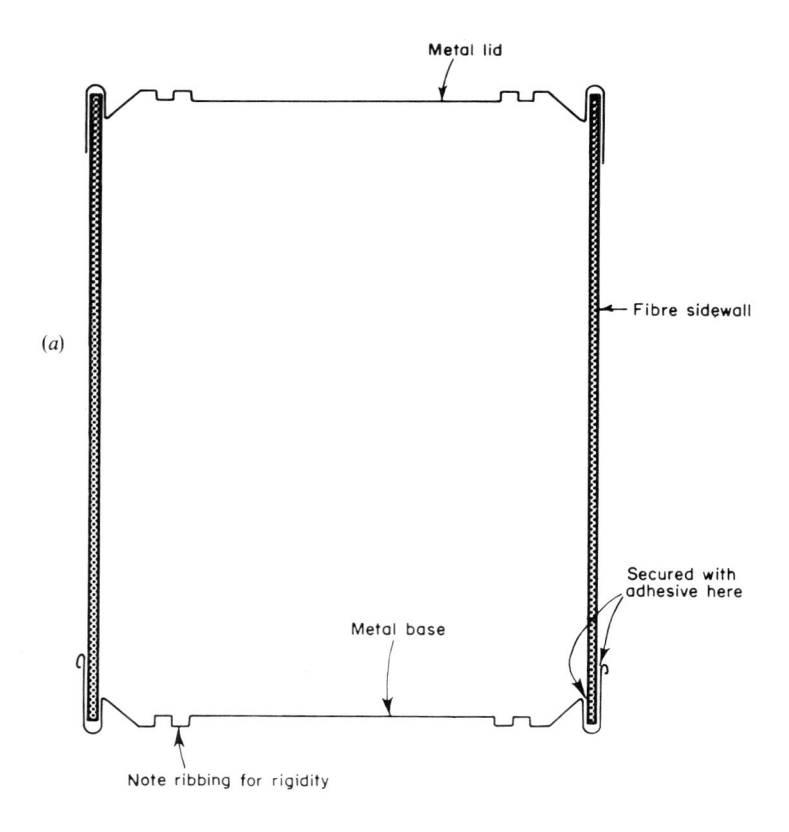

Figure 7.2 (a) drum with metal base and lid

1.109

PAPER-BASED PACKAGING

of a short collar attached to the top end of the drum inside the side-wall, so that it protrudes a few centimetres. A pillbox lid then fits snugly over this spigot so that the side of the lid and the side-wall of the drum give a flush joint. The second method is to insert a snug-fitting sleeve into the full length of the drum (figure 7.1e), which again protrudes a few centimetres. This method is more expensive, since more material

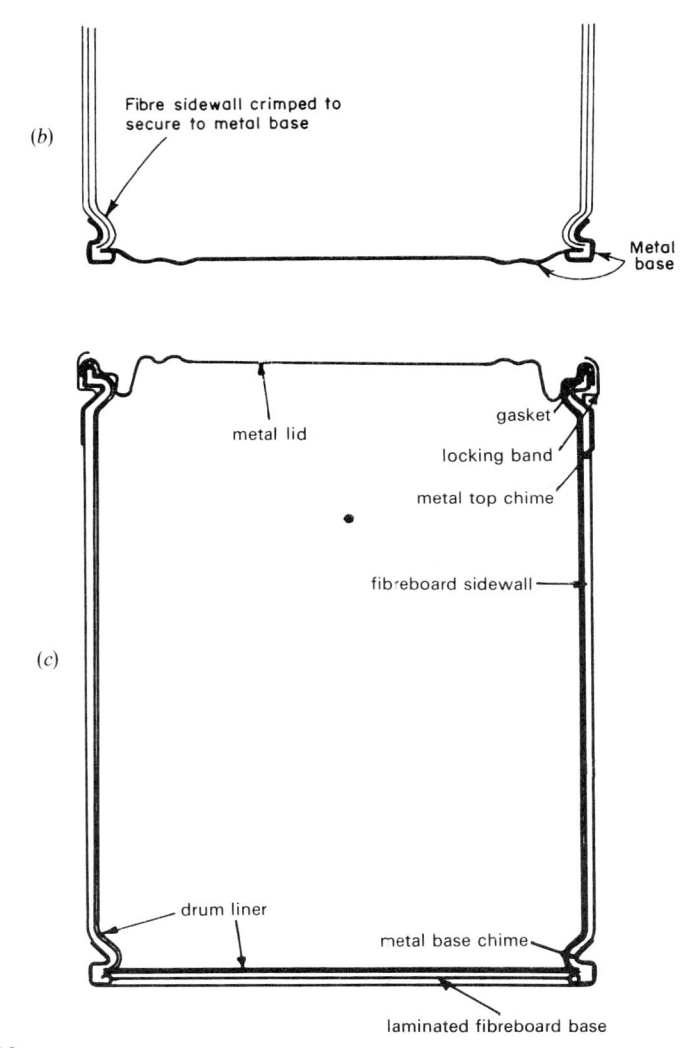

Figure 7.2
(b) fixing a metal base by crimping
(c) fibre drum with laminated base secured with metal hoop (chimb or chime)

is used, but at the same time it gives additional strength to the complete side-wall. The third method, which is the least common, is to use a plug-fitting lid (figure 7.1*f*). This, as the name implies, refers to a lid which pushes inside the drum rather than fits over it. This method does not afford such a strong closure, and is usually limited to drums of small capacity.

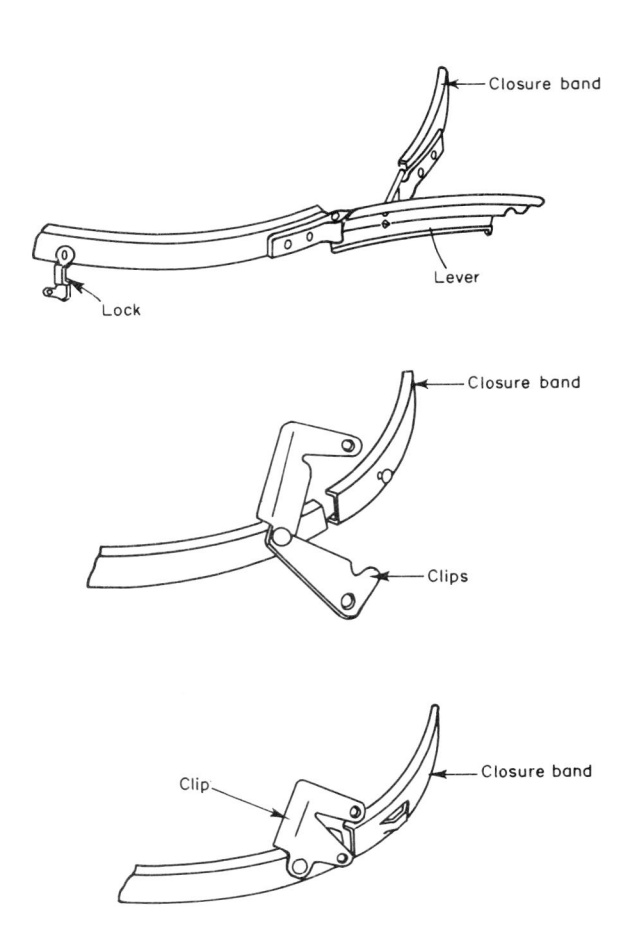

Figure 7.3 Types of catch for closure bands

1.111

Use of fibre drums

Originally fibre drums were used only for the carriage of dry goods, including chemicals, pharmaceuticals and foodstuffs. It was soon found that they are ideal for solid cylindrical objects, e.g. heavy rolls of such material as plastics sheeting, linoleum, expensive papers, aluminium foil, etc. Usually fibre drums with recessed ends are used for these materials, since this decreases the risk of damage to the edges of the rolls.

With improved methods of manufacture, and more styles of drum available, it was found that various refinements can be incorporated to extend the range of products for which drums would be suitable. The first improvement was the introduction of a moisture vapour barrier into the side-wall to permit the packing of hygroscopic, deliquescent or other extremely sensitive products. These barriers may be of bitumen-laminated kraft paper or polyethylene-laminated kraft paper or, for very sensitive products, aluminium foil laminated with bitumen to a kraft paper. The barrier is "buried" in the side-wall of the drum, i.e. it is wound in the side-wall so that it is not immediately adjacent to the inside surface or the outside of the drum. Its position within the side-wall can be varied, and will depend upon the purpose for which the barrier is used.

The second improvement consisted of extending the use of the drum to the carriage of semi-solids, which are usually unsatisfactory in the plain fibre drums.

Fibre drums can also be lined with various materials, e.g. polyethylene aluminium foil and vegetable parchment. This permits drums to be used without the separate loose polythene liner for the carriage of semi-liquid materials. In the United States, a semi-liquid is defined as a material having a minimum viscosity of 5000 cP at temperatures up to 38 °C, exclusive of material normally shipped under refrigeration, in which case the viscosity is measured at the shipping temperatures. If the material contains solids susceptible to phase separation, the viscosity measurement is determined on the liquid component when this is greater than 10% by weight. So far no agreement has been reached in the United Kingdom concerning a final definition of a semi-liquid, but the tendency is to follow the American one. Excellent examples of materials which fall within this definition are lard and shortening, paste adhesives, certain printing inks and thick syrups (see figure 7.4).

The inside surfaces can also be treated by spraying with paraffin wax or special polyethylene wax blends, or with a group of materials loosely known as *release coatings*, which include silicones, alginates and polyvinyl acetate emulsions. These are used if drums are to carry materials having a melting-point of 50°C or higher, e.g. bitumen, waxes, resins and lanolin. They are all filled after liquefying by heat, and the special interior treatment

FIBRE DRUMS

Figure 7.4 Fibre drums being filled with hot pickles. These drums, made from a combination of fibre-board, steel and plastics, can be used to transport pastes and semi-liquids
The Bowater Organization

prevents them from bonding to the fibre side-walls when the liquid sets solid on cooling. The solidified contents are easily removed by cutting away the container with a sharp knife.

As well as semi-liquids, certain liquids can also be carried if a separate container is used inside the fibre drum, or if the side-wall and base are lined with special duplex linings (laminates of polyethylene-fibre-polyethylene-fibre). The inner container can be of thin-walled plastics, e.g. blow-moulded polyethylene or closed-bag liners made from welded flexible PVC sheet. If duplex linings are used, the lining joint must be secured with a suitable heat-sealed tape, and the base sealed with an inert caulking material. Liquids packed in this way include sodium hypochlorite, liquid detergents, liquid fertilizer and insecticides, and concentrated fruit juices.

A further use of fibre drums is for irregularly shaped solid objects by the use of internal fitments and cushioning materials. Users are finding that a cylindrical package is easier to handle than a square one. The cylindrical drums can be rolled "milk churn" fashion by one person, whereas a square box or wooden case requires at least two persons to lift it, or some form of mechanical handling.

The external surfaces of fibre drums are decorated when required by spray painting or varnishing, and by printing in various colours. The printing method is usually silk screen. The painted surface not only decorates but also provides a certain resistance to moisture.

Specifications

A British Standard for fibreboard drums for export (BS 1596:1961) provides a minimum specification and a series of performance tests for the carriage of up to 250 kg.

The British Railways Board require that fibre drums to carry dangerous goods be approved by the British Rail Research Department and that approved drums are stamped with a copy of the appropriate certificate of approval. Fibre drums to carry non-dangerous goods by British Rail should be constructed to BS 1133 (Packaging Code) section 7.

Two points to be remembered when using fibre drums:

(a) It is extremely important that drums should always be stored (both when filled or empty) in an upright position, and not on their sides.
(b) Fibre drums should not be stored in the open air or exposed to the elements, unless the manufacturer has deliberately incorporated special features to make them waterproof.

CHAPTER 8

Moulded Pulp Containers

Gordon J. P. Porteous

Moulded pulp containers may be defined as articles moulded from a mixture of water and any type of fibrous material capable of being treated by normal paper-making processes. In practice these raw materials may be divided into two main classes as follows:

1 Mixtures in varying proportions of virgin mechanical and chemical wood pulp, either with or without the addition of other waterproofing, hardening, colouring or other materials.
2 Wastepaper pulps, either with or without the addition of materials as above.

The forming process used in the production of moulded pulp containers is, in many respects, similar to paper-making but with the important difference that, whereas paper is formed on a travelling wire screen, moulded pulp containers are made in a mould fitted with a screen formed to the shape and profile of the container to be produced.

Manufacture

There are two distinct moulding processes employed for the production of pulp containers. These are known as *pressure injection* and *suction moulding* respectively. Although these two processes produce articles having different characteristics, the preparation of the raw material is basically similar for both, and follows the same lines as that used in the paper-making industry.

In the pressure injection process the mouldings are produced on semiautomatic machines. Each machine is fitted with a mould normally having six sections of which five are movable. A measured amount of pulp and water mixture is admitted into the mould, and the article is formed by blowing in air at a pressure of approximately 4 kg/cm^2 (50 lb/in^2) and at a temperature of approximately 480°C. The moulded articles leave the mould containing 45–50% of moisture, and are subjected to an after-drying operation which also sterilizes the mouldings.

Pulp mouldings produced by the pressure injection process combine great strength and rigidity with light weight. The thickness of the walls is normally approximately $2\frac{1}{2}$mm, but this may be increased or decreased at will. It is possible to produce mouldings of almost any shape, and this point is illustrated by the fact that large quantities of novelty articles such as toys, animals, hats, etc., are produced by this method.

Mouldings can also be produced with special characteristics, such as waterproofness or wet strength, by adding suitable materials to the water-pulp mixture.

Before an article can be produced by any type of moulding process, it is first of all necessary to make a suitable mould. In the pressure injection process one mould is the minimum. If very large quantities of mouldings are to be produced, multiple pocket moulds are employed; but the process is also well adapted to the production of relatively small quantities of not less than 5000 of any one article. It will be appreciated that the cost of the mould has to be spread over the number of articles produced, and hence the greater the quantity the lower the tooling charge per unit.

In the suction process the pulp mixture is deposited and formed to the shape of the mould by the application of a partial vacuum to one side of the mould screen. As compared with the pressure injection process, the formed article as it leaves the mould contains a very much higher percentage of moisture, generally of the order of 85%. This residual moisture has to be removed by an after-drying process. For certain applications an after-pressing process, using heated dies, gives a stiffer and smoother article which can more readily be colour-printed with the customer's name, trademark, etc.

The suction process is normally operated by fully-automatic moulding machines which require a minimum of two forming moulds and one transfer mould. It is general, however, that several moulds are used, anything up to 24 per machine, dependent on machine size and article to be manufactured. The tool costs involved are therefore relatively high, and the process is normally employed only for the production of articles required in very large numbers. Moreover, by virtue of the design of the machines, the process is mainly used for the production of open-top tray-type containers.

Choice of manufacturing process

The choice of moulding process for any given article is largely determined by the considerations set out above. Some articles can be made to best advantage by pressure injection and others by the suction process.

Pressure-injection moulded-pulp articles are particularly well suited to the packing of glass bottles and jars. Examples of successful use are the packing of whisky for export. This form of packing consists of a pulp sleeve moulded to the profile of the bottles which are packed head to tail in a wood or fibreboard case. This method enables a considerable amount of space to be saved.

Another successful type of pulp container is designed to hold and protect Winchester quart and other bottles of various capacities from one ounce to one gallon (30 ml to 5 litres), used by the essence, perfumery and similar trades. These containers have a square or rectangular cross-section and the bottles are supported on specially designed panels which prevent the transmission of shock to the bottle.

Pressure-moulded containers are also used in very large quantities for packing and protecting electrical and engineering components, for the protection of highly finished machined parts, and for inter-departmental transit during manufacture.

Manufacturers' names, decorative designs, trade marks and other features can be embossed on the article at no extra cost. The containers may also be labelled, coloured by dyeing the pulp, or finished in any desired style by means of a spray gun or other decorative process.

A more recent application of the suction-moulded process is the manufacture of disposable bedpan liners, kidney bowls, etc., for hospital use.

The same method of production is used as in other suction mouldings and, with the addition of an emulsion wax, a satisfactory degree of waterproofing is achieved.

After use, the soiled articles are macerated in specially designed disposal machines situated at various points in the hospital, and the slurry fed into the sewage system.

There is also an application in the pressure-moulding process for the production of disposable hospital equipment. This takes the form of moulded urine bottles. Here the same additives are used as in the suction process, and the ultimate use and disposal are the same.

Other types of moulded-pulp article are also produced in the form of novelties, such as animals, dolls and other complicated shapes. Further use for the process is the manufacture of specially-shaped collecting boxes for charitable societies. Well-known forms of moulded pulp articles made by the suction-moulding process are egg flats, food trays, and many other forms of tray-shaped articles for packing fruit and other commodities.

In summing up, for specialized shapes requiring a high degree of protection and a low tooling cost in relation to quantity, the pressure-moulding process is most suitable, but for large quantities of mouldings with a low unit cost the suction process would be selected.

PART TWO
METAL-BASED PACKAGING

CHAPTER 1

Metal Packaging— The Basic Materials

F. A. Paine

The main metals used in packaging are mild steel sheet, tinplate, terne plate, galvanized mild steel sheet, stainless steel, aluminium alloys and aluminium.

Mild steel plate, usually cold-reduced strip-mill steel, is the principal metal-drum-making material. Terne plate, a mild steel coated with a tin/lead alloy, and galvanized steel, with an electrolytic deposit of zinc, are also employed. For special contents aluminium alloy sheets, commercially pure aluminium sheet, and stainless steel (normally an 18–8 nickel chrome steel) are used for drums and kegs.

Tinplate

Tinplate is the principal material for metal boxes and cans.

Tinplate is mild steel (i.e. low-carbon steel) coated on both sides with tin. The base steel plate or strip is manufactured by rolling hot steel ingots down to a strip with a thickness of 0·07 in (1·8 mm). The strip is then pickled continuously in a bath of hot dilute sulphuric acid to a finished gauge of 0·006 to 0·020 in (0·15–0·50 mm). The sheet is finally annealed and temper-rolled to impart the required hardness and surface finish.

Since the early 1960s economies have been produced by reducing the thickness of the steel base. These economies depend on the fact that further cold reduction of the sheet produces a material having a greater intrinsic stiffness; hence a thinner sheet can be used for some applications. The plate is then known as 2CR (double cold reduced) or DCR (double reduced).

The tin coating was originally applied by running the steel plate through a bath of molten tin (hot dipping process). Although this method is still employed, the greater bulk is now made by a continuous electro-plating process (giving electrolytic tinplate). Hot dipping gives a comparatively thick coating, with a lower limit of about 22 grams per square metre (11 g/m^2 on each side). This figure corresponds to the old figure of 1 lb per basis box (1 basis box = 31,360 square inches of tinplate, i.e. 62,720 in^2 total surface area).

The electrolytic method is more flexible than hot dipping and produces coatings down to 5·6 g/m^2 (2·8 g/m^2 on each side). In addition, it is possible with the electrolytic process to produce differentially coated tinplate, i.e. tinplate with a different coating weight on either side. Such material is used when the protective requirements for the outside and inside of a can differ. The lowest feasible tin coating can then be used for each surface, leading to overall reduction in tin coating costs. Manufacture of electrolytic tinplate is a highly technical operation, requiring extensive initial capital investment. In spite of this, it has almost entirely ousted hot-dipped tinplate for can manufacture, because of the savings in tin coating costs.

Hot-dipped tinplate possesses a naturally bright finish, whereas electrolytic tinning produces a rather dull coating. Electrolytic tinplate can, however, be brightened by heating, momentarily, either in a bath of hot oil or by electrical induction. This process is known as *flow brightening*. Flow brightening not only improves appearance but also the resistance of the plate to corrosion. The dull mat finish of the as-plated sheet is due to a porous finish, and the protective properties of the coating are much improved by melting the tin to give a more coherent finish. After flow brightening, electrolytic tin coatings are treated to remove any tin oxide formed during the process. They are then treated in chromic acid, dichromate or chromate/phosphate solutions to stabilize the finish. This is because tin oxide is often affected by subsequent storage or baking of the tinplate.

The high-speed equipment now used for can making necessitates the use of a film of lubricant on the tinplate surface. Cotton seed oil, or synthetic oils such as di-octyl sebacate or di-butyl sebacate are normally used. The oil film is extremely thin—of the order of a few tenths of a millionth of a millimetre—and the lubricant must be compatible with any subsequent lacquer coating used.

Tin-free steel (TFS)

The final step in tin economy is the elimination of tin entirely, hence the name *tin-free steel*. This material is produced by electrolytically coating mild steel plate with a chromium/chromium oxide film. The process was developed in Japan and one version, known as Hi-Top, is being made in the United Kingdom. The chromium/chromium oxide layer is even thinner than the thinnest tin coatings normally used, and must be lacquered before it can be used in the manufacture of containers. It is, however, satisfactory for protecting the steel from rusting during transit and storage prior to can manufacture.

The cost of TFS is lower than that of tinplate, but is increased by the necessity for lacquer coating. If plain tinplate can be used for packaging a particular product, therefore, it will normally be cheaper to use tinplate. On the other hand, if the tinplate has to be lacquered, then lacquered TFS will normally be cheaper.

Blackplate

Blackplate is the name given to the base plate of mild steel before tinning. It can be used for container manufacture, using the techniques of welding or cementing for the production of the side seam. Both surfaces have to be protected with some sort of coating, otherwise rusting easily occurs. In general, blackplate is very little used today.

Aluminium

Aluminium and its alloys have long been used for the manufacture of rigid containers, although not to the same extent as tinplate. Aluminium is also used in the form of foil (both alone and laminated to other materials), and for collapsible tubes.

Rigid cans

The basic fact about aluminium that conditions its use in container manufacture is that it cannot be soldered by any commercially acceptable technique. Rigid containers are made by two basic techniques, therefore, neither of which involves the use of soldered joints. These are:

(a) impact extrusion
(b) drawing or stamping from sheet.

Impact extrusion

A circular disc (or slug) of metal, about 2·5 to 5·0 mm thick and of a diameter approximately equal to that of the required finished container, is held in a shallow die, and struck by a punch. Some of the material flows up and around the outside of the punch, forming the wall of the container, while the metal remaining in the die forms the base of the container. After the can body has been pressed, the base is given an inward dome, by a forming operation, in order to give a stable base. The container wall is then cut to the correct length, and either flanged for subsequent seaming, given a threaded profile to take a screw cap, or given an external bead profile to take a plastics snap-on cap. The containers are either lacquered or anodized after manufacture in order to protect them from corrosion.

If a ring slug is substituted for the circular disc in the production method, then by using an appropriately shaped die, a one-inch aperture surrounded by a solid bead can be obtained, instead of the solid base. This type of container is used for aerosol manufacture. The open-bottom end is flanged as before, and a tinplate or aluminium end is seamed on. Aluminium is the preferred material because of the possibility of electrolytic corrosion at the junction of the tinplate and aluminium.

Drawing or stamping from sheet

For fairly shallow draws, the sheet or strip can be pre-lacquered. This pre-lacquered strip is fed into a press (or a series of presses) where a metal blank is cut out, formed to the shape of the can, and finally trimmed. This process is used for such cans as the rectangular fish can, circular paste cans, or for the containers used to hold 50 cigarettes.

Deeper-drawn cans are manufactured in stages. One example of multiple-stage manufacture is the drawing and ironing process. The aluminium alloy sheet is fed into a press which blanks it out and shapes it into cups. Each cup is then drawn to reduce its diameter and increase the height. The thickness of the cup wall is reduced by about 70%, whereas the bottom of the cup remains substantially the same. The container is trimmed, washed (to remove lubricant), lacquered internally, and finally printed by a rotary process. The drawing and ironing process is shown schematically in figure 2.7. Before leaving the subject of drawn and ironed cans, it should be noted that this method has also been adapted to the manufacture of tinplate cans.

Foil

Aluminium foil is manufactured by rolling slabs of pure aluminium (99·2 to 99·5% aluminium) down to a thin foil. This rolling process starts by taking an ingot of several tonnes in weight. This is rolled down at an elevated temperature to a slab about 60 metres long. Cold rolling is then carried out by passing the slab through a line of rolling stands until the desired thickness is attained. The thinnest foil used in flexible packaging is around 0·009 mm. Thicker gauges are used for milk bottle tops or for certain semi-rigid containers. The latter may utilize foil of a thickness between 0·1 and 0·2 mm.

The high deformation caused by cold rolling makes the aluminium brittle, reduces its elongation at break, but increases its tensile strength. Aluminium foil in this form is referred to as *hard-temper foil* and is unsuitable for normal wrapping application or for the manufacture of pouches because of its tendency to break at creases. A hard-temper foil is required, however, for the so-called *push-through blister packs* for tablets, in order that the foil should break when the tablet is pushed from the plastics film side of the blister.

Corrosion of tinplate

Billions of tinplate containers are successfully used each year for packaging a wide range of products. Nevertheless, many different sorts of attack on the containers by the packaged product can occur. Many products can attack tin itself, but detinning is preferable, since dissolution of the tin by the product delays attack on the underlying steel. This is important, since the tin coating is a very thin one and is not 100% complete, particularly after the rigours of fabrication into a can on high-speed equipment. Attack on exposed areas of steel, unchecked by preferential attack on the tin, could lead to rapid pinholing of the container. Attack on the tin is not beneficial, however, when it leads to discoloration of the product, or to off-flavours due to metal pickup. Here, the tinplate must be lacquered.

The product may also attack the lead solder where this is present. Highly alkaline products can slowly attack and dissolve lead solder, with consequent leakage at the side seam. Some organic compounds also react chemically with lead solder, resulting in blackening of the solder and the product. In aerosol shaving creams, for example, the presence of stearic acid in the formulation can result in the precipitation of lead stearate along the side seam. When the can is shaken, this precipitate may clog the aerosol valve. The problem may be reduced by lacquering the inside seam. However, perfect coverage of the seam is difficult to obtain under commercial conditions, so that high tin-content solders are often used where such problems occur with lead solders. Another answer to the problem is to use a corrosion inhibitor. Testing is essential, in such instances, because many corrosion inhibitors work only for particular products.

Attack on the base steel normally occurs only at imperfections in the tin coating, at cut edges of the tinplate, or when the tin has been completely dissolved by the product. The use of lacquer coatings, while protecting the sufaces, cannot prevent attack at cut edges. Open-top cans do not present problems of cut edge exposure, but lever lid tins (such as paint tins) have an exposed cut edge at the ring. This is the component seamed to the top of the paint tin body to provide the seating for the push-in lid. Although the cut edge is curled under to keep it out of contact with the product, corrosion can still occur with aqueous products from water vapour in the head space.

Mechanism of corrosion

Corrosion is an electro-chemical phenomenon involving a transfer or displacement of electrons. If two dissimilar metals are immersed in water or some other conducting solution (electrolyte) and then connected, a current will flow and we have a galvanic cell. The electric potential set up between the two metals provides for the transfer of electrons through the wire. (If the water was absolutely pure, only a very small current would be produced.) The two metals in such a cell are called *electrodes*. One is called the *anode*, the other the *cathode*. The flow of current is caused by the flow of electrons from the anode to the cathode.

The loss of electrons from a metal anode leaves the metal below the liquid surface as a positive ion which is soluble. The anode, therefore, slowly dissolves in the electrolyte. With a tinplate container, the two metals in contact with each other are immersed in the packaged product which acts as the electrolyte. If the tin is the anode and the steel is the cathode, then the tin is dissolved while the steel remains untouched. The tin is referred to as a "sacrificial" anode, because by its dissolution it protects the steel. Tin, however, can behave either as the anode or the cathode, depending on the conditions, and this provides a means of controlling the behaviour of the tin.

One of the important factors controlling the electro-chemical behaviour of tin is oxygen. Normally, the concentration of oxygen in the product in a tinplate container is low. Under these conditions the tin is anodic and the steel cathodic—the desirable condition, because the tin is dissolved preferentially and thus the tinplate protects the base steel, even if the coating contains pinholes. At higher oxygen concentration, however, the tin becomes the cathode and the steel the anode, when rapid corrosion of any exposed steel occurs, usually causing perforation of the container.

Polarization

The passage of an electric current through an aqueous electrolyte leads to the formation of hydrogen gas at the negative electrode. Most of the hydrogen escapes, but some gas bubbles cling to the cathode and cause a current to flow in the opposite direction to that originally produced. The effect reduces the overall cell current. This phenomenon is known as *polarization* and in commercial galvanic cells would obviously be a drawback. In a corrosion situation, however, anything which reduces the corrosion current reduces the corrosion rate. Conversely, any factor which reduces polarization will accelerate corrosion. This gives the clue to an explanation of the effect of oxygen dissolved in the product on the corrosion of steel in contact with tin. High oxygen contents in the product react with the hydrogen deposited at the cathode, reduce the polarization, and restore the normal corrosion current.

Oxygen and moisture are the two factors necessary for corrosion of iron, and both must be present simultaneously. The corrosion product of iron (rust) is an oxide of iron. Moisture is necessary for the formation of metallic ions, which then dissolve with subsequent corrosion. Neither dry oxygen nor oxygen-free water cause appreciable corrosion of iron.

Any factor which interferes with the normal function of the electrodes (the components of the container—tin and iron) or the electrolyte (the packaged product) may slow down or even stop the corrosion process. Conversely, any factor which intensifies their function will accelerate the corrosion process.

The materials from which the electrodes are made have an important influence on the rate of corrosion, because they determine the electric potential of the corrosion cell. The common metals can be arranged in order of their tendency to go into solution, and to form ions. This series is known as the *electro-motive series*; metals high up the series (such as sodium) go into solution very readily, while metals at the bottom of the series, such as gold and platinum, do not go into solution at all, and hence do not corrode. Two metals which are far apart produce a larger voltage when they form the electrodes in a galvanic cell than two metals close together, which produce a much smaller one.

Aluminium and iron are far apart and, therefore, the use of aluminium ends on tinplate containers is to be recommended only when contact of the aluminium and iron is prevented by some coating. Otherwise rapid corrosion of aluminium will occur. This holds true only if the product (electrolyte) does not change.

As might be expected, the acidity or alkalinity (pH) of the environment is an important factor in corrosion. As a rule, acid solutions (low pH) are more corrosive than neutral or alkaline ones. However, the actual rate at which a metal corrodes in a solution of given pH depends on several other factors, including oxygen concentration and polarization, which have already been discussed. Three other factors are important from a practical standpoint.

(a) Stress corrosion

Stress corrosion is the term applied to many patterns of corrosive attack in which stress is believed to accelerate corrosion. It often occurs when normal corrosion is almost negligible. Many theories to account for stress corrosion have been suggested, but the most acceptable is that electrochemical action occurs between stress-caused anodic areas and the more cathodic unstressed areas. Since the areas of stress are anodic, corrosion takes place here, and the tensile stresses present open up crevices, thus exposing fresh metal to further attack.

The stresses can be caused by cold working of the metal, or because of a too-severe drawing operation in fabrication.

(b) Presence of inhibitors

Corrosion inhibitors are often added to an environment or electrolyte to reduce corrosion attack. Such materials work by interfering with the functions of the electrolyte or the electrodes, and thus slow down or stop the corrosion process. Inhibitors added to a canned product are designed to reduce or stop the electric current between the tin and steel in the case of tinplate containers.

(c) Passivity

Under certain conditions a metal electrode may cease to dissolve, although it appears unchanged. The metal is said to be in a *passive state*. The attainment of passivity depends on the electrolyte. With metals like iron it is favoured by alkalinity, but other metals such as tungsten become passive more easily in acid solutions.

Passivity can also be produced without the action of an electric current. For example, if iron is dipped into concentrated nitric acid, there is a brief reaction, which rapidly ceases, and the metal becomes passive. Nitric acid is an oxidizing agent, and it is significant that iron can also be rendered passive by heating it in air. The mechanism of passivity seems, therefore, to be allied in some way to oxidation, and it has been found that metals in the passive state have an extremely fine oxide film on the surface. This film acts as a protective barrier and prevents further solution of the metal.

Corrosion testing

One obvious method of assessing the possibility of corrosion is to carry out storage tests on containers and the products concerned. If the conditions of storage are also equivalent to those expected in use, then this method will give good results. The only snag is the time-scale involved. If a shelf-life of two years is envisaged, then a test period of much less than 4–6 months is unlikely to give meaningful results.

Another method is electro-chemical testing, commonly known as *corrosivity testing*. This is based on one of Faraday's laws of electrolysis which states that the amount of decomposition is proportional to the current and to the time for which it flows (i.e. the total quantity of electricity). Conversely, when a given weight of metal dissolves, it gives rise to an equivalent quantity of electricity which flows as an electric current. The magnitude of this electric current can be measured, as can its directional flow. This tells us which of the two metals is dissolving (or corroding) and how rapidly the corrosion is taking place.

The method is applicable to products packed in containers constructed of at least two metals (such as tinplate—one electrode is tin and the other is steel). By this method some very valuable information can be gained, even after a period of only 24 hours, especially when it is evaluated in conjunction with practical experience. Thus, at the end of 24 hours we know which of the two metals is anodic (and hence will corrode) and which is cathodic and being protected. We also know the magnitude of the current, and hence the magnitude of the corrosion to be expected. This method must not be used in isolation, and storage tests should also be carried out in order to verify the results and to determine whether elevated temperatures can reverse the polarity or break down the product in some way.

Lacquer coatings

The basic function of an interior lacquer is protection of the product rather than protection of the container. Lacquers can be beneficial and many prolong the effective shelf-life. No lacquer is known, however, that is effective in all circumstances, and the product must always be taken into consideration, as well as the likely storage conditions. Since lacquer coatings are usually applied to the sheet tinplate prior to can manufacture, the lacquer must be able to withstand the mechanical shocks associated with can-making. In addition, the lacquer must be easy to apply and cure, resistant to the product, provide the barrier required, and be economical in use.

For food or pharmaceuticals, can lacquers must not give any toxic hazard, and must be free from odours and flavours. For products which are particularly susceptible to contact with metals, one additional lacquer coating is given after the can has been manufactured. This second coating is applied by spraying, and serves to cover imperfections in the side seam, or those caused by mechanical damage during manufacture.

There is a wide variety of can lacquers available, including oleoresins, phenolformaldehyde, epoxy resins and vinyls. There are also specialized types, such as one containing zinc oxide. This is used for products containing sulphur-bearing proteins (processed peas, for example). The sulphur would normally cause blackening of the can and its contents, due to iron sulphide formed by attack on the steel at lacquer imperfections. The zinc oxide in the lacquer removes the sulphur, with consequent formation of zinc sulphide, and prevents the blackening reaction.

Lacquers are also applied to the outside of tinplate containers to improve corrosion resistance, particularly if the cans are to be exported to tropical areas, or are likely to be stored under damp conditions. With printed cans, the external decoration acts as a protection against corrosion. It should be noted, however, that external decoration does not cover the soldered side seam, so that a base lacquer may be needed in addition to the decoration in extremely corrosive conditions. Alternatively, the can may be given an external side-stripe of lacquer along the side seams.

CHAPTER 2

Metal Cans

D. W. Price

Almost every food product one can name has, at some time or another, been packed in a metal can. It is accepted as commonplace that tins of paint are found in the shops and cans of oil at the petrol stations. World wide, the metal can is used to convey and distribute the necessities and luxuries of life.

Not many years ago there were some who forecast the demise of the tinplate container, claiming it would be displaced by newer plastics materials or lightweight glass, and that new ways of processing food would no longer require the can in the form we know it.

The metal food can continues to be made in ever-increasing quantities, and the new materials and processes have found their places alongside it for those products for which they are best suited.

Some metal containers have, however, passed out of common use. Cigarette tins are now a novelty or a rarity in the United Kingdom, and instant coffee is now almost entirely in glass jars for retail distribution.

These paragraphs have used the words *tin* and *can* almost indiscriminately to refer to various metal containers. The words have come to mean much the same, and the word *can* is more widely used than in the past. The more subtle differences in meaning may be ignored, and the words are used almost synonymously in the rest of this chapter. Almost but not quite. A hinged-lid cigar tin is surely never a *can* and an all-aluminium beer can is difficult to accept as a *tin*.

Tinplate is still the most common raw material for tins and cans. This is mild steel sheet with a very thin layer of tin on each of its surfaces. Blackplate is not often used, but it is most easily described as tinplate without the tin, i.e. mild steel sheet, available in the same range of thicknesses as tinplate.

METAL-BASED PACKAGING

Tin-free steel is like tinplate, but the tin has been replaced by other corrosion-resistant metals such as chromium.

Aluminium is being used in increasing quantities.

The built-up body construction

Built-up body construction is the form of construction in which a metal container is produced by using more than one piece of material and joining the several pieces together with suitable seams. This distinguishes the method from that which produces a solid-drawn body having no seams, and clearly having been produced from a single piece of material.

The open-top food can is the most familiar use of the built-up form of construction. It is called *open top* because it is supplied to the canner with the top of the can open to receive the product. Top end components are supplied separately to be attached by the canner.

To manufacture this style of can, a rectangular blank of tinplate is mechanically wrapped around a mandrel to form a cylinder. Where the two ends of the blank meet, an interlocked seam is normally made, and this is soldered to make it leak proof, before the bottom end is secured by a double seam. Figure 2.1 shows a section through an interlocked side seam (*a*) when the side seam "hooks" are first brought together around the mandrel and (*b*) after they are hammered flat to secure the seam just prior to soldering.

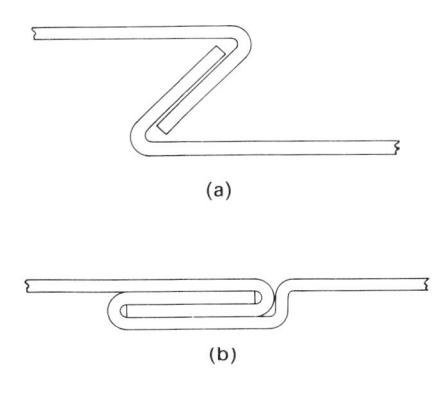

Figure 2.1 Interlocked side seam
(*a*) before hammering   (*b*) finished seam

Figure 2.2 illustrates the stages in the formation of the so-called *double seam* which secures the bottom end to the body. An identical method is used by the canner to secure the top end to the can after filling. Reference to figure 2.1 shows that an interlocked side seam has four

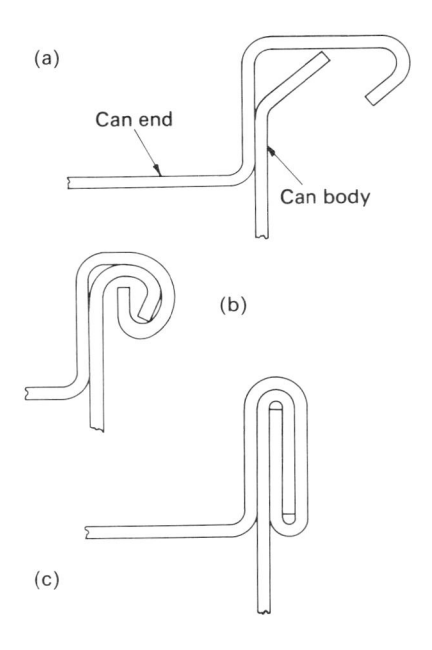

Figure 2.2 Formation of a double seam
(a) end placed on body (b) seam part-formed (c) finished seam

thicknesses of material, and these would need to be bent over into no less than eight thicknesses before the can body is seamed to the bottom (or top) end. This would be quite unacceptable, not only because of the unsightly bump it would leave in the double seam at its junction with the side seam, but because the massive local distortion in the end seam at that point would be a source of leakage.

To avoid this, the body blank is prepared for seaming by having specially shaped notches cut in each of its corners so that, instead of the interlocked seam being formed over the whole length of the can body, the extreme ends are simply overlapped.

Figure 2.3(a) shows what is meant by notches, and at (b) is shown a section along the length of an interlocked side seam, near one end where the flange has been formed to prepare for double-seaming an end. The transition from four to two thicknesses of material can be seen.

Where the two thicknesses of material simply overlap, they are secured by soldering, this being done at the same time as solder is applied to the interlocked part of the seam. The area is called the *lap section* of the seam.

METAL-BASED PACKAGING

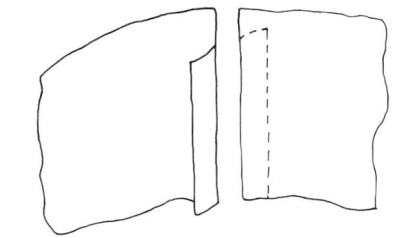

a) Top of can body notched away to make lap.

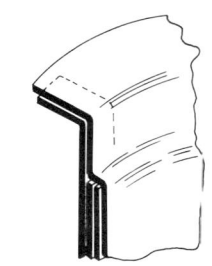

b) Longitudinal section through interlocked seam showing lap joint at top.

Figure 2.3   Transition from four thicknesses to two at the end of the interlocked side seam

Despite this way of reducing the thickness of the side seam at each end, the intersection between the side seam and the end double seam remains a potential source of leakage. To eliminate this possibility, a resilient compound is applied to the whole peripheral area of the can end before seaming. This *lining compound* fills any small interstices which might otherwise occur when the end double seam is made. Figure 2.4 shows where this compound is applied to a can end. Strictly speaking, the lining compound is hardly necessary at any position except where the side seam intersects, but it is easier in high-speed production to apply the compound to the whole periphery of the end.

Treatment of can seams

End double seams

The introduction of end lining compound to assist in sealing double seams is referred to above. The constituents of can-end lining compounds can be varied to suit various circumstances and methods of application, but are mostly natural or synthetic rubbers which are either dispersed in water or dissolved in a suitable solvent. The compound employed is squirted through a small nozzle into the annular area of the end which

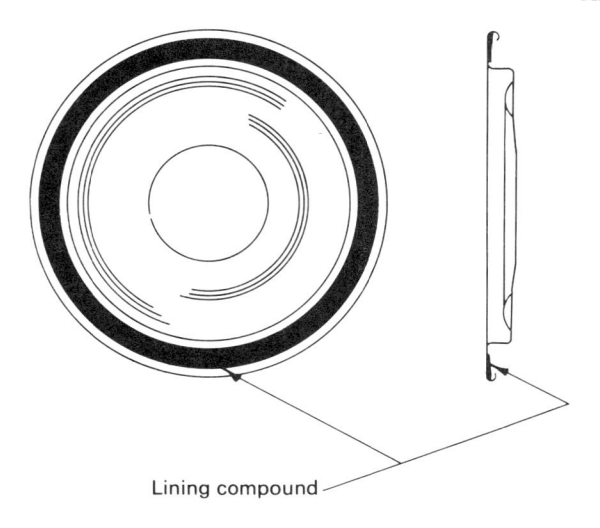

Figure 2.4 Placement of lining compound in can end

will subsequently become part of the double seam (figure 2.4). The end is rotated beneath the nozzle so as to ensure an even spread of compound, which is quite fluid when applied. If water-based compounds are used, the ends are passed through an oven which cures the compound by vulcanizing it after the water has been driven off. The process is irreversible, i.e. the compound cannot be rendered fluid again by warming or by the addition of common solvents once it has passed through the curing operation.

Solvent-based compounds can be allowed to cure in the air.

Can double seams may also be soldered, but this is a practice which is no longer common. It is to be seen on some containers which serve unusual purposes in engineering, such as oil or brake fluid tanks or filter bodies, some of which are required in sufficient quantities to be able to enjoy the economies of can manufacturing techniques. Some large tins, of 5 litres capacity and above, have their end seams soldered, particularly if they may be called upon to hold penetrating or volatile fluids for long periods in an unpredictable or hostile environment.

Interlocked side seam

The most common means of sealing against leakage is to use solder. Means have been developed over many years to undertake side-seam soldering reliably and at very high speed on automatic equipment. The interlocked seam can be made so that the surfaces of the tinplate in the folds are not quite in contact with each other. Hot solder is therefore drawn into the small space by capillary action and, once there, is solidified by being cooled under air blasts or by contact with the atmosphere.

The high costs of tin and lead which go to make solder have stimulated a search for other ways of sealing side seams. Soldering also calls for the use of energy to pre-heat the cans and melt the solder.

An early alternative was rubber solution. If applied to the body blank before the interlocked side seam is hammered flat, it can be effective in filling the joint. Many alternatives and more elaborate materials have been tried, all being generally referred to as *solutions*.

Other materials, which can also be applied to the edges of the body blank just before (or just after) the hooks are formed, are designed to adhere to the metal much more strongly than does solution, and to set harder than solution. Some of these materials, commonly referred to as *cements*, are thermoplastic and can be applied through nozzles as jets of hot liquid. Once placed on the metal, they are immediately trapped and hammered in the interlocked side seam.

Although there is a variety of solutions and cements available to the can maker, the distinction between the two material categories should be stated. Solutions fill the interstices of the interlocked side seam, adhering only strongly enough to stay in place. Cements not only fill the gaps which might cause leakage, they add something to the mechanical strength of the seam by the strength of their bond to the metal surface.

Other types of seams and joints

End seams

The previous section has referred at some length to the double seam and figure 2.5 shows two other ways of securing ends to bodies on built-up tins, compared with the double seam. The double seam is by far the most common way of securing ends to bodies, not only because it is fundamentally easier than other techniques, but because high-speed machinery is available for its production as a result of the exacting technical demands of the food canning industry.

The single seam has its uses and is employed mainly where mechanical strength and liquid-tightness are not vital. Its simpler form makes it attractive for use in tins of unusual cross-section, and it is often to be found on talcum powder tins.

The capped-on end produces a lap-joint around the end. It is always necessary to secure such ends by something other than their frictional fit, and solder is the most usual medium. This type of joint is useful

METAL CANS

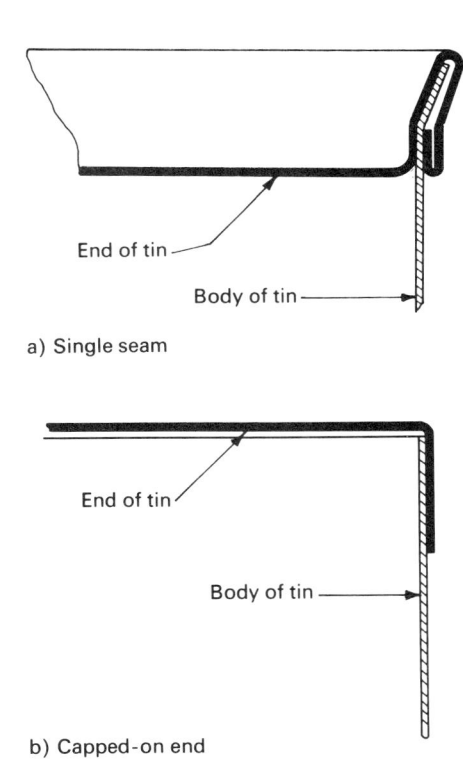

a) Single seam

b) Capped-on end

Figure 2.5   Other ways of securing ends to tins

when a very thin end component has to be fitted to a thicker body material. The technique is not now, however, common in the United Kingdom.

Side seams

Figure 2.6 shows two other types of side seam to be compared with the conventional interlocked seam already discussed at some length. At (*a*) is a variation of the interlocked seam, and this is known as a "Mennen" seam or powder seam. The usual interlocked seam is strong in tension, i.e. it is well able to resist the effects of internal pressure within the can, or any other mechanical forces tending to expand the can from within. It is not as strong in resisting those forces which might tend to crush a tin from the outside. That is where the Mennen seam becomes useful. It is often used on talcum powder tins which have shoulder components tightly fitting on the outside which tend to crush the tin.

METAL-BASED PACKAGING

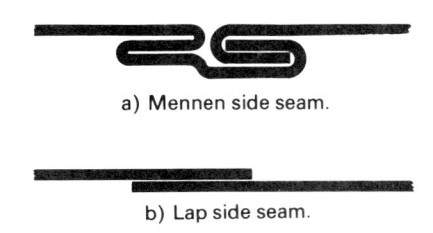

a) Mennen side seam.

b) Lap side seam.

Figure 2.6 Other types of side seam

The lap seam shown in figure 2.6(b) looks deceptively simple, but becomes complicated by the ways in which the two overlapping edges are secured to each other. Soldering is probably the oldest technique, but solder is not a strong material and is not well able to withstand high stresses continuously applied for long periods, such as occur when a can is pressurized or vacuumized. The can maker has, therefore, resorted to other materials; in particular, modern polymeric materials which are able to bond metal to metal and which are often thermoplastic. This makes them capable of hot application to the edges of the body blank by jet or roller, but they solidify by rapid cooling as soon as the overlapped edges are pressed together. This method of side seam formation can be employed on metals other than tinplate, such as tin-free steel or aluminium, both of which defy soldering by traditional techniques.

Lap joints also lend themselves to welding. Resistance welding is used and, although such joints may appear to be continuously welded throughout their length, X-ray examination shows that they are formed by a very large number of overlapping stitch-like welds which give the outward appearance and practical effects of a single continuous weld. Such joints are mechanically strong in tension and compression.

The question of material economy is of ever-increasing importance, and an examination of the various diagrams will show that the sheet metal required to produce a tin of a given peripheral length will vary with the type of seam chosen. The hooks of an interlocked or Mennen side seam are between 3 mm ($\frac{1}{8}''$) and 2·5 mm ($\frac{3}{32}''$) long, and the overlap in an adhesively bonded or welded-lap side seam is 3 mm to 4 mm. In order of material economy, the seams fall in the following order:

1 Lap seam—most economical
2 Interlock seam
3 Mennen seam—least economical

Cans without side seams

Much of the space in earlier sections has been taken up in discussing

2.20

the various ways in which seams may be produced and prevented from leaking. Clearly there would be some advantage if seams could be partly or wholly dispensed with.

Because of the advanced manufacturing techniques which have been developed for high-speed can making, labour costs are far less significant than material costs. Any way of saving material is worth exploring. A can of a particular height and diameter will require a minimum area of sheet metal for its construction. The area can be minimized by reducing the number and complexity of the joints and, once that is achieved, attention falls upon the thickness.

Many tins and cans could have certain of their parts reduced in thickness without undue risk to their main purpose of conveying and preserving their contents. This is particularly true of the bodies of cylindrical cans. Two difficulties have stood in the way of this. Firstly, it is not so easy to produce very thin tinplate with the same assurances of quality as can be given for thicker sheet. Secondly, such thin sheet as we are now contemplating, would have the "feel" of thick paper and would behave in unacceptable ways on high-speed machinery, mainly because of aerodynamic effects which are seldom noticeable on conventional thick material. It is not uncommon to find can-making processes which require tinplate sheets or flat blanks to be pushed through the air at speeds exceeding 350 miles per hour (560 kph).

These and other reasons led to the development of the Drawn and Wall-Ironed Can, often abbreviated to DWI. Although the tooling and machinery to make such cans calls for high precision and advanced technology, the process is fairly simple to understand. It is illustrated in figure 2.7. A shallow tinplate cup is first produced by the long-established technique of drawing from a metal blank. This cup is then forced through several rings or dies, each slightly smaller than the previous one. The punch which does this supports the inside of the cup to stop it collapsing, and consequently the cylindrical side wall is progressively reduced in thickness. After passing through the last dies, the can is removed from the punch and is automatically carried away to be cleaned and printed. Because the material is reduced in thickness, it is increased in area, and the can is therefore made taller as it passes through the die rings. A trimming operation follows the wall-ironing, and this ensures that every can is the same height, with a well-finished open top which will eventually receive the top end component.

Although the process takes a little time to describe, it occurs so rapidly as to appear instantaneous. The high speed and the considerable reduction in wall thickness achieved, demand that copious amounts of lubricant be used in the press operations, and these must be removed by thorough cleaning. Tin on the surface of tinplate acts as a lubricant itself and

2.21

METAL-BASED PACKAGING

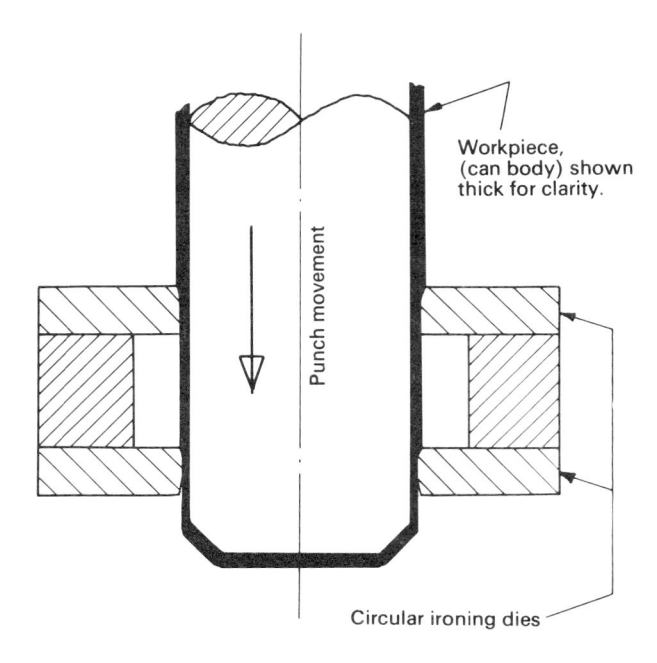

Figure 2.7 The DWI process (drawn and wall-ironed can). A cup-shaped component is first drawn from a round blank. The cup is forced through the ironing rings by the punch. Only two rings are shown, but three or more may be used in practice. Each ring reduces the wall thickness by between 20% and 35%.

assists the operation but, by using tools which have been properly prepared, the process will also handle aluminium.

DWI cans have very thin body walls where the material has been ironed, but the bases remain substantially the same thickness as the stock material fed to the process. Whilst much is gained in the way of material economy by using this process, the can maker is faced with the difficulty of printing on the cylindrical surface of the can after manufacture, whereas conventional can making allows printing to be carried out on flat sheets which are cut and formed into cans after printing.

General-line built-up tins

The term *general line* is used to refer to all of the many types and styles of tinplate container which cannot be described as open-top cans. Factories tend to specialize in one or the other. Although some manufacturing techniques are common to both, the management style required to run a general-line business is different to that which is appropriate to open-top can making.

To attempt to catalogue all of the tins available would be space-consuming and is unnecessary. Reference to trade literature will assist the reader, as will a study of those British Standards which relate to tin box making. Some types of tin are, however, of fundamental importance, and it is these we now describe and discuss.

The slip-lid tin

This is the typical "tin". It serves the housewife as cake tin or biscuit tin, and has aided the retail sale of many dry commodities. It is seldom employed for liquid products.

The body and bottom end are secured to each other by a conventional double seam, and the body side seam is most often an interlocked seam which may be soldered, doped or cemented. There is no technical reason why the side seam should not be a welded or bonded lap seam. The bottom end seam may contain compound or, like the side seam, be soldered. Consequently, the tin without its lid is able to hold water or the finest dry products without leaking or sifting. The lid, by definition, slips on the body (see figure 2.8). It is not secured by solder or any other sealing medium, because it is meant to be put on and taken off with ease, and without resorting to any tools or mechanical aids. The lid is *solid drawn*, i.e. it is made from a single piece of sheet metal formed by pressing through a steel ring or die by a punch. The process is examined more thoroughly in the section dealing with Solid Drawn Tins. The lid has no seam or joints, and is thus well able to prevent the passage of liquids, moisture vapour or finely divided products.

What tends to let down the slip-lid tin is the simple friction fit between lid and body. This is unsealed so that liquids, gases and fine solid particles can find their way in and out of the tin. Nevertheless, for some products the slip-lid tin is ideal. An example is individually wrapped sweets, which need both atmospheric and mechanical protection. The individual wrapping provides the protection from the atmosphere, and the tin augments this and provides protection from mechanical damage. Perhaps most important, a tin presents an opportunity to use lithographic printing to great advantage for display and advertising.

Many ways can be found to improve the protective capability of the slip-lid tin. The problem is usually a matter of preventing oxygen and moisture vapour from entering the tin. A very common approach is to use self-adhesive tape to cover the joint between lid and body, and machines are available which enable this to be done rapidly.

Another method is to secure an adhesive-coated impermeable dia-

METAL-BASED PACKAGING

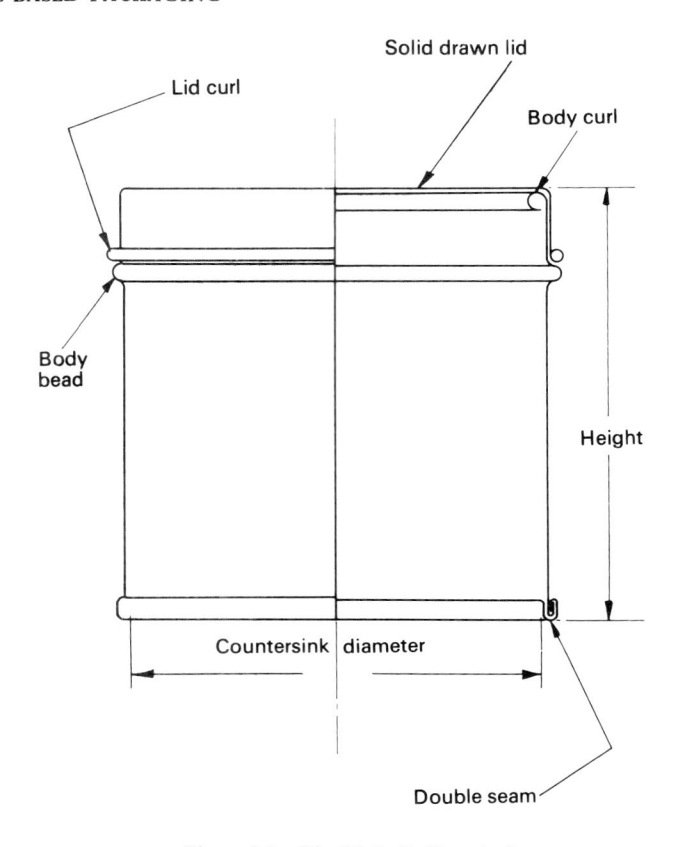

Figure 2.8 Slip-lid tin (half section)

phragm to the top edge of the tin body by heat. Aluminium foil is a suitable barrier, and so is paper coated to make it gas and moisture-vapour-resistant.

It is not uncommon to find slip-lid tins completely over-wrapped with transparent film. This, too, contributes to protecting the contents from moisture during transport, storage and display. It is assumed that with all these measures the removal of the sealing material by the user presents little hazard to the contents, which are consumed in a relatively short time following initial opening.

These remarks about slip-lid tins serve to underline certain principles which can be applied to other types of metal container.

1   In choosing which type of tin to use, the total package must be considered. This means that account should be taken of

(a) any primary wrapping, actually in contact with the product
(b) any lining or cushioning material within the tin
(c) the tin itself
(d) any sealing tape or over-wrapping material outside the tin
(e) any other outer packaging, such as a fibreboard case with or without cushioning material, used to transport one or several tins to the point of sale.
2 The tin as a protective package can often be adapted to new uses by the employment of materials and methods evolved in other areas of technology.
3 It is important to define the packaging requirement. Leaving aside matters of display for the moment, the question is often a matter of how much protection the cost budget will allow.

The lever-lid tin

When discussing the slip-lid tin it was said that a difficulty exists in making a highly efficient seal between lid and body and that it is common to use measures such as over-wrapping to provide improved protection to moisture or oxygen-sensitive products. The reason for the inadequate seal lies in the method of body construction. This must be rolled or bent to its cylindrical shape, and is impossible to achieve with a guarantee of absolute smoothness to the exterior of the body. Instead of bending in to a smooth cylinder, the material will often develop numerous small flats producing, instead of a cylinder, a polygon with a large number of sides. Shortcomings of the process and local differences in raw material will produce larger irregularities. In addition there is the side seam which, however it is made, will present an irregularity which cannot be sealed by a slip lid well enough to contain liquid.

To achieve the quality of seal necessary to contain liquids or exclude moisture and gases, a further component must be added. This component is the *lever ring* (figure 2.9). This is a seamless component made by drawing the metal from a flat sheet. It is seamed to the top of the can in the same way as the bottom end is fitted. Because it is seamless, the lever ring enables another drawn seamless component—the lever lid— to be fitted to it with good assurance of an effective seal.

Many styles of lever ring and lever lid have appeared on the market, some having particular technical merit and others being different just for the sake of a change. No lever ring is really easy to make. The aperture in the ring and the diameter of the lid must both be made to close engineering limits, if a satisfactory and consistent fit is to be achieved between any ring and any lid out of many thousands. In fact, the tolerances on the diameters of the ring and lid for a 1-litre paint tin are typically as follows:

Ring diameter: 100·18 mm   Tolerance +0·02 mm
                                       −0·08 mm
Lid diameter: 100·41 mm   Tolerance +0·05 mm
                                       −0·05 mm

2.25

METAL-BASED PACKAGING

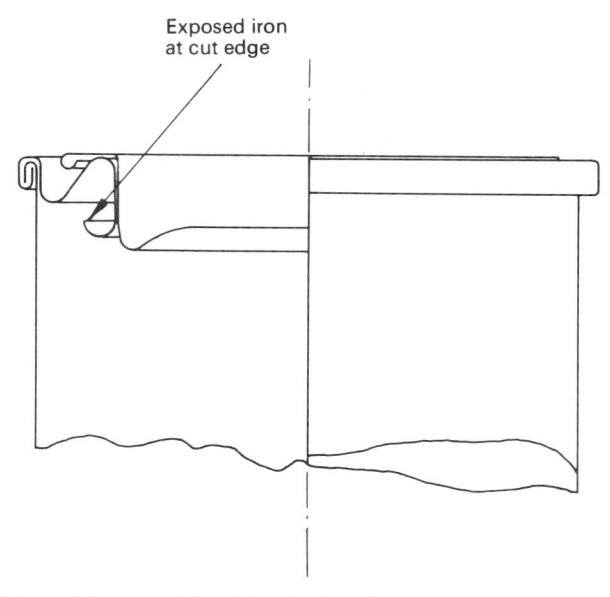

Figure 2.9 Lever lid and ring (half section) showing cut edge of ring

No lever ring is easy to make, but it is easier to finish with the cut edge of the tinplate around the aperture on the inside of the can rather than the outside (see figure 2.9). This is unacceptable for water-based products which will attack the iron exposed at the edge and not covered by tin in the same way as the rest of the surface. This explains the justification of one variation in design of the lever rings where the exposed cut edge of the metal finishes on the outside of the can and, although exposed to the atmosphere, will be attacked only very slowly compared with the rate of corrosion it would suffer if within the container.

Some manufacturers consider that the lever-lid paint can is much more difficult to make than the conventional food can. Certainly it has four components against the food can's three, and the precision necessary to make the closure components has already been noted. The argument is, however, based on the packaging requirements of the lever-lid paint can which must be:

1 easy to fill (through a large aperture)
2 easy to close at high speed
3 easy to open without specially designed tools
4 easy to reclose and seal after use.

When in storage and on display, the can must prevent evaporation of volatile solvents. Effective reclosure after use has to be achieved in the presence of paint or other products which will have contaminated the sealing surfaces. Quite a demanding set of requirements.

The lever-lid paint can is sometimes criticized for its lack of security. Because of the demanding requirements on the can and, particularly because it must be easy to open, it will not withstand unlimited rough handling. If a 5-litre can of emulsion paint is dropped on its side from a height of 1 metre, the lid is almost certain to be forced off by the sudden surge of pressure caused by impact with the ground.

Ways of preventing or reducing the risk of lid removal are available. Spring clips can be fitted to help retain the lid, or an overall cover or capsule may be spun under the double seam. Spots of solder applied to three of four points on the periphery of the lid do the same job as spring clips. All of these are more or less effective in adding to security, and more or less effective in irritating the user who wants to use the contents. It may be significant that the vast majority of paint cans are sold without any added security devices, and very little paint seems to be spilled in retail shops or the pavements outside.

Oblong pourer tins

Before leaving the subject of built-up tins, some mention must be made of the oblong pourer tin. It is another important example of a difficult task being attempted because it is justified by non-technical considerations. Figure 2.10 shows the general style of such tins which are commonly available in capacities from 125 millilitres to 5 litres, the smaller capacities being supplied without handles.

Round tins would be stronger and easier to make, and could well be cheaper, and yet this general style of oblong tin enjoys world-wide popularity.

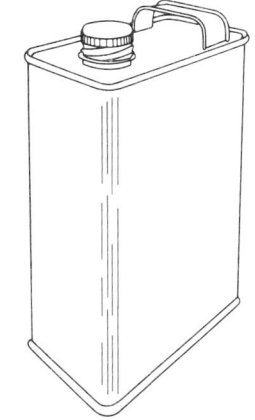

Figure 2.10   5-litre oblong pourer tin

METAL-BASED PACKAGING

The flat display area on the sides and easy storage afforded by the rectangular shape seem to outweigh other factors. This type of tin is often chosen for products which are easily able to penetrate the most minute leaks, and the consistency of mechanized production enables the tin to meet these demands.

The difficulty in making an oblong built-up tin is concerned with the corners. When making a double-seamed round tin, the manufacturer employs what might be described as a *metal spinning operation*. Although the extent to which material is redisposed in making a double seam is limited, the process is similar to that employed by silversmiths and others in making deep cup-shaped articles from flat sheet. Such techniques benefit from speed and smoothness of operation. The smoothness is lost when an oblong tin is made. At the radius connecting one straight side to another, mechanisms must undergo violent changes of direction in forming the seam. This causes uneven wear of bearings and gears, and results in loss of quality which must be countered or forestalled by maintenance activity.

This example is used simply to indicate the nature of the technical difficulty. It must be accepted that round tool and machine components are easier to make with precision than are oblong items, and this is the root of the difficulty presented in the manufacture of an oblong tin.

An in-between method—the locked-corner tin

The previous section discussed built-in tins which are made of several pieces of material put together in such a way as to form a hollow container. A later section explores the features of so-called Solid Drawn Tins.

Between these two methods of construction there lies a type of tin known usually as a Locked Corner Tin or sometimes as a Sunk and Seamed Tin. In tin box making "sinking" refers to the making of a depression within a component, or to the manufacture of a dish or cup-shaped article from a flat sheet or blank.

The locked-corner tin is a rectangular tin, the body of which is made from a single blank. It is distinguished by the fact that the body has sharp corners, not rounded or radiused like oblong built-up tins, and each corner has an interlocked seam. Figure 2.11 shows both the blank and the finished body. The body is made by pushing the blank through a die which is itself rectangular and of a size to match the required outside dimensions of the body. The four corners of the die are, however, of a very special form, and they trap the edges of the blank as it is pushed through by a punch, in such a way as to form an interlocked seam at each corner.

Examination of the sketch of the blank shown in figure 2.11 should be sufficient to suggest that, in terms of material use, this manufacturing

METAL CANS

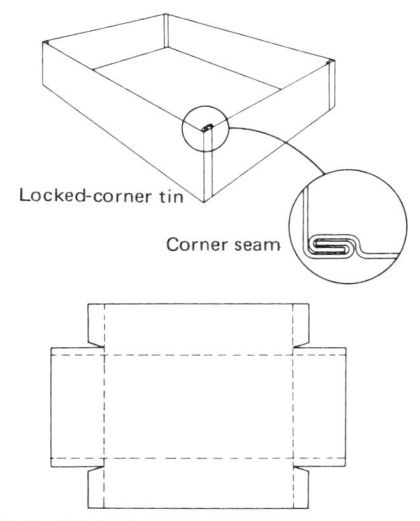

Figure 2.11 Locked-corner tin

technique is most economical for shallow tins, and grows progressively less economical as the body depth is increased. This is simply because deep tins require the blank to have a large, almost square, piece cut away from each corner. The material cut away is called a "notch", and the deeper the tin the bigger it has to be. Material cut away in the notch is discarded and therefore wasted.

Despite the requirements for carefully made tools and the loss of material from body notches, locked-corner tins are used when sharp square corners are required within the tin, as when packing cigarettes and some types of biscuit. They may be fitted with a solid-drawn slip-lid or hinged lids as circumstances demand.

Solid-drawn tins

Solid-drawn tins are commonly used for packing tobacco under vacuum, and in other shapes for packing fish, such as sardines. They have many other uses, and in the last year or two a variation of the solid-drawn tin has become of major importance in the packaging of food and drink.

They have been referred to once or twice already when comparisons of cost or performance were discussed. The lid of a round slip-lid tin is solid-drawn from a single piece of material, without the use of seams or joints. The first stage in the manufacture of a DWI can requires that

METAL-BASED PACKAGING

a plain round cup should be drawn from a circular disc or blank of raw material.

In general engineering, the process of drawing cup-shaped components from flat material is common, and consequently the fundamentals of the process are well described in engineering textbooks. For a detailed exposition of the subject or knowledge of the mathematics involved, these should be consulted. For our present purpose it is sufficient to examine those matters which bear upon the solid-drawn component as a tin or part of a tin.

The vacuum tobacco tin

This is the most appropriate use of the solid-drawn tin today, and it would be difficult to find a better example of fitness for purpose.

Figure 2.12 illustrates the main features of a round vacuum tin. When discussing slip-lid tins it was noted that a perfect seal between lid and body is almost impossible to achieve, because

(a) A rolled body is not smoothly rounded and thus prevents a good seal being made on the diameter.
(b) The presence of a vertical side-seam creates a major irregularity preventing a good seal.

Both of these shortcomings are avoided in a solid-drawn tin. Even so, when the top edge of the body is rolled over (or *curled* to use the tin boxmakers' term), minute irregularities can still exist which would stop the tin holding a vacuum. To overcome this, a small amount of resilient

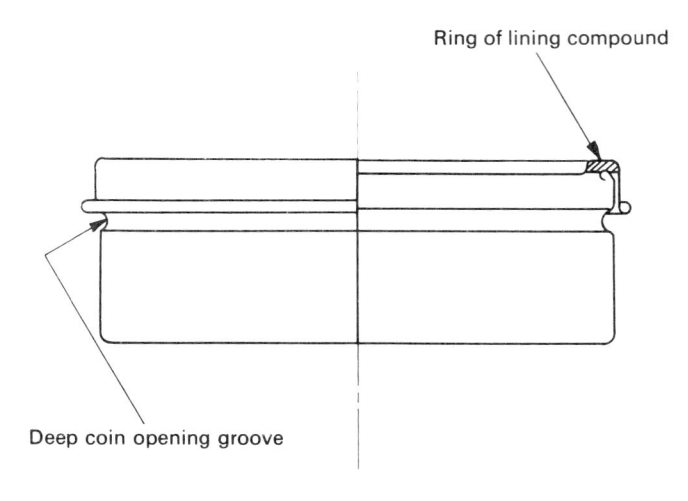

Figure 2.12   Round vacuum tobacco tin (half section)

2.30

lining compound is introduced into the lid of the tin to form a gasket near the periphery, into which the top edge of the body can become embedded. After this precaution, the tin becomes a highly satisfactory vacuum pack which can be opened with a large coin.

Twisting a coin in the groove provided merely lifts the lid from the body locally (near the coin), and allows air to enter until the internal pressure equals that of the atmosphere, and the lid can then be lifted off. This point is worth making because it is easy to overlook the magnitude of the force acting to keep the lid closely on a vacuum-packed tin. The following calculation illustrates the point.

Given that:
The diameter of the tin is 68 mm and the pressure difference between the inside of the tin and the outside atmosphere is 0·9 bar (1 bar = 10^5 N/m^2 or roughly 1 atmosphere).

Note that the achievement of a perfect vacuum within the tin would give a pressure difference exactly equal to atmospheric pressure. In practice, a near perfect vacuum is difficult to achieve and also unnecessary.

The area of the lid which is subjected to the pressure difference is

$$\pi \times (34)^2 \text{ mm}^2 \quad \text{or} \quad \frac{\pi \times 34^2}{10^6} \text{ m}^2$$

The total force acting on the lid because of atmospheric pressure and tending to keep it shut is therefore

$$\frac{0·9 \times 10^5 \times \pi \times 34^2}{10^6} = 328 \text{ newtons approximately}$$

This force would be produced by standing a weight of 33·4 kg (74 lb) on top of the lid.

Such a force, acting on the lid of the tin to keep it closed, makes for a most secure closure, in the mechanical sense. The air-tight seal prevents the contents from deteriorating by drying out or by losing flavour or aroma.

Some popular tobaccos sell steadily and, with proper retail stock rotation and frequent replenishments, the vacuum tin seems hardly necessary and might be replaced by a less costly and less air-tight container. Other brands sell more slowly or sales are erratic because of advertising or other promotional devices. Even with popular brands, it is not always possible to guarantee perfect stock rotation, and the tin provides a valuable protection against deterioration brought about by chance or by accidental long storage. There is also much to be said for a standardized container which allows a tidy approach to retail display and demands no special precautions by the shopkeeper.

Manufacturing limitations of deep-drawn tins

When a deep-drawn tin is produced, the metal from which it is formed

METAL-BASED PACKAGING

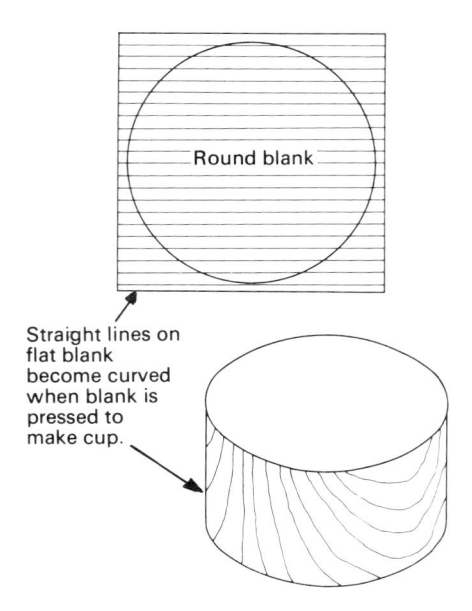

Figure 2.13 Distortion which occurs on drawing a metal cup

undergoes great strain because of the way in which it is redistributed. Figure 2.13 illustrates this for a round tin. If a piece of tinplate is printed or marked with a rectangular grid, when in the flat state, the grid will take up the new pattern, as shown, when a round deep-drawn cup-shaped component is made from it.

It may be of interest to explain that, despite the major redistribution of material which takes place in the drawing operation, the area of sheet metal remains substantially unchanged, i.e. the metal blank from which a drawn component can just be made will have the same total area as the finished component. There are minute differences which occur in practice, but for most purposes the assumption of area equality is a safe one. This distinguishes drawn components from the component or can body made by the DWI process. With DWI there is a very substantial increase in the total surface area of the metal during the process.

Returning to the drawing process, an examination of figure 2.13 should suggest that, for a given diameter of finished component, the deeper the draw the more severe will be the distortion of the top edge. Also as the ratio of depth to diameter increases, the forces involved in pressing the article to shape become greater and may exceed what the sheet metal will stand, thus leading to fracture during manufacture. Distortion of the surface affects printing and must be allowed for in advance. Possibly more important is the fact that surface coatings, applied to the metal to produce a uniform coloured effect or to supplement the protection from corrosion given by the coating, are less pliable than the metal. Surface coatings tend to craze under severe distortion, and their appearance becomes more mat than in those areas where the strains are less. In extreme circumstances the adhesion between coating and metal can break down altogether and leave plain metal showing through.

There are then two ways in which the depth to diameter ratio of a drawn component is limited; by severe strain which can cause fracture, and by surface disturbance which can make the appearance visually unacceptable. The strains may be minimized by drawing the cup-shaped component in more than one operation, i.e. by starting off with a shallow cup, and then subjecting it to more drawing operations, each of which will increase its depth and reduce its diameter. This can produce extremely-deep small-diameter components by working through several easy stages, and the technique is quite common in general engineering. It is, however, expensive both because of the manufacturing time it takes to perform the numerous operations, and because of the amount of manufacturing plant it employs. In tin box making it is seldom possible to accept such costs.

The manufacturing limitations for deep-drawn tins are really concerned with what can be done in a single drawing operation. For round tins, a safe limit is a depth to diameter ratio of 1:4 which means a depth only $\frac{1}{4}$ of the diameter. Having made that statement, it must be noted that ratios as extreme as 1:1 have sometimes been successfully attempted, but this is with material the thickness and physical properties of which have been carefully selected, and where the manufacturing operation has been specially designed to achieve the required result.

Rectangular solid-drawn tins present a special case which is illustrated in figure 2.14. If the rectangular body is divided into its geometric parts, it is seen that the straight sides are formed by simply folding them through an angle of 90° to the base. No drawing takes place along the straight sides—there is no redistribution of material as there is in the side wall of a round tin.

The rounded corners are drawn, and each can be considered as one fourth of a small round tin. Because the corners are drawn, they impose the acceptable limitation on depth. The depth of a rectangular drawn box with rounded corners is normally limited to:

$$\text{Depth} = 2 \times \text{corner radius}.$$

If the four corners could be brought together as in figure 2.14, it is possible to talk in terms of the diameter of the tin so formed, and the

METAL-BASED PACKAGING

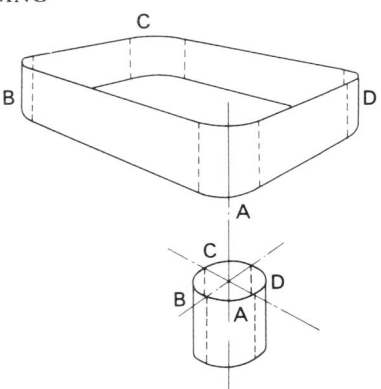

Figure 2.14 Significance of corner radii on a rectangular drawn box. The straight sides of a drawn rectangular box are simply bent into position. Drawing takes place on the radiused corners. The severity of the operation may be judged by assuming the four corners are brought together to form a small round tin.

depth limitation would become:

depth = diameter, because diameter = 2 × corner radius.

It is interesting to note that, even for rectangular tins, the 1:1 depth to diameter ratio previously suggested as the absolute limit for deep-drawn work, using a single drawing operation, is again observed.

The whole subject of solid-drawn tins is complex and important, not only because of the usefulness of tins made by this method, but because of the extent to which the manufacturing technique is employed in can making and tin box making generally. The references to solid-drawn tins could just as easily have been to solid-drawn components. The lid of a slip-lid tin is solid-drawn; so is the shoulder of an oval talcum powder tin. Even the end for a conventional food can is solid-drawn, although the draw depth is so shallow as to be almost insignificant. More significant is the drawing of metal involved in making a cone-shaped top for an aerosol can, and a particularly complex example is a screw neck made as an integral part of the top of an oblong pourer-type tin.

Much of what has been said regarding solid-drawn tins is therefore of importance to the manufacture of separate components which form parts of tins and cans.

Economics of making tins

As with other manufactured articles, the cost of making a can is made up of:

Direct material costs
Direct labour costs
Overheads including indirect materials, indirect labour, depreciation, factory rent, rates and cost of energy, etc.

In general-line tin box making, direct labour costs may be between 5% and 10%, with material 30% to 60% of total selling price. The remainder is overhead and profit. In high-speed can making, direct labour is an even lower percentage, but overheads tend to be higher because of the cost of the complex machinery involved.

Assuming that, as a rough guide, material will approximate to 50% of the cost of any tin or can, it is thus the largest identifiable single cost. Much ingenuity has been displayed by engineers in trying to achieve the greatest possible economy in the consumption of tinplate or other box-making material, and considerable sums of money have been invested to minimize waste in making cans by high-speed methods. The same pressure for economy has not had much effect on the selection of tins for various products. These are determined by the market needs—what will sell, or what is acceptable as distinct from what is most economical.

The ideal geometry

The problem seems to be a matter of containing the greatest volume by the least area of sheet material. If that were all there were to it, every tin would be a sphere, because a spherical shape is the most economical way of using sheet material to contain a given volume. In practice, of course, we are confined to using tins which are either cylinders or cubes (or in the more general case, rectangular prisms).

Assuming that the tin has no seams, no joints and no overlapping edges, in the way that a slip lid overlaps the body, then if cylindrical, the tin is like a short-section cut from a seamless tube and fitted with two ends, each of which is a simple disc of diameter just equal to the diameter of the tube. For such a container, it can be shown that the maximum volume is contained by the minimum area when (*a*) for a cylindrical container, height equals diameter; and (*b*) for a rectangular container, all sides are equal.

As in many situations, such mathematical perfection is upset by practical considerations. Not the least of these is that circular components are cut for the cylinder ends from square pieces of tinplate (see figure 2.15). This is wasteful because more than 22% of the square piece will be thrown away as shred, and less than 78% used to make a lid or bottom end.

METAL-BASED PACKAGING

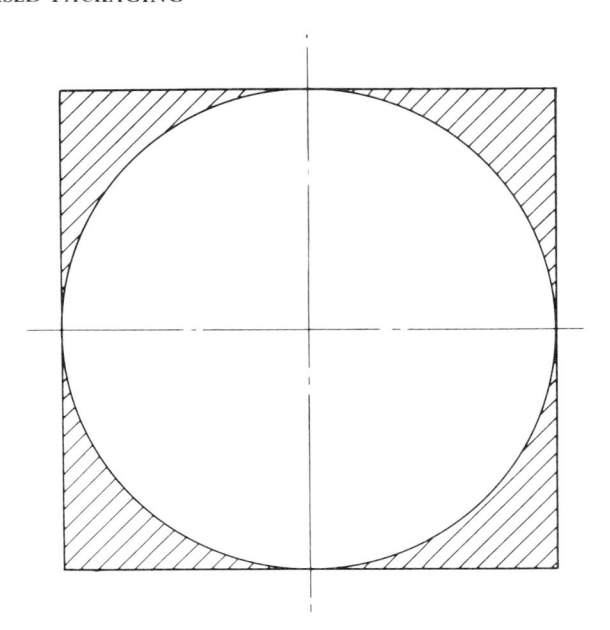

Figure 2.15 Round disc taken from square

It is also necessary to allow for a seam in the body of the tin and, with a slip-lid tin, there is an overlap of the lid on the body to consider, as well as the material in the bottom end seam. Taking account of all of this, the most economical proportions for a slip-lid tin are such that the height exceeds the diameter slightly. As a rough guide, a height to diameter ratio of 1·2:1 might be about right. The exact proportions can, of course, be calculated in each instance; they will vary slightly from small tins to large ones, and be determined by the exact amount of material employed in seams and the overlap of the slip lid on the body.

Contrast these ideal proportions for economy with tins used for retail sale. Some tins do have proportions approaching the ideal for material economy, others are far less economical. Typical of the latter are large round biscuit tins which are popular because of their re-use value in the home. Here, considerations such as display value and utility outweigh material economy in the designer's order of importance.

It is almost impossible to find a cubical container in use today, and the popular oblong pourer-type tins used for oils, solvents and many other liquid products, are a long way from being of economic proportions. The reader might like to make his own list of reasons for this, noting in passing that a cube is also uncommon in architecture.

(a) Single row from plain strip.

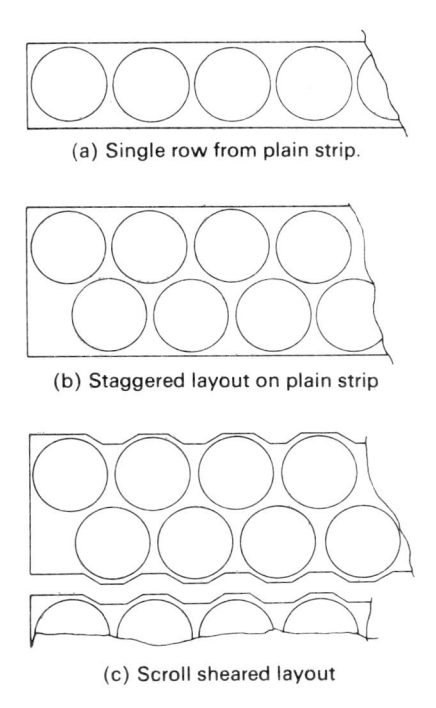

(b) Staggered layout on plain strip

(c) Scroll sheared layout

Figure 2.16 Progressively economical ways of cutting discs from strips

Getting the most out of available material

A large part of the can makers' work is concerned with cutting round discs from sheets of tinplate or aluminium. The obvious way to do this is shown in figure 2.16a. The strip, cut from a sheet of tinplate, is fed to a stamping press and the discs are each cut out in quick succession.

In practice, the tool fitted to the press is designed to produce, say, can ends, and one almost-finished end would be formed at each stroke of the press; but it is convenient to talk in terms of discs, because we are only concerned with the plane geometry of the problem for the moment.

It is not practicable to arrange cutting so that the edges of the discs just break through the edge of the strip. Instead, a small margin must be left along the edges of the strip and between each disc. The material which remains after the discs have been cut out is called *shred*. In figure 2.16a the shred which remains is about 25% of the original sheet. In other words only 75% of the original strip finishes as discs.

Such a wasteful arrangement is to be avoided, if at all possible. Figure 2.16b shows a rather better arrangement, where the discs have been staggered, and two rows are cut from a single wider strip. The result here is that about 75·5% of the material goes into the discs.

The principle of staggering the pattern of discs is a sound one, but most of the waste arises along the edges of the parallel strip. Figure 2.16c shows how the can maker reduces this. A large sheet of tinplate is cut into strips, with zig-zag edges as shown, and the strip is then fed longitudinally into the stamping press. This system achieves the use of 82% of the strip. This is a much better result, and its only weakness is that because squared sheets of tinplate are usually involved there is still waste along the straight edges. Even that has now been overcome by using coils of tinplate instead of sheets and cutting the zig-zag pattern strip from continuous strip, fed from a coil. In this way, an overall 82% usage is very nearly achieved.

The zig-zag edge is not as easy to cut as a simple straight edge, and requires a carefully made pair of shear blades and a particular type of shearing machine to accept them. The machine is called a Scroll Shear, and the blades it uses are expensive to produce and maintain. Equipment of this kind can only be justified by the long and continuous use which stems from standardization of containers. Food cans, beverage cans, aerosol cans and other such containers, for which there is massive demand, are made only in certain heights and diameters which are the subject of national and international standards. Because of the UK membership of the EEC and our metrication programme, the various standards are undergoing changes. Such changes could be expensive in new tooling or changes to tooling, and cannot be entered into hastily or without proper consideration of their merits. To the manufacturer, a standard is only of value if it remains substantially unchanged for long periods.

Measuring and describing tins

This has undergone substantial change recently for three inter-related reasons. These are the UK entry into the EEC, metrication, and increased activity by the International Standards Organisation (ISO). Can users and other readers may still meet the obsolete terminology, however, and it is necessary therefore to make a brief reference to the old method.

The unit of measurement was $\frac{1}{16}$ in and this unit was used in two ways. A simple slip-lid tin with the bottom end seamed on, could be described as $700 \times 3\frac{5}{16}$ in. The diameter was (and still is) stated first and consequently the 700 refers to the diameter. It means 7 in exactly, i.e. that the diameter is neither $6\frac{15}{16}$ in nor $7\frac{1}{16}$ in. Other sizes stated in this way might be 307 ($3\frac{7}{16}$ in) or 404 ($4\frac{1}{4}$ in). Although the word *exactly* has been used, what is meant is "closer to 7 in than to $6\frac{15}{16}$ in or $7\frac{1}{16}$ in". But *what* is closer to 7 in? In this instance the diameter is measured outside the double seam of the bottom end. When the three-digit notation is used, e.g. 700 or 404,

METAL CANS

etc., it is understood that a diameter accepted as standard is referred to. If the diameter was $7\frac{1}{4}$ in, then it would have been written as $7\frac{1}{4}$ in because the trade had not accepted such a diameter as standard.

Whereas the digital method refers to the diameter outside the double seam, a size shown as inches and fractions ($7\frac{1}{4}$ in) refers to the diameter inside the recess of the end—within the *countersink*, to use the trade term. On any tin, the actual difference between the countersink diameter and the diameter outside the seam is approximately $\frac{1}{8}$ in (see figure 2.17).

To return to our slip-lid tin, then, it measures 7 in diameter over the double seam and it is $3\frac{5}{16}$ in tall, the convention here being that the height is the overall height of the tin, without the lid fitted.

The digital notation is only employed to demonstrate the height of tins and cans when they are established as standard sizes.

So much for the past. Now measurements are in millimetres and, although the descriptions look similar, they mean something different, e.g. a large rectangular tin might be described as:

$$231 \times 217 \times 118$$

This means that the tin has base dimensions of 231 mm × 217 mm and is 118 mm tall.

A round lever-lid paint tin is described as:

176 diameter, 5-litre lever-lid tin.

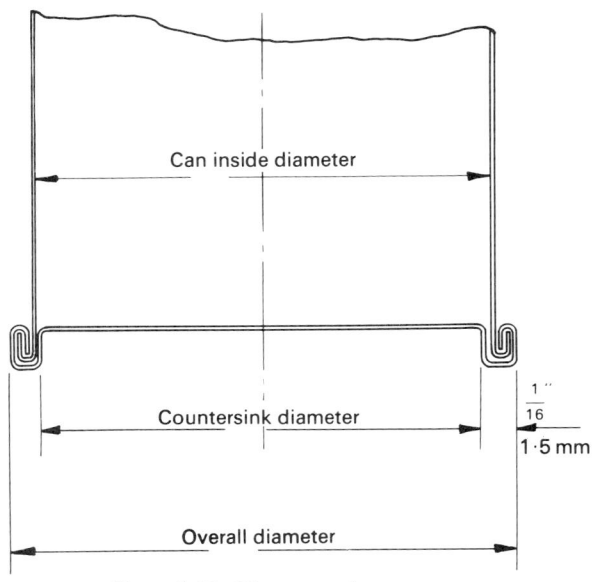

Figure 2.17 Diameters of a can

2.39

METAL-BASED PACKAGING

Generally, the diameter in millimetres is stated, followed by the nominal volume. If the reference is to an oblong pourer-type tin, forming part of a standard range, the base dimensions are given followed by the nominal volume. As with the rectangular tin of the first example above, the larger of the two base dimensions is given first.

Tins of non-standard height or non-standard volume, give both the diameter and the height in millimetres:

$$e.g.\ 176 \times 120$$

This is the method which has also been adopted in the United Kingdom for describing cans with both ends seamed on, such as food, beer and beverage cans. There is, however, a move to have such cans described by their diameter and volume, and the outcome of current discussions is not yet known.

The use of metric units is becoming commonplace, but the dimensions do not always refer to the same features as with the old digital method, using a unit of one sixteenth of an inch. For height, there is not much change. The height of a food can is the external height, measured over both top and bottom seams. This is rounded up or down to the nearest millimetre.

With cross-sections, an attempt is now made to use dimensions as close as possible to the inside diameter of round tins, or the inside length and width of rectangular tins. This is easy for the manufacturer, who knows what size of can he is trying to make, and can obtain precise measurements from the mandrels or other tools over which the can bodies are formed. It is not so easy for the can user who may not be equipped to measure the inside diameter of a can, even if it happened to be rigid enough and uniform enough to permit this. His problem might be a matter of identifying the size of can from two or three possible standard sizes. If the can is filled and has both ends seamed on, access to the interior in order to measure its diameter is not even possible.

A method is available which overcomes this difficulty. For round cans with a double seamed end, it is sufficient to measure the diameter inside the countersink of the end and round-up the measurement to the nearest whole millimetre. The same procedure will also serve to establish the size of a rectangular container.

When other types of tin or can are encountered, it will usually be sufficient to accept the nominal dimension of a cross-section as the actual inside dimension to the nearest millimetre. A full exposition of the subject is given in the British Tin Box Makers' Federation publication entitled *Metric Information*, which also describes how the tin box maker deals with the problem.

Closures

Compromise in design must always be accepted, no matter whether the object is a motor-car, a central-heating system, or a suit of clothes. More often than not, the extent to which our wishes are achieved depends upon cost. The useful life of a motor-car, the reliability of a central-heating system, or the workmanship in a suit, are limited by cost. The compromise, in all instances, is dependent on how much we can afford.

With closures, cost is one of the limiting factors, but there are others, and the perfect closure possibly does not exist. If it did, it might easily pass unnoticed, because no-one would complain about it. The lever lid and ring, used on paint cans, is a good example of compromise. The lid must enter the ring easily, but it must also remain there, even when the can receives a degree of rough handling. None of the product must leak from the closure, but the lid must come off without difficulty, with only the twist of a blunt screwdriver. The lid must also go back again and make a seal, even when the ring is covered with paint, and these functions must be achieved at the least possible cost.

Had the lever lid and ring not been developed over a long period of time, it is difficult to believe that it could be introduced today. No packaging engineer or designer would be in any hurry to accept such a demanding set of requirements.

Because of this need for compromise, it is worth while to analyse some of the basic properties of closures. Such analysis leads to an understanding which helps in selecting or specifying what is needed for a product or for a new packaging requirement. There are two broad categories of closures:

1 The "once only" closure with no reclosing facility.
2 The reclosable closure.

The once-only closure

Some typical examples are:

The open-top food can opened by a can opener.
A beverage can with an easy-opening end or pierced by a special opener.
The engine-oil can with a tear-off strip covering both a pouring hole and a vent hole.

Such closures are cheap to produce, provide good resistance to leakage, and are suitable for their particular purposes. They include the "frangible diaphragm" beloved of patent specification writers. Some require an opening tool—typically the can opener.

A common factor is that this type of closure is useful only for those products which will be dispensed and completely used from the can within a short time of its being opened.

2.41

The reclosable closure

The lever lid and ring is clearly in this category, and so is the screw cap and neck. Many other closures are now available involving plastics spouts, pourers and other dispensing devices, and the development of these continues at such a pace that to list them will hardly be of service to the reader. What does seem to be important are the principles involved. All reclosable closures have three properties in common:

1. Friction
2. Resilience
3. Amplification of effort.

Friction

It is friction which keeps a cork in a glass bottle—friction between the cork and inside of the neck of the glass bottle. Friction also retains a lever lid in its ring against the internal pressure which may build up because of heat from the outside, or against a sudden surge of internal pressure resulting from dropping the tin.

Note that the friction between two tinplate surfaces can be markedly changed by the application of lacquer to one or both surfaces.

With lever-lid tins it is sometimes necessary to have the inside of the tin protected from corrosion by a surface lacquer. The lacquer also has to be applied to the lever ring and the inside surface of the lever lid. Care must be taken to ensure that the application of such lacquer has not reduced the friction between the surfaces to an unacceptable level, which will limit the security of the closure.

Only friction holds a screw cap tight on its associated neck. The way in which the friction is used best becomes clearer as we examine the other two properties involved.

Resilience

When a cork, a stopper or a lever lid is forced into the aperture designed to receive it, some resistance is met, initially, which is greater than that which opposes further entry of the plug. This initial resistance is produced by the compression of the plug, the expansion of the aperture, or both. With a cork in a glass bottle neck, the change of size is almost all due to the cork. The change in size of a tinplate lever lid and ring is due to what the engineer describes as the interference between the lid and the ring. The lid is slightly reduced in diameter and the ring is slightly enlarged and, if the lid is removed, the two components will tend to revert to their original size. That is what *resilience* means—the ability

METAL CANS

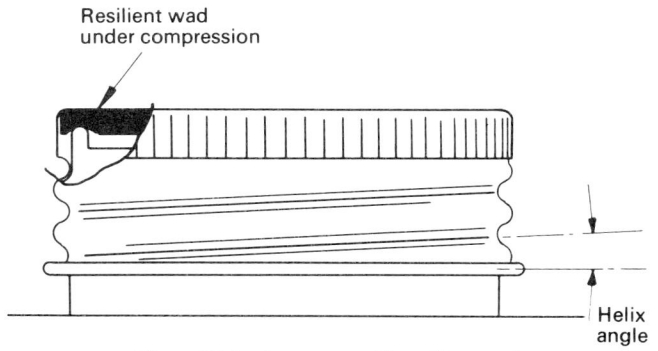

Figure 2.18 Screw cap with resilient wad

to change diameter in contact, and subsequently revert to the original dimensions when the two parts of the closure are freed from each other.

Resilience is found in the wad of a screw cap, which thus acts as a spring between the inside of a cap and the top of a screw neck, as shown in figure 2.18. It is the wad that makes the necessary seal, of course, but it has the additional function to provide a resilient member between the neck and the cap. Without it, the cap would be either tight or loose. When tight, the slightest rotation in the unscrewing direction would make it loose, and it would be insecure because of this. Any slight shock or continuous vibration in transit, or even sudden changes in temperature, could cause the cap to loosen on its neck.

The presence of a resilient wad aids security. The cap must unscrew through a reasonable angle from the fully tightened condition to reach a point where the previously compressed wad expands to something like its original uncompressed condition. Because of this range of movement, a cap with a wad is better able to withstand transit hazards than one without. Perhaps more significant, in practice, is that the most resilient or most elastic wad will be the one which provides the best security, other factors being equal.

Amplification of effort

This property is easier to appreciate in some closures than in others. With screw caps, the helix angle of the thread provides the amplification (see figure 2.18). The *helix angle* is the angle the thread appears to make with the horizontal plane when the cap is held with the axis vertical— the way it normally appears on the top of the tin. The smaller this angle, the greater the amplification of effort, i.e. the greater the force produced to compress the wad for a given effort applied to tighten the cap. Practical considerations in the manufacture of screw threads in thin

METAL-BASED PACKAGING

sheet metal prevent this angle becoming very small. Within the range of angles available, however, the force compressing the wad is substantially greater than the force applied by hand or by machine to the periphery of the cap when it is tightened.

A note for engineering readers is necessary at this point. The way in which the rotation of a screw cap on a neck or a nut on a bolt produces a higher axial force than is applied to the periphery of the cap or nut is strictly due to what the engineer calls *mechanical advantage*. The reason we do not use this established term is that with other forms of closure, such as the lever lid and ring, there is amplification of effort which is not strictly mechanical advantage.

It was noted when discussing resilience that a lever lid pressed into its associated ring will cause the ring to expand, and the lid to be compressed. When in the closed condition, therefore, the forces are held in balance, the ring exerting radial forces tending to crush the cap, and the cap exerting equal outward forces tending to expand the ring.

These forces exist because of the resilience of the two metal components. It happens that the total radial forces acting in this way are greater than the force required to push the cap into the ring. It is evident that the greater the radial forces the better is the closure able to prevent leakage. Amplification of effort exists, therefore, because the total of the radial forces produced is greater than the force required to press the cap vertically into the ring.

Materials of construction

These can be listed as:

Tinplate
Blackplate
Tin-free steel
Aluminium

Other materials, such as brass, are used in small quantities for things like talcum powder closures. Plastics, too, have become accepted as useful for pourers and closures, but it is the main container-making materials with which we are now concerned.

Tinplate

Tinplate is the most important can-making material in the sense that can-making consumes more of it than of any other single material. This situation is likely to continue for some time to come.

Tinplate is thin mild-steel sheet with a very thin coating of tin on each

2.44

surface. The material is available in thicknesses ranging from 0·17 mm to 0·31 mm in 0·01 mm increments.

The thickness of tin on each surface is included in the thicknesses stated above. Not only is the tin coating very thin, it is difficult to measure in terms of millimetres or other linear units. This is not only because of the smallness of the measurements involved. There is no clear dividing line between the tin and the steel. At the interface between the two metals, a tin/iron alloy is formed. Towards the surface this is rich in tin and towards the steel it is rich in iron. In other words, the tin and the iron interpenetrate to some extent and blur the dividing line between each other.

Instead of trying to measure and express the thickness of tin coating in linear units, it is customary to talk in terms of the weight of tin per unit area. Grams per square metre are the units now employed.

The tin may be applied either by passing the steel sheets through a bath of molten tin (Hot Dip Tinning) or by electrolytic deposition (Electro-Tinning). Hot dip tinning is used where a heavy coating of tin is required, and electro-tinning is employed for thin coatings. So as to distinguish between the two in specifications and on orders the prefix "H" is used for hot dip material and "E" for electrolytic.

The designation of electrolytic tinplate with $11·2 \text{ g/m}^2$ of tin on each surface of the sheet is shown as E.11·2/11·2, and this represents a thickness of tin of approximately 0·001 54 mm or 0·000 060 6 in, on each surface. With electrolytic tinplate it is possible to achieve a different weight of coating on each side of the sheet, and the material is then known as differentially coated tinplate.

Blackplate

Put quite simply, blackplate is tinplate without the tin. It does not have many applications in tin box making, mainly because the absence of tin makes it vulnerable to rust in the presence of even the slightest moisture. This can be countered by applying synthetic lacquer to the sheets before they are exposed to storage or other hazards.

Blackplate is not black at all, today. When new and clean, it is almost indistinguishable from tinplate to a passing observer. It is called *blackplate* because, in the days when it was rolled by hand, it acquired a thin coating of black oxide when it was left on the surface and not dissolved away in acid, as it would be today, before the material left the mill.

Tin-free steel

Tin-free steel is mild-steel sheet (tinplate base) with some other corrosion-resistant material applied to the surfaces instead of tin. The substitute metal coating is usually chromium, or more specifically chromium and chromic oxide together.

Steel with this form of coating is sold in the United Kingdom under the trade name of Hi-Top. The development of this material has been justified by the high and rapidly increasing cost of tin.

Chromium is not cheap and is itself increasing in cost as years pass. However, the thickness of the chromium coating required is substantially less than the thickness of tin necessary to achieve about the same corrosion protection for the underlying steel.

There seems no doubt that the industry will learn to adapt itself to the use of this and other tin-free steels. Elsewhere in this chapter it has been stressed that can-making is largely a matter of making good joints and seams. Chromium-coated steel is almost impossible to solder and difficult to weld. This has held back its use as a direct alternative to tinplate, although recent developments of adhesively-bonded side seams are now proving useful in this area, and welding techniques are likely to be available in due course.

Aluminium

This metal is becoming increasingly useful to the can maker, particularly for drawn and wall-ironed (DWI) applications. Small solid-drawn cans with easy-opening ends are now being offered in aluminium, and aluminium ends with easy-opening aids have become commonplace on beer and beverage cans. There is something to be said for having the whole of the can of the same metal, because scrap reclamation and re-cycling become easier. This is stimulating the movement towards the all-aluminium can with an easy-opening end, which is simpler to make in aluminium than in tinplate.

Like chromium, aluminium becomes protected from corrosion by a thin layer of oxide which forms on the surface as soon as the metal is exposed to air. Also, like chromium-coated steel, aluminium is almost impossible to solder.

Decoration of cans and tins

There are four methods in use for the external decoration of cans. These are:

(a) offset lithography
(b) dry offset printing
(c) silk screen printing
(d) paper labelling.

In addition to and in support of offset lithography and silk screen printing, roller coating is extensively used. This is the method which is also employed to apply protective coatings to that side of the metal which will form the inside of the can.

Offset lithography

Lithography has its roots in artists' methods of reproducing their work. For hand methods of reproduction, the picture to be reproduced is drawn on the surface of fine-grained stone slab using wax crayons. When the stone is moistened, water is absorbed by the unwaxed areas and repelled by the lines and marks drawn by the artist. Subsequent application of greasy ink to the whole stone by means of rollers, pads, or other means deposits ink only in the waxed areas and completely repels it in the wet areas. If a sheet of paper, board, or other suitable material is then pressed hard against the stone surface, the ink image will be transferred to the paper or board. Repeating the procedure with several stones provides the means to achieve multi-colour reproduction. This technique is still in use by artists to achieve small-quantity reproduction of their work.

Stones were originally used in tin printing. Because tinplate is non-absorbent, and because its stiffness prevents its being brought into continuous close contact with the stone, some way had to be found to transfer the coloured image from the stone to the metal. Paperboard of a particular grade and thickness was found to be successful. The board was pressed against the stone, from which it collected ink in the shape of the required image, and the inked surface was then pressed against the surface of the tinplate so that the image was once again transferred, this time from the board to the tinplate.

When an intermediate web or blanket, such as the paperboard sheet, is used in this way, it is said to "offset" the image. Thus the process obtains its name, *offset lithography*. Many years ago, the paperboard blanket was replaced by a fine-grain rubber sheet, and the stone slab by a metal plate which could be bent and stretched around the surface of a cylinder. These and other innovations permitted the full mechanization of the process, so that in the latest modern equipment, sheets more than 1 metre square are printed in multi-colour machines at the rate of 3000 to 5000 sheets per hour.

Drying the ink

Because tinplate, aluminium and other metal sheets are non-absorbent, the inks must be dried on the surface rather than allowed to soak into it as they do on paper. Consequently, freshly printed sheets must be held out of contact with each other and passed through an oven before they can be stacked, or handled through the making-up operations. The ovens are normally gas-fired, and the products of combustion (mixed with air) are forced over the printed sheets to carry away the solvents and harden the resins which carry the coloured pigments.

Other special inks are being introduced which may make the space-consuming ovens unnecessary in due course. These inks are of quite different formulation, and contain no great quantity of volatile solvents. The resin in which the pigments are supported is capable of being polymerized by exposure to ultra-violet light. If the intensity of the light is strong enough, the exposure time is only a fraction of a second; the ink is then dry, and ready to receive a subsequent colour, superimposed where necessary. Alternatively, a clear varnish can be applied immediately following the exposure to radiation.

Limitations and pitfalls

Offset lithography is a versatile process, and it can achieve many useful and striking effects, if properly used. Like other processes it has its limitations, but these can usually be avoided with thought and planning.

The process does not like fine detail. If a choice of type face is possible, it is best to select one which has an absence of serifs and contains clear open letters. Such a choice will ensure pleasing results over long printing runs. Another typical difficulty is that concerned with the crest displayed by Royal Warrant Holders. Obviously those entitled to use this device want to show it as often as possible, and insist that it appears at least once on each tin. Unfortunately manufacturers are seldom willing to devote much space to it in the designs, and printers are frequently faced with having to reproduce it in a strong colour in the design with a heavy weight of ink and within a space, say, 10 mm square. The detail is almost entirely lost, and the crest is barely recognizable.

Many designs can be achieved by printing three colours and black on a white background. There is a commercial pressure to minimize the number of colours employed, because the cost of printing is almost entirely proportional to the number of colours used in the printing run—the number of passes. However, it sometimes happens that, regardless of cost, an effect is required which necessitates six or even more printing passes. From the earlier explanation it will be evident that to print six colours may require six journeys through a hot oven. This means that the initial background coating, which is often white, will have to pass through the oven eight times in all, because one pass is necessary to dry the white itself and, when all else is done, an overall varnish will be applied which again requires a pass through the oven. It is not surprising that the white background may become yellowed or discoloured because of all the cooking it receives. Generally it is best to make a genuine effort to minimize the number of colours employed, cost considerations apart.

Other things to avoid are connected with the interaction between printing and tin box making. The final appearance of the printed surface depends upon a great many things. The surface of the metal itself is the first of these. The tin coating can vary in its reflectivity and there is, for example, a considerable difference between the surface of a tinplate sheet and the surface of an aluminium sheet. To achieve an identical appearance on both is almost impossible.

More frequently this change in gloss or reflectivity occurs because of the making-up process itself. For example, consider a simple slip-lid tin. The body has been made by bending it into a circular shape. Its reflectivity has been wholly preserved and might even have been enhanced by the curvature of the metal. Fitting over this we have the vertical wall of the slip lid; it has been subjected to a drawing operation in a press tool which has redistributed the metal. Such working always tends to dull the appearance, if only slightly, and it generally tends to weaken the strong colours. The perfect match between lid and body required by the designer is not obtained, and probably cannot be obtained, no matter what precautions are taken. The deterioration due to drawing is greatest on deep small-diameter components. One of the most demanding requirements is to match the drawn shoulder on a talcum powder tin with the body wall.

Registration between the printed design and the physical profile of the finished tinplate component is worth a word or two. Most of the time this creates no problem and, if a design appears misplaced in the lid of a tin, it is more likely to be due to an error in making up than in printing. However, designers often succumb to the temptation to surround their design with some sort of frame or margin. It would be ridiculous to debate the aesthetics of this in general terms, but examination of actual examples suggests that the trouble this causes on drawn lids and other press worked items is not justified by the objective sought. A design which is allowed to run across the lid surface and down the edge in one continuous pattern or colour will permit some inaccuracy in stamping the lids—even up to 3 mm on a large lid, before the error is noticeable. If the design on the top surface is surrounded by a sharp line in some contrasting colour, the slightest misplacement in stamping the lid is made glaringly obvious.

METAL-BASED PACKAGING

Roller coating

Roller coating is not a printing operation in the sense that it allows the reproduction of an image. Roller coating provides the means to roll a continuous coating of solid colour on to the surface of a metal sheet. Many lithographed designs are printed on a roller-coated surface. Most often, the background "colour" used is white.

The process is what its name suggests. Coating material is flooded over the surface of a rotating roller surfaced with synthetic rubber or some other elastic material. A sheet of metal is passed under the roller, against which it is held by another roller, and picks up a quantity of coating material. The application is followed by oven drying.

This technique requires less skill than printing. The machines run faster than printing machines, and the weight of coating applied is considerably greater than can be achieved by printing. It is the method used to apply protective lacquer to that side of sheet which forms the inside of a can.

Dry offset printing

The meaning of "offset" was explained under the heading Offset Lithography. It was seen that lithography requires the use of water and grease in cooperation to confine the ink to the image areas. In dry offset printing, the water is eliminated and a printing plate with a raised image, like a letterpress plate, is used instead of the substantially flat printing plate used for lithography. Dry offset printing can be achieved on a lithographic printing machine with little more effort than the use of a letterpress plate and leaving out the water. The results are not as good as can be achieved by lithography but benefits of speed and reduced stoppages can follow from the use of the dry offset technique in simpler designs especially adapted for the process.

However, the new importance of dry offset printing arises because of the development of the drawn and wall-ironed can. This has already been discussed earlier in the chapter and it is evident that cans made in this way must be printed after manufacture. The process as applied to cylindrical can bodies requires the can to be on a mandrel and able to rotate against a rubber blanket stretched over a cylinder. To avoid the complications involved in getting precise registration between one colour and another on a featureless can body, all colours are printed simultaneously. With offset lithography of flat sheets, colour to colour registration is achieved by working from the front edge of the sheet and one side of it, i.e. the machine has a way of detecting these two edges and placing the coloured image the correct distance from each. On a DWI cylinder, the perfectly smooth cylindrical surface provides no feature which could enable colour to colour registration to be achieved around the cylinder. Consequently each colour is printed in turn on to a rubber blanket from which all are offset simultaneously, on to the can surface.

To allow this, it is preferable to arrange matters so that no two colours overlap each other in the design. This means that multi-colour half-tone printing, requiring dots of one colour to be printed over dots of another colour, is also to be avoided. It can be done, with difficulty, if frequent stoppages for cleaning up can be tolerated; but this is seldom the case.

Silk screen printing

Silk screen printing involves the use of a fine silk (or more commonly nylon) screen mesh, mounted in a supporting frame. Ink can pass through the mesh, except where the holes have been blocked by resin or by impermeable patches. This affords a system for producing a coloured image, and multi-colour work can be done if the ink is dried on the printed object between each pass.

To print a design on a finished can, the can must be rolled against the silk screen whilst ink is forced through the mesh by a squeegee.

It is easier to screen-print metal in the flat and make up the tin afterwards as with lithography, but other considerations sometimes require the tin to be printed after manufacture.

Silk screen printing is used on work where the quantity required is too small to justify the costs of the preparation required for lithography. It is economically acceptable on short runs where a good standard of decoration is essential.

Paper labelling

The majority of food cans still carry paper labels. There are two main reasons—the first is that the thermal processing and automatic handling a food can must suffer, can damage a printed decoration, unless special precautions are taken in the design and management of the processing equipment.

Secondly, printed cans reduce the options. Each design may be used only for the product advertised on the outside and, if printed cans are used by a factory dealing with a variety of fruit and vegetable products, careful planning is necessary to ensure that the right cans are available at the right time and none need to be discarded as surplus.

Paper labels, on the other hand, are cheaper than cans and easier to store. Modern labelling machinery holds no terrors for the plant manager and does a thoroughly effective job.

CHAPTER 3

Composite Containers

B. Lindop

The manufacture of composite containers started in this country towards the end of the 1920s to supply the need for a cheap package to provide mechanical protection. Gradually this attitude towards composites has changed, and they now enjoy a distinct position in the packaging spectrum.

A *composite container* is, as implied by the name, a container made from more than one constituent material, generally consisting of a paperboard body with metal or plastics ends.

In the post-war period, major developments have taken place in the manufacture of composite containers, and this may be illustrated by the varied uses to which a composite can be put to meet the large number of marketing needs. This chapter endeavours to assist a potential user to make the correct choice of pack for his particular requirements. To achieve this, a knowledge of the characteristics and limitations of certain composites is invaluable.

Types of composite container

There are basically two types of composite container available: spirally wound, and convolutely or straight wound (see figures 3.1 *a* and *b*).

(a) Spirally wound composites

Only cylindrical shapes can be produced by this method, yet in a multitude of sizes. To form the tubular body, two or more plies of board are superimposed and glued together around a stationary cylindrical

COMPOSITE CONTAINERS

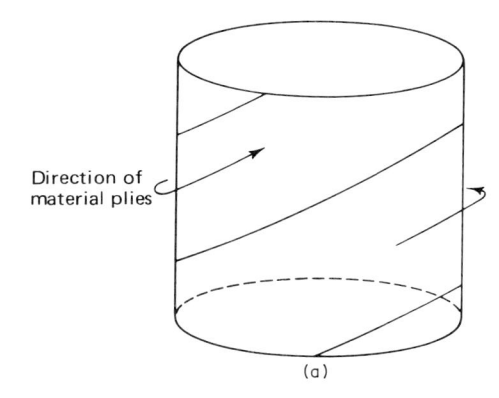

Figure 3.1a Spiral winding

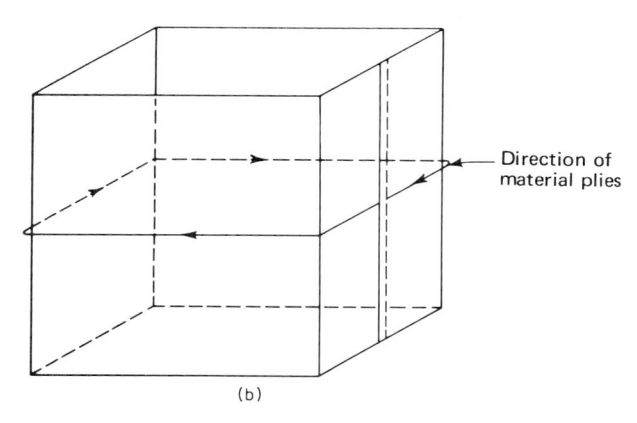

Figure 3.1b Convolute or straight winding

mandrel in a spiral manner. Each board ply is applied at an angle, which varies according to the width of the board used and the diameter of container required, and it is conveyed along the mandrel by means of a driven rubber belt which is wrapped spirally around the mandrel. When the requisite length of tube is achieved, it is cut off automatically by a revolving saw and ejected from the mandrel.

Naturally the weakest part of the container is where the plies meet and, to reduce the possibility of damage when external pressures are exerted, it is essential to ensure that all joints are well butted and the various plies overlapped in the correct position.

A cylindrical tube can only be formed by having all the plies in the wall construction overlapping. This interlacing of adjacent layers of board provides an adequate bond, and the two-ply constructions of this type are suitable for many purposes.

(b) Convolutely wound composites

In contrast to the spiral winding technique, a wider variety of shapes, including round, square, triangular, rectangular and oval, can be produced by straight winding. In this process the board from the reel is passed over gluing rollers and fed into a gripper in the winding mandrel. This mandrel then rotates, pulling the required number of wraps of material around itself to form a tube. After each rotational cycle of the machine, the board web is guillotined to separate the tube from the parent reel. An ejector pushes the partly formed tubes along the mandrel for the subsequent operations of rolling, labelling and reforming. The reel of board used is usually of a width equal to the height of several individual bodies, and for non-cylindrical work is slit down to the correct individual box height on the feed table of the machine. These separate webs are wound together and produce multiple bodies for each cycle of the machine. For round work, the full web width can be used throughout the winding operation, so producing a long tube which can be gang-labelled instead of individually labelled, and finally cut into individual bodies at the reforming stage.

The strength of a composite container is an important factor to consider, and can be improved by increasing the number of plies or by using thicker board. Where wall thickness limitations apply, to provide more strength it is better to use more plies of thinner material to achieve the required thickness. The standard board calipers used range from 250 micrometres to 500 µm, but occasionally use is made of material as thick as 750 µm.

In addition to the two basic types of composites, there are also certain modifications which add to the variety and range available. One variation of the spiral-wound composite is the sleeve-type container.

Sleeve-type composites

Included in this section are insecticide puffer packs, balloon inflators, ammunition cases, pharmaceutical calculators, and telescope sleeves. This type of container, as the name implies, consists of two, three or more separate cylindrical tubes which are assembled together to form one complete sleeved unit. The inner portions have an overall external diameter marginally less than the internal bore of the outer tubes, so that a compact unit is formed when all the integral parts are united.

A sleeve-type container is both practical and necessary, yet serves two

COMPOSITE CONTAINERS

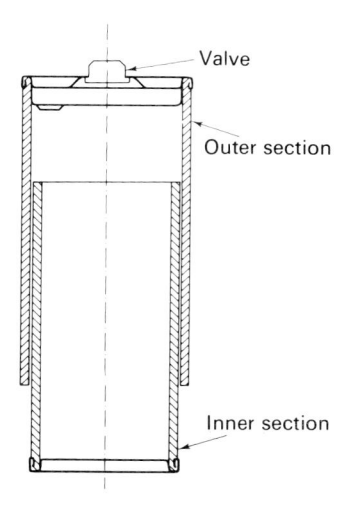

Figure 3.2 Inflator pack: sleeve-type composite

completely different purposes, protective and operative.

Generally composites have an average body wall of three plies of material, which is quite adequate to withstand normal handling and fulfil most requirements, but as always, there are exceptions to the rule. Certain containers must endure considerable stress, to provide suitable protection to their contents, and in some instances sleeved containers are advantageous. Where durability of the pack is critical, a sleeved container (which is reinforced by the interlacing sections forming a double or treble-thickness wall) will suffice.

Exceptionally strong containers can be fabricated in this basic style. By assembly of component parts, it is possible to make containers with walls over 35 mm thick. Such constructions, fitted with special board collars, metal discs and injection-moulded retaining rings are used by the Services for ammunition cases.

From an operative viewpoint, it is essential to have sleeve containers for puffer, inflator and many other kinds of novelty packs. An inflator pack is merely one open-ended container inverted into another which is fitted with a valve assembly; it operates in the same way as an ordinary pump (figure 3.2).

Body and functional barrier materials

It has now been established that there are two basic types of composites, spiral and convolute, but each of these can be further subdivided. Dependent upon the purposes for which a composite is intended, it can be classified as *functional* or *non-functional*. The latter category includes all containers which are required to offer mechanical protection only.

When considering the potential of a composite pack, the question of function is of first importance, and it is the material construction which determines this.

Every packaged article at some time is subjected to the risk of mechanical damage, or contact with environments that can cause deterioration. The prime function of any container is to protect the contents, during distribution, from the manufacturer's premises to the consumer's. Since it may take several months to complete the distribution cycle, it is important that the container should withstand this prolonged handling. The physical characteristics of the product, the design of closure, the shelf life required and the price are all relevant factors which influence the correct construction.

Various materials are employed in the manufacture of composites, and the next section describes those in common use.

Body materials

1. *Chipboard*

Most composites are fabricated from this type of paperboard which is used where high strength is not paramount. Chipboard has no special functional properties; it is not a barrier material and will not protect against atmospheric influences, but it does provide mechanical protection. It is suitable for containers which hold non-foodstuffs, such as small engineering parts, scouring powders and insecticides. It is possible to use white lined chipboard for some foodstuffs where flavour contamination could occur in contact with unlined board. The white lining also gives a clean appearance at low cost, but does not give any protection against moisture. For decorative purposes chipboard can be lined in a variety of colours.

2. *Kraft paper*

Probably the most dramatic change has been the use of kraft paper as a protective packaging material. Although this material is light, it still imparts sufficient strength to composites with thin walls. Kraft paper readily absorbs moisture from the surrounding atmosphere, but this can be prevented by treating it with compounds and various laminations which increase its protective properties. It is heat-resistant up to 150°C.

Lining materials

During recent years several lining materials have been developed as a result of technical research, and it is now possible for a composite to provide a greater degree of product protection than one might expect: for example, motor lubricating oil can be packed in composites. Some of the principal functional linings used are described below.

1. *Pure vegetable parchment*

Pure vegetable parchment is a tough treated translucent pure cellulosic paper which will withstand immersion in boiling water and is especially resistant to oils, greases, mild acids and alkalis. It is free from odour or taste, and has the great advantage of high strength when wet, advantages which have created a good market for its use in composites for liquids and fresh produce.

2. *Wax laminates*

Paper/wax/tissue laminates have a low water-vapour transmission rate. They are suitable for products requiring protection from moisture loss or gain, such as biscuits, sweets, and certain pharmaceutical tablets and powders. They provide limited protection from the flavour and odour aspect, and very little resistance to grease.

3. *Aluminium foil*

Aluminium foil is the best and most economic protective liner for composite containers in terms of cost and performance. Two standard gauges are usually used for composites but it is available in other gauges. The foil is often laminated to paper, which reinforces it, and in some instances other "carrier" materials impart heat sealability. Coatings and print can also be applied to foil and, where necessary, it can be suitably lacquered to prevent corrosion. Its main advantages are non-toxicity, opacity, impermeability to water vapour, freedom from flavour and odour contamination, and resistance to oil and grease. In most instances, some form of laminated aluminium foil is used in composite manufacture, and plain unsupported foil is not used for the body, only in connection with the top closure as will be seen later.

4. *Polyethylene coated paper*

Polyethylene coated paper is used in various gauges, but mainly 150 gauge polyethylene (low-density) extrusion coated on paper. It is suitable for products with high moisture content. Containers with a polyethylene/paper liner ply can be made by utilizing a patented process of extrusion which seals the join or overlap whilst the tube is being wound. Such containers are adequate for packing products like dairy cream, frozen liquid eggs, and various water-based emulsions.

5. *Polyethylene coated aluminium foil*

Polyethylene coated aluminium foil can be used in conjunction with the extrusion sealing process. Hot-melt adhesive lap sealing can also produce composite containers which are liquid-tight and which have extremely low moisture-vapour transmission rates. For very hygroscopic products this is an ideal barrier lamination, and it is particularly practical for such products as dried milk and powdered drinking chocolate.

6. *Glassine*

Glassine is a supercalendered smooth dense highly-beaten paper which was developed for purely functional purposes for a wide variety of greasy and oily products. It is made primarily from chemical wood pulps and, when waxed, lacquered or laminated, is highly resistant to transmission of water vapour. The fibrous nature of ordinary paper is almost removed by the processing, leaving a non-porous sheet which provides, in addition to grease resistance, a degree of odour resistance and protection from bacterial infiltration. Glassine, made in various colours, is consequently used as a protective liner for cereals, tea, coffee, bread, grated cheese, butter, soap. tobacco products, chemicals and oily metal parts.

7. *Glassine/foil/glassine*

This is a triple lamination of pure aluminium foil sandwiched between two layers of glassine. Both materials used separately have good barrier characteristics, so naturally in laminated form they provide a material of exceptional quality. Highly-volatile solvent-based products such as adhesives and sealants are best packed in composites with this liner.

8. *Silicone release-coated paper*

This makes an internal lining material which is particularly useful for hot-filled sticky bituminous and resinous products and hot-melt adhesives. Any product of an adherent nature can be packed in these papers and, when required, freely extracted without any attachment to the body wall.

If necessary, the composite can be torn down the spiral butt joint in order to strip the body away from the product. In certain instances the contents shrink on cooling, and can therefore be tipped out as a solid block.

9. *Vapour corrosion inhibitor (VCI) paper*

This paper can be incorporated as an internal ply in composite containers used for steel and iron products, such as roller bearings, small precision instruments and gear shafts, where there is a risk of rusting.

Other barriers

Other materials which provide some form of barrier include paraffin wax, with which the containers can be impregnated, and bitumen adhesives applied between the layers of board. Paraffin wax provides a high degree of water-vapour resistance and increases rigidity, while bitumen adhesive provides both liquid water and water-vapour resistance.

In some containers, instead of applying a barrier lining as an extra ply of material, it can be sprayed onto the inner wall. During the final stages of making the body, as the end component is seamed on, a fine pressurized jet of quick-drying "flushing compound" is sprayed inside. As the container rotates, centrifugal force ensures that the compound is evenly distributed on the wall. By this method, a coating is also applied to the component seam, giving extra protection against seepage where the metal is attached to the board.

Virtually any mechanical or functional barrier material which can be laminated to or coated on paper can be used in the manufacture of composite containers, but this potential has not been extensively developed because of the restrictive cost factor and limited demand.

Types of closure

By definition, a composite container is any fabricated pack with a board body and possibly one or more tinplate or plastics components. The first component part having been dealt with, let us now examine the functions of the second.

Composites almost always require some type of reclosure and/or a dispensing device. In general, closures which are applied to metal cans are commonly used for composites too, but there are certain additional methods which are peculiar to composites.

Every closure used has some definable feature which, if expressed correctly by the customer, can help to convey his exact requirements in

METAL-BASED PACKAGING

terms which the supplier understands.

The three basic types of closures are similar to those on metal cans, but apart from these, there are others which are applicable only to composites.

Generally composites are supplied with the bottom component already in place and with the top closure loose—to be applied after the container has been filled. It is important to differentiate between closures which require no seaming by the customer and those which do. First of all, let us consider those which require no additional seaming operation, where the customer is only required to place on the lid.

1. *Frictional engagement*

In this category, there are three different types of closure:

(a) Slip (on) lid.—This is a rather loose type of closure, the rim of the lid fitting over the outside of the body. Since the closure is not air-tight, where there is a possibility of deterioration due to ingress of water vapour, it is advisable to seal the container after filling by securing the lid with tape; but this detracts from the appearance. An alternative method of firmly securing the lid is to over-label a plain composite after filling with the lid in position. As the rim or "skirt" of a slip lid is usually about 13 mm deep, the label encircles this part of the lid and fixes it securely. This operation is only achieved successfully with a raw edged lid. It is appreciably more difficult if a curled-edge slip lid is used. The advantage of a curled-edge lid is that it enables easier removal because of improved grip on the lip of the curl.

A refinement of the ordinary slip lid is the "captive" slip lid which is peculiar to composites. It is an ordinary curled-edge slip lid which is reshaped by the customer after filling, whereby the component is clenched into the body wall during the closing operation. Since the lid is beaded on to the body, rendering a tight fit, this obviates the need for taping, and after first removal the lid can be replaced by "clicking" back over the body head (see figures 3.3*a*, *b* and *c*).

(b) Plug (in) lid (composites only).—As the name implies, this component is inserted into the aperture at one end of the body and completely closes the container. The flush engagement of the sides of the lid with the internal container wall produces a tight-fitting closure which can, however, be opened with moderate finger leverage. To accomplish this, plug closures have curled flanges, which also prevent them being pushed wholly into the container. If an exact cubic capacity is required, eliminating ullage, it is essential that the countersunk depth of the lid is taken into

COMPOSITE CONTAINERS

Figure 3.3 Slip (on) lids

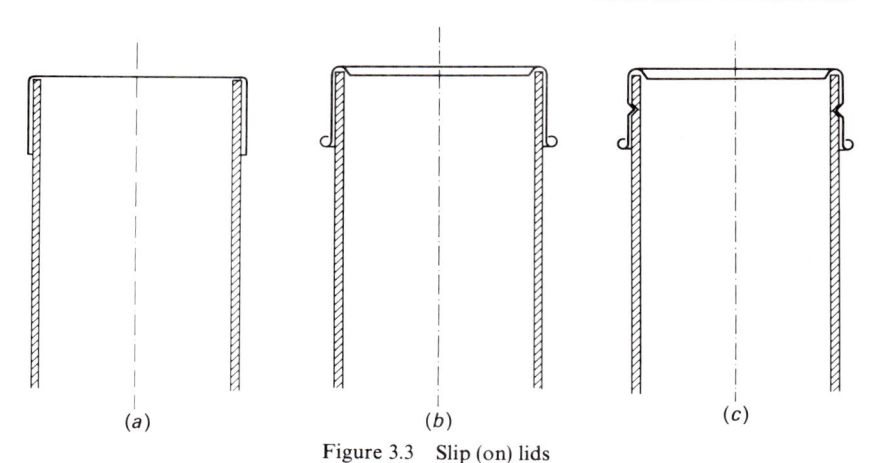

consideration, otherwise the overall effective depth will be insufficient (see figure 3.4).

Both plug (in) and slip (on) lids have certain characteristics which make it very difficult to decide which type to adopt. They are both fitted by assembly on to the composite after filling, and are not sufficiently air-tight to offer any functional protection. There is very little advantage in using one type as opposed to the other, and choice can really be decided upon only after considering the conditions the composite has to withstand. Under normal conditions both function equally well, but probably the plug lid is slightly more secure, as it is held in position by the body wall. As a general guide, slip lids are usually cheaper than plug lids, although this may not always be true, depending upon the methods of manufacture.

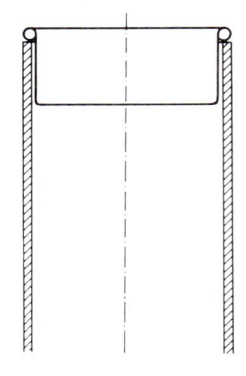

Figure 3.4 Plug (in) lid

METAL-BASED PACKAGING

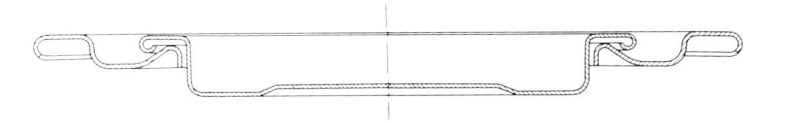

Figure 3.5 Ring and cap assembly

(c) Lever lid (as for tins).—The lever lid is always used in conjunction with a ring which is seamed on to the top end of the container (figure 3.5). The composite can be supplied with the bottom-end component and the top-end ring seamed on and, after filling, the customer merely has to insert the lid into the ring. Where this type of closure is used on food containers, particularly those which are dispensing packs, it is advisable to incorporate a safety rim on the ring. Bulk catering packs, for example, where the contents may be dispensed by hand, invariably have this type of "safety" ring to prevent injury when extracting one's hand from within the ring aperture (figure 3.6).

One optional feature of the lever lid and ring closure is a diaphragm which may be incorporated to act as an extra seal. Here, the container is supplied with the lever lid, ring and diaphragm assembly seamed on, and the bottom end open. The customer fills through the bottom, and then seams the bottom end on. In order to remove the contents, it is necessary not only to lever off the lid, but then to puncture the diaphragm. The diaphragm, which may be paper, parchment, or aluminium foil, operates as a functional barrier, the tamper-proof feature being a secondary characteristic. An example of this is a custard powder container.

These closures have an inherent disadvantage, however, which can be an inconvenience at times. Everyone will know how difficult it can be to lever open a paint can unless a suitable implement is available. Fortunately for the housewife, food-can lever closures do not require the same degree of pressure to release the lids. A recent development has now considerably reduced this problem, and the new modified-profile "lever lid" closure (figure 3.7) can be easily opened by slight twist leverage with a coin under the bead of the lid. Equally, it can be re-closed simply by exerting hand pressure on the lid to snap it over the bead contour of the ring to give an air-tight seal.

Figure 3.6 Safety ring and cap assembly

Figure 3.7 New profile ring and cap assembly

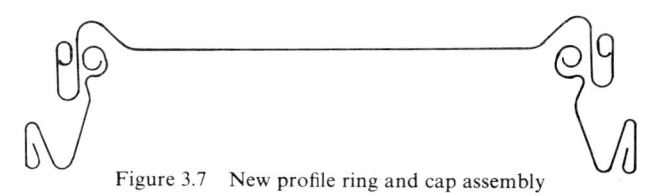

2. *Screw thread engagement of the lid*

This type of closure is not now extensively used on composite containers, as its importance has gradually diminished with the introduction of plastics dispenser devices. There is one design of screw thread closure which is used, however, and this is the threaded sprinkler neck and screw cap with assembled wad which is used on talcum powder composites. This type of closure is similar to the ring used in the lever lid and ring assembly, but with a much smaller aperture, through which the customer fills the container. As the composite already has both ends seamed on prior to filling, all the packer has to do is to insert the neck and cap unit into the ring, and the pack is complete.

No more need be said on screw thread components in this context as the matter has been discussed in the section on metal cans (page 2.42).

One significant characteristic of the closures mentioned so far is that the packer is not involved in any seaming operation once the container is filled, with the exception of the lever lid assembly, which after all is an optional feature depending upon the packers' circumstances.

3. *Non-detachable closures*

The third category of closure is the "seamed on" type, which necessitates a further seaming operation after filling, by the customer, to complete the pack. Basically, there is only one closure accomplished by this permanent clenching method, but there are variations on this type unique to composites which are slight modifications of standard ends.

(a) Standard end (see figure 3.8).—An example of this is the base components of any composite. This *end*, as the component is called, is clenched on to the body wall and not interlocked with the body as in

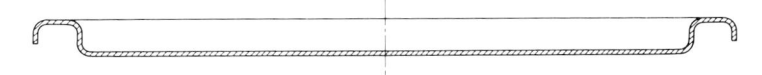

Figure 3.8 Cross-section of end

double seaming. The flanges of the component are initially flat or slightly curled, with sharp edges. During the seaming operation, these edges are turned inwards and compressed against the board wall, whereby the end is attached to the body. The resultant joint or "seam" is quite firm and will withstand considerable internal and external pressure.

(b) Semi-perforated top.—This is an end component which has been partially perforated in a number of positions to facilitate puncturing by the user. No metal is removed initially; it is only scored to prevent leakage when filled prior to use. This top is not commonly used and is gradually being replaced by the fully perforated top (*c*).

(c) Fully-perforated top.—This is an end component which is completely pierced in a number of positions, i.e. metal is totally removed. The holes are then sealed by a removable board disc or piece of adhesive tape. Sometimes an additional dispensing feature is inserted into the fully-perforated top, such as the plastics sprinkler top used for talcum powder.

(d) String opening containers.—This closure is a development peculiar to convolute containers only, where the opening device is inserted automatically during the winding operation. (A circumferential thread of string cannot be introduced into a spirally manufactured body because of the angle at which the board is applied on to the mandrel.) Once the container has been filled and finally sealed with a standard end by the customer, it can be opened by pulling the circumferential thread which perforates the label, whereby the two sections of the container can be separated. Since this style cannot possibly be opened without affecting the external appearance, it possesses an inherent tamper-proof feature.

(e) Double seamed end.—As explained earlier, the normal method of seaming an end on a composite container is to turn in the component flange and compress it against the body wall. With composites, this has always been the general technique, but the latest development now being employed is the *double seam* which is, in fact, the standard seam for a metal can.

To produce a double seam it is necessary for the base of the body to be "flanged" (figure 3.9*a*) so that the engaging component can grip the body and fold it over upon itself during the seaming operation. Now, instead of the metal merely pressing against the body, the two are interlocked together in a tight curl.

A distinct advantage of the double seam is that it is much more rigid and difficult to separate from the body. Under normal conditions an

COMPOSITE CONTAINERS

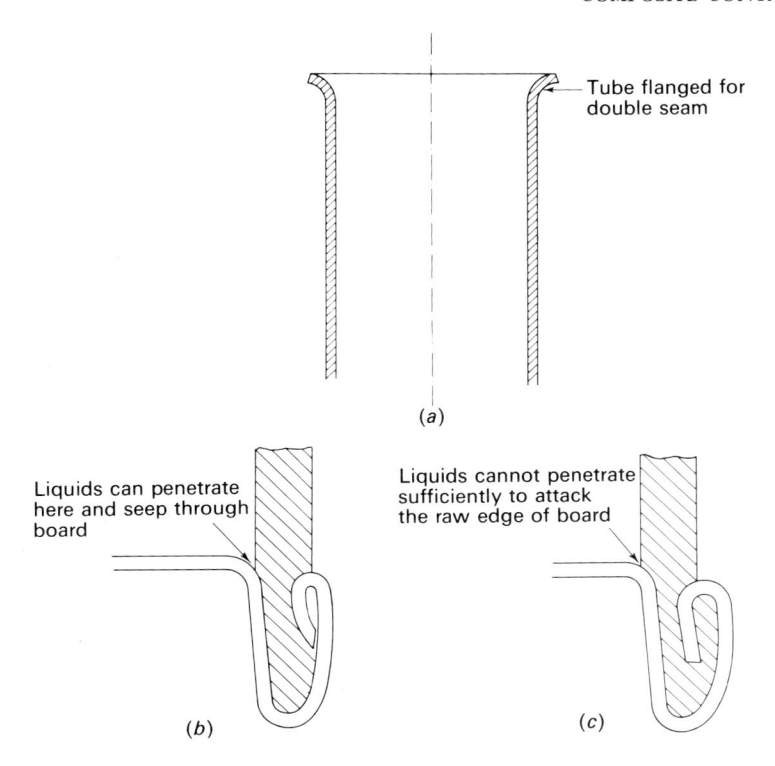

Figure 3.9 Double seaming

ordinary seam is quite adequate for most purposes, but where composites have to withstand severe external pressure, then it is better to use a double-seam container, as there is less chance of the end becoming separated from the body. Another feature of the double seam is that it prevents seepage of liquid contents, since the raw (cut) edge of the board is protected within the enclosed seam. With normal seams it is possible for liquid contents to penetrate between the metal base and body, and then seep through the body wall, thus spoiling the external appearance (compare figures b and c).

4. *Membrane closures*

(a) A similar closure to a lever ring component with a foil or paper membrane and plastics reclosable snap-on lid has been developed in recent years. When the membrane has been pierced and removed, the pack can be closed by replacing the lid over the protruding ring seam. It is only styling and easier opening facility that differentiates this from a conventional diaphragmed ring and lid closure.

(b) Some membranes are heat-sealed to a polyethylene coated liner, thereby eliminating the need for a metal rim under which a seal is attached to the body. When the membrane has been removed, a plastics plug-in lid closure reseals the full aperture. This form of closure has a limited application, as it can be used only when a heat-sealable inner lining is incorporated in the body construction.

Generally these closures are fitted to the container by the supplier, and the customer is required to assemble the base component after filling, thus giving a tamper-proof unit.

Usage

There is a wide variety of components available, but we are limited to some extent by the functional performance of the individual types. Actually the product determines the most suitable closure. Composites with two seamed-on components are more practical for small engineering spares and industrial components. Both types of perforated closure are suitable for scouring powders and insecticides. Plug-lid closures are used for sports goods, toys and novelties. Slip-lid closures are used for some confectionery, and ring-and-cap closures for custard and milk powders.

In addition to the many tinplate closures, there is also a wide range of plastics dispensing and pouring devices which can either be inserted into or glued on to the metal components.

Basic cost factors

The cost of a composite container is determined, to a large extent, by the requirements of the commodity packed. A thorough investigation is conducted on every new project to determine the correct liner material, adhesive, tinplate lacquer and coating, required to fulfil these requirements. After all, performance is of first importance, size and shape being secondary. The product governs the cost of all the material used; the size influences the production speed; and sales appeal imposes the cost factor of decoration. To arrive at a satisfactory situation, we must relate performance, size, style and shape with cost.

(a) *Materials and construction*

At present, of the two different types of composite, it is cheaper to produce spiral than convolute or straight-wound composites on a comparable basis. Convolute-winding machinery has much greater bearing on size limitation and cost than spiral-winding equipment, since it can only accommodate a narrow web width. Although all non-cylindrical composites are more costly to produce than cylindrical spiral ones, this premium may be counterbalanced by the savings in transportation costs and storage space.

With the many varied operations and special processes involved in production, it is very difficult to lay down any definite principles upon which we could base the cost structure. Since the material used represents a major part of the total, its cost serves adequately as a general guide.

The board forming the body is applied from a standard-width reel, which itself is slit from a larger reel. As the board for any particular composite is cut to a uniform width, there is no shred or trim allowance necessary, and the total board allocated forms the bodies without much wastage.

The size and shape of a composite has an indirect influence on cost in connection with the tinplate closure. On most cylindrical composites there are one or two closure components. Each of these circular components is formed from square or rectangular sheets of tinplate, and consequently there is a percentage of waste material which becomes increasingly more expensive as the diameter becomes greater. In theory, square or rectangular components are more economic propositions, but this advantage is offset by the present technical limitations of non-cylindrical composites.

The most economic situation to aim for is one where the cost of the pack does not outweigh its usefulness, and in this respect it is essential to arrive at the most suitable balance between material cost and performance. As tinplate is more expensive than board, it is better to use more board and less metal, provided circumstances will allow it. For example, the board cost in a tall narrow composite could be almost the same as or slightly more than, the tinplate cost, whereas in a squat composite the board cost could represent as little as a quarter of the tinplate cost. As a general rule, it is cheaper to have a tall composite, where there is more board used than tinplate, rather than have a squat composite where the tinplate cost is disproportionate in relation to the board cost. By careful correlation it is possible to assess an equilibrium cost, whereby the two factors determine the optimum size of a composite relative to material cost.

(b) *Labelling*

A plain board container is not particularly attractive, and almost invariably the body is labelled to give it the appropriate customer appeal. There are two ways of labelling composite containers: during the winding process, or afterwards as a separate operation.

The first method is operatively cheaper, but the initial gravure cylinder cost for printing the labels is high, and therefore it is only economic for large quantities. However, flexographic printing, as opposed to the more expensive rotogravure process, has overcome this problem to some extent, and shorter runs have become viable propositions where relatively straightforward graphics are involved. Intricate half-tone designs with fine registration tolerances could not be achieved. Dimensions and angles are extremely important on spirally applied labels. When originating designs, the layout angle must coincide with the angle at which the label is wound on to the tube, so that the print finishes in the horizontal plane.

Spirally applied labels can be printed either with a random or with a registered design. Random printed spiral labels present little difficulty, whereas registered designs have to be carefully prepared and laid out, and consequently tend to be more expensive.

A more significant advantage of the random printed spiral label is that, as the design is repetitive around the body, the print can be read from any angle in a horizontal plane. Random printed labels may not appear at first sight to be very attractive, but attractiveness is not the only selling point; labels are made to be read, and this is an outstanding characteristic of this type of label.

The second method, known as flat or "gang" labelling, necessitates an additional operation where the "gang" of several individual label units is fed into a machine, coated with adhesive, and applied automatically to a length of tubing. This type of labelling is more expensive because of this extra operation, although origination costs are considerably less, and relatively small quantities can be economically produced. Most paper labels for this application are printed letterpress or litho, but gravure-printed foil labels can create really impressive embossed designs.

Irrespective of the method employed, the governing factor of cost is the detail of the actual design, the number of colours involved, and the quantity. A basic principle is that the greater the quantity, the lower is the unit cost. This is particularly so with labelled composites, where there can be a considerable price change effected by quantity variation.

Measurement of composite containers

Once a customer has decided upon a particular item, it is essential that the necessary information is conveyed to the supplier in precise terms. To express the dimensions in standard units of measurement is not sufficient, because the composite maker's interpretation of this information may be completely different from that which was intended. In

COMPOSITE CONTAINERS

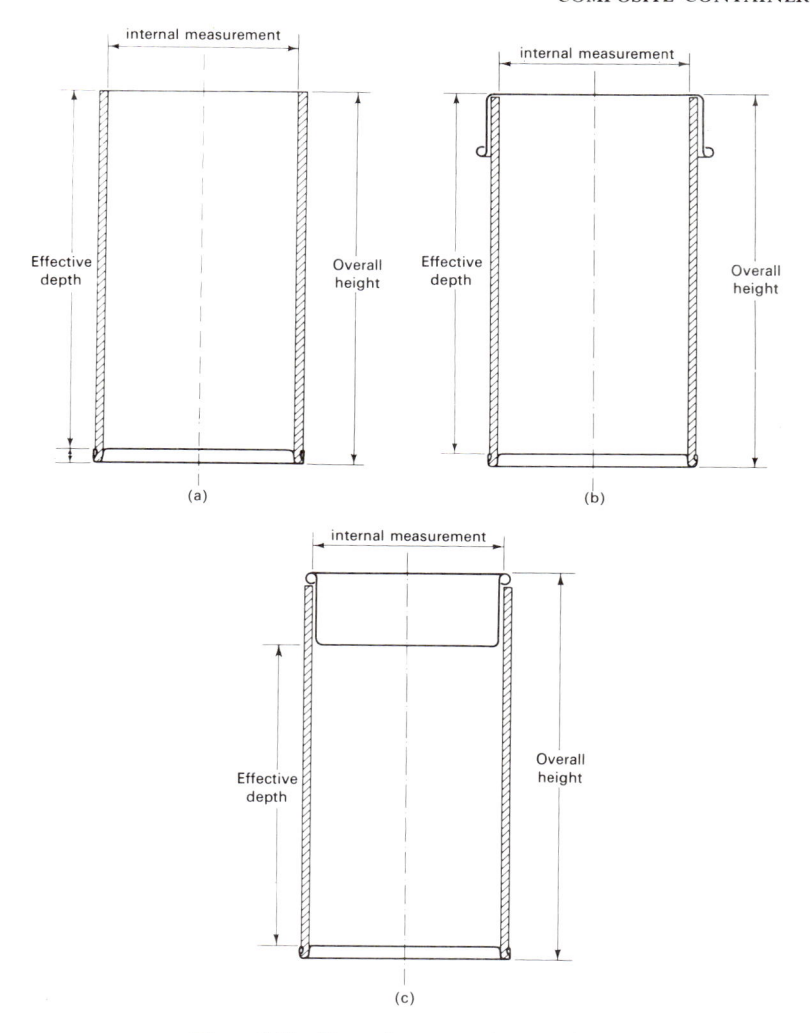

Figure 3.10 Measuring composite containers

general, composites are defined by the base and height dimensions only, with the base details given first (see figures 3.10a, b and c).

The basic diameter is taken as an internal measurement, excluding the wall and seam thicknesses. This exclusion is merely for the supplier's benefit, as to include various wall thicknesses would make measurement extremely tedious, since sizes would have to be defined to the nearest thousandth of an inch. Most base dimensions are expressed in fractions of an inch as an internal measurement, but sometimes dimensions are described in the trade by the standard digit system. If the diameter of a composite is expressed as a series of numbers, then it can be assumed that this is a digit system measurement, and the diameter is taken as an overall measurement across the seams. Under this system the last two digits denote sixteenths of an inch, and the preceding digits denote the whole inches; e.g. 211 means $2\frac{11}{16}$ in, 502 means $5\frac{1}{8}$ in, 700 means 7 in.

The body height dimension is taken over the top and bottom seams, and described as a deep external measurement. Component protrusions beyond these seams are not taken into account, and this is a very important point to consider when calculating packing-case sizes, storage space and distribution costs. For example, a talcum powder composite may have an external depth of $4\frac{1}{2}$ in, but an overall depth of $5\frac{3}{8}$ in from the base joint to the top of the plastics sprinkler cap.

To express the exact depth of a composite container can be confusing, especially where the cubic capacity is critical, so the following principles should be borne in mind. The diameter or base dimensions are always taken as internal measurements, but the depth can be expressed in two ways according to the style of container. A composite container, i.e. a container with a metal end seamed on, has an external depth measured over the component seams, whereas a board tube (core) without any components is described by its overall body length.

Another peculiar characteristic of composite measurement to observe is the countersink depth of certain components. Slip (on) lid closures do not affect the internal depth of a container, but the plug (in) lid does reduce the effective internal depth by as much as 15 mm in some instances. Other components do not make such internal depth alterations, but some standard ends are panelled, and this does affect the packing of certain products because of the uneven surface.

Testing

The function of a good package is to ensure that the product to be marketed reaches the consumer in perfect condition, i.e. no deterioration of the product should occur during distribution or storage prior to use. Inadequate formulation or construction can cause instability and decomposition of many products. It is necessary, therefore, to check thoroughly all materials to ensure that optimum protection is achieved and maintained. Tinplate can be treated with phenolic, polyvinyl, epoxy or polyurethane resin solutions to provide protection against product contamination. Obviously such coatings will function properly only if the base metal is stable. All components are shaped, some with pronounced contours, under considerable pressure, so the materials used must resist such stress without fracture.

Specimen pieces of tinplate are therefore subjected to bend and fracture tests to determine their ductility.

It can take several different press operations to make some components, consequently they have to withstand greater stress.

The Jenkins Test

To decide the most suitable metal, a sample is tested by placing it in a machine and bending alternately through 90° in one direction and then the other until it breaks. Its resistance is measured by the number of bends it takes before fracture.

The Erichson Test

Another method of testing tinplate is to place a piece in a clamp exerting a steadily increasing force, using pneumatic pumping until the metal splits; at this point an indicator registers the pressure applied.

The body section of the composite is generally more susceptible to damage or failure. Several tests can be carried out on paper and board materials to determine their suitability.

Some of the more important tests relative to this section are:

(i) Cobb absorbency.—A conditioned 100 cm^2 specimen of material is weighed, then wetted with clean water for a given time, blotted and reweighed. The resultant weight increase is expressed as a gain in weight in grams per square metre. The less absorbant material is obviously better where moisture resistance is important.

(ii) Moisture content.—A Marconi Moisture Meter measures the change in electrical conductivity with moisture content on the surface of the material. The moisture contained is defined to be the weight loss as a percentage of the original weight after a specimen has been dried in an oven at 102–105°C to a state of equilibrium.

(iii) Mullen burst.—A fixed area of material is subjected to a steadily increasing force from one side. A rubber diaphragm expands in contact against a given area of the sample by hydraulic pumping and the pressure at which the material bursts is recorded. Many composite cartridges are used under pressure and therefore must be made from high-burst-strength materials, e.g. kraft.

(iv) Puncture.—This is a rather simple test to determine the force required to pierce a piece of board. A swinging pendulum device is used, increasing the angle of swing until the sample punctures.

(v) Compression. — A completely assembled container is placed between two press platens, one of which is fixed. Steady pressure is applied by one platen being driven against the other. The compressive force required to produce failure in the side walls of the container is measured and recorded, plotting pounds force against amount compressed in inches (newtons against millimetres).

Composite or metal container?—the choice

A composite is a versatile container, strong and rigid, yet lightweight and easily disposable. It is relatively inexpensive, attractive and can be used on orthodox can-filling equipment. This type of pack has many advantages to offer, but it cannot be categorically distinguished from an all-tinplate container. In some instances the two are complementary, yet for other products one would be incompatible with the other.

Generally speaking, the performance of metal cans against composites cannot be determined, as there are so many incomparable aspects. Each has its attributes and, in the final analysis, the decision will be based upon which of these qualities has the greater influence on what is considered the most important factor in marketing the product.

CHAPTER 4

Aerosol Packaging

A. Simpson

Over the last 20 years or so, a new method of packing and dispensing products known as *aerosols* has become very important. The system, first applied commercially in the United States during the latter part of World War 2, was used for some 1000 million packs in America in 1963, and 3000 million in 1973. In the United Kingdom the growth has been from 70 million in 1963 to 450 million in 1973.

The feature that started this packaging system was the realization that a liquefied gas inside a pressure vessel can provide a constant pressure in that vessel even though it is being emptied. For instance, by having a solution of insecticide mixed with the liquefied gas, it is possible, by controlling the ejection of the mixture using a small orifice, to fill a room with finely divided particles of the insecticide. From such a beginning, a system has been developed by the introduction of new combinations of liquefied gases, different containers, sophistication of the valve mechanism, etc., into an accepted method of dispensing many other products.

Even over the comparatively short time that this method has been in existence, the pressurized container has become so important that it is essential to consider it, like the collapsible tube, as a way of doing a job.

Definitions

Aerosol.—An integral ready-to-use package incorporating a valve and a product which is dispensed by prestored pressure in a controlled manner when the valve is operated.

Valve.—A mechanical device, the operation of which permits the controlled emission of the product from the aerosol in a predetermined manner.

Cap.—A removable protective cover over the valve actuator, located in such a manner as to prevent accidental operation of the valve.

Propellant.—A material which provides the power to eject the contents.

Types

A number of different types of aerosols exist, and it is convenient to divide them into three classes according to the way they function.

(a) Surface sprays: paints, hair dressings, moth proofers, lubricants, polishes.
(b) Space sprays: insecticides, air fresheners.
(c) Foams: shave creams, hand creams.

By considering the method of producing the spray we get:

1 High propellant content.
2 Low propellant content.
3 Special valves which themselves break up the spray, either by mechanical means or by allowing an excess of the propellant to be mixed with the spray.

Components

There are five essential parts. Figure 4.1 shows these and each of them is dealt with separately below.

Caps

The cap performs two functions, that of enhancing the appearance of the container and protecting the valve from accidental operation.

A cap often has the orifice of the valve built into it together with either a locking device to prevent accidental operation or a shielded plunger. Such "actuator caps" have the advantage that the user does not have to remove them to operate the unit. Whilst making the whole item more attractive, they are more expensive than a separate button and cap.

Valves

The final orifice performs a vital function in the successful operation of the aerosol. It is incorporated into an actuator button and may take a variety of forms. It can consist of a single hole of closely controlled dimensions, in a plastics moulding; or of a multiplicity of holes leading to

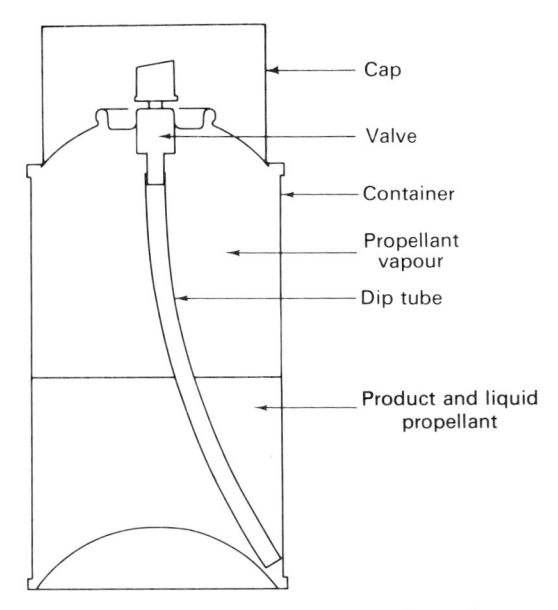

Figure 4.1 Section through an aerosol container

one final opening and, in such instances, often several plastics mouldings, or mouldings and machined parts, are locked together. Even a simple button actuator may still have a complicated maze of fine orifices and expansion chambers moulded into it. The button is mounted on a stem which is usually spring-loaded, so that a fixed rubber gasket closes over a hole in the stem when the unit is in the rest position (figure 4.2).

Depressing the button uncovers the hole and connects the interior of the container to the atmosphere. The pressurized material inside is thus released. There are a number of ways of making this seal. For instance, some have a solid rubber diaphragm that is pushed away from the boss of the valve cup by the stem which has a slit in it. Others have a stem in two parts, but still have a hole in the gasket so that the product flows underneath the gasket to the stem, and thence to the button.

The housing (see figure 4.2) holds the spring and gasket in place and enables the cup boss to be deformed after assembling the valve so that the gasket is compressed. The housing may have a spigot for a dip tube, so that the container can be held upright whilst the valve is being operated. The spigot orifice is often specially sized to restrict the flow of product.

A vapour-phase tap valve is one in which vapour from the gas phase in the pack is introduced into the housing using a second hole. Here a stream of product and liquefied gas and a stream of vapour meet each other and alter the spray characteristics. The flow rates of these two, as

2.75

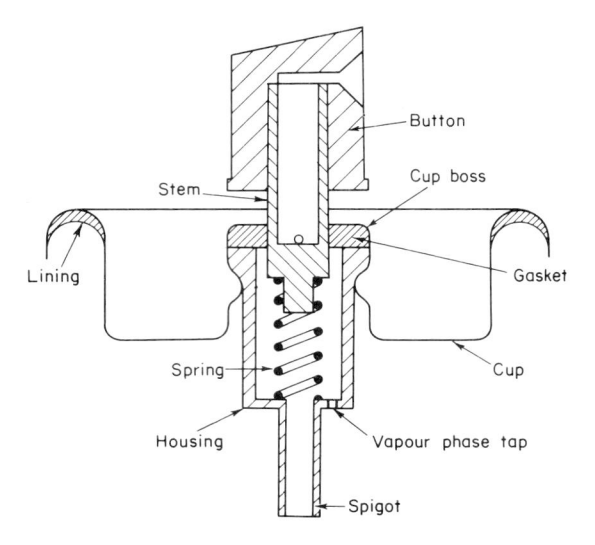

Figure 4.2 Parts of a typical valve

determined by the relative sizes of the holes, must be carefully selected to ensure a satisfactory performance of the valve.

The cup can be made of tinplate or aluminium. The lining, which is used to make the seal between the valve and the container, may be of the "flowed in" type as used for vacuum seals or consist merely of a rubber washer. The latter is usually used when either the product is very searching or the container is less uniform than the ideal.

The various components of the valve can be made from a number of different materials. For instance, the gasket can be soft or hard, resistant to some solvents and not to others. The spring can be of stainless steel or mild steel, and the housing of Delrin, nylon or metal.

The selection of a valve involves checking that the spray characteristics (spray rate, angle of discharge, particle size, distribution of particles, etc.) are suitable for the particular product, as well as seeing that it has a satisfactory operation throughout the anticipated life of the pack.

Containers

Metal, glass and plastics have been used for pressurized containers, and each has its place. Aluminium cans, while more expensive than tinplate cans, have the advantage of seamless construction, and hence the possibility of all-round decoration. They are manufactured by extrusion and can be of monoblock construction, in which instance the opening for the valve (which has been standardized) is made by a multiple forming operation. If the container is of two-piece construction, then the valve opening is manufactured as part of the forming process and merely trimmed. The end, which may be of tinplate or aluminium, is then double-seamed into place, using a sealing compound to ensure a satisfactory joint.

Tinplate containers available in the United Kingdom are all "built-up", i.e. a sheet of tinplate is bent around a former and, either by welding the edges, or by interlocking them and soldering, a cylinder is made. Two ends are then double-seamed into place to form a container. One of these ends has the standardized valve opening, and the other is concave in order to withstand the internal pressure. Such containers are not usually as strong as their aluminium counterparts, and in some countries drawn-steel containers are available at a premium cost.

Glass and plastics containers are used, either where the product is too corrosive for metal, or where the product is expensive enough to justify the extra cost. With glass containers there is always a risk of shattering, and a general limitation on the size and internal pressure is necessary. To a certain extent this can be overcome by coating the glass with a suitable plastics material, or even by inserting the bottle into a steel, aluminium, or spirally-wound paperboard tube. This increases the cost, although the coating or the outer cover can be suitably decorated, thus saving the expense of a label. With plastics containers there are problems in that many plastics are permeable to some common ingredients of aerosol formulations, e.g. certain perfume oils migrate through them, and water vapour often has an appreciable transmission rate.

Labels

No matter whether the label is printed on to the container body during manufacture or as an additional operation after filling, it should have a warning to the effect that:

> **This container is pressurized. Keep away from heat, including direct sunlight. Do not puncture or incinerate, even when empty.**

Propellants

Various gases have been used, and in fact the soda-water siphon is an example of a pressurized pack, although is not normally considered to be an aerosol package. Compressed gases have the disadvantage that whilst they provide an internal pressure in the container, it tends to diminish as the container empties. This means that the spray characteristics alter throughout the life of the container and may become unacceptable. Proper selection of the valve, product carrier and propellant, can minimize this effect, but in general aerosols are not suitable for those materials which require a uniform spray pattern. The most usual compressed gases used are:

> carbon dioxide
> nitrous oxide
> nitrogen

Table 4.1 lists properties of the common propellants.

It is very common practice to use a mixture of Propellant 11 and Propellant 12 and such a mixture has a vapour pressure of 240 kPa at 21°C. This is very suitable for use with tinplate containers which commonly have bursting strengths up to 1000 kPa. The use of the butanes as propellants is becoming widespread, and blends of these with propane are made in order to control the vapour pressure. There is some hazard with the use of propane and butane, particularly in the filling operation, and suitable precautions must be taken. However, once the pack has been sealed satisfactorily there is very little difference in the hazard between properly formulated packs with flammable and non-flammable propellants.

The selection of a propellant depends upon a number of factors such as:

> spray characteristics
> valve
> viscosity of product
> nature of product
> cost

The performance of a final pack is affected by several variables and, since these are not independent, the work must be carried out by experienced people.

Products

In theory, anything that can be made up into a liquid or paste form can be dispensed by this method. In practice, only those products which have some practical advantage in using this method of dispensing are so packed.

In general terms, it is important to ensure that the consistency (viscosity) is correctly adjusted to give satisfactory performance. Whilst it is possible to check that this has been achieved by suitable laboratory testing, in view of the wide range of valves and propellants available, it is essential that the formulator has sufficient experience to ensure that the most economical as well as the most advantageous combinations are used.

Having produced a satisfactory product, the physical characteristics

Table 4.1. Properties of common propellants

Common name	Chemical name	Chemical formula	Boiling-point (°C)	Density (g/ml at 21°C)	Vapour pressure* (kPa at 21°C)
Propellant 11	Trichlorofluoro-methane	CCl_3F	24	1·48	92
Propellant 12	Dichlorodifluoro-methane	CCl_2F_2	−29	1·32	585
Propellant 114	Dichlorotetra fluoro-methane	$CClF_2CClF_2$	4	1·47	190
Butane	n-butane	$CH_3CH_2CH_2CH_3$	−1	0·56	212
Butane	iso-butane	$(CH_3)_2CHCH_3$	−10	0·58	315
Propane	Propane	$CH_3CH_2CH_3$	−42	0·50	860

* $1\,lb/in^2 = 6·895\,kPa = 6·895\,kN/m^2 = 68·95\,mbar$.

must be recorded, and then a suitable series of test containers are packed. These containers are then stored and opened at intervals to check that their performance and the physical characteristics of the product are satisfactory. It is common practice to extend such tests over a period of eight months, although occasionally shorter times are used. Special techniques, usually available only in the larger laboratories, can reduce the initial screening for corrosivity to as little as one month.

The physical properties that the product must have and the behaviour of the finished pack should be noted. Typical factors requiring investigation might be:

        The product: density
                                    viscosity
                                    pH
                                    colour
                                    low-temperature stability
        The finished pack: moisture content
                                    spray characteristics
                                    residue
                                    internal pressure

Filling

There are two methods of filling that can be employed: one of these, commonly called the *cold method*, involves the cooling of the product and the propellant below the boiling-point of the propellant. The second method (*pressure filling*) involves handling the product at ambient temperatures and filling it into an open container.

Cold filling (figure 4.3)

It is essential that the product is cold stable, and that it does not become too viscous or precipitate at low temperatures (say $-20°C$). The system requires special filling equipment that must not only be capable of working at these temperatures but must also maintain the low temperature should the production line stop for any reason. Because of the capital cost involved and the large amount of energy required to heat the can and contents back to room temperature for testing, labelling and packing, the method has largely been superseded except for specific applications.

Pressure filling (figure 4.4)

Alternative filling sequences are illustrated in figure 4.4. After the product has been placed in the container, the air has to be removed; this is done either by replacing it with vapour propellant (purging) or by

AEROSOL PACKAGING

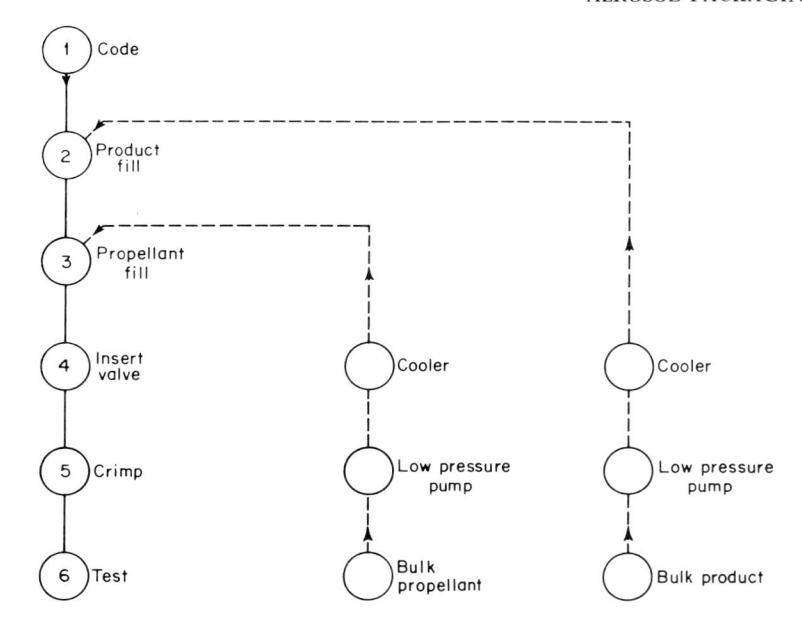

Figure 4.3 Cold-filling sequence

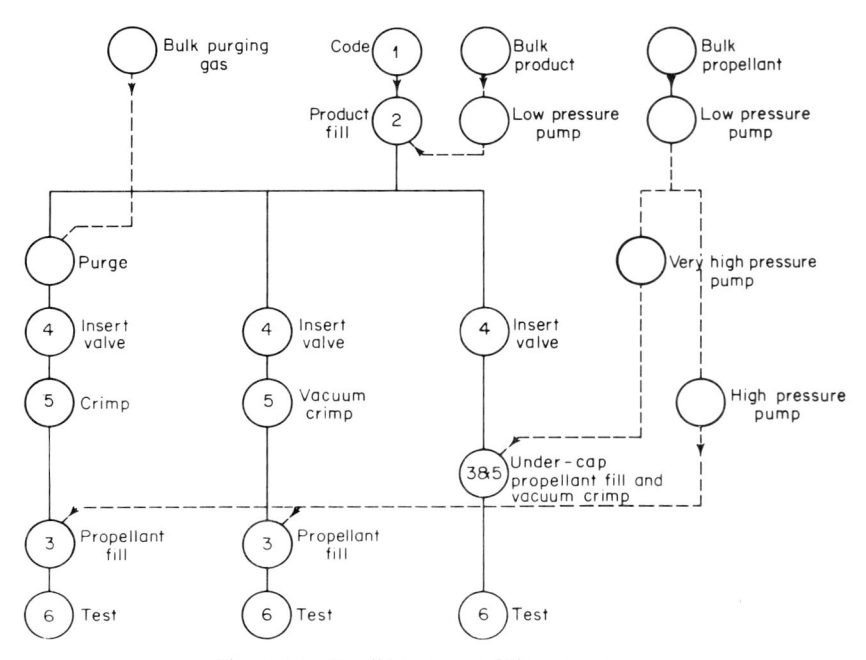

Figure 4.4 Possible pressure-filling sequences

2.81

drawing a vacuum immediately prior to sealing the valve in place. The vapour purging method has the advantage of simplicity, but is more expensive to run since it involves the dropping of several grams of Propellant 12 vapour into the can in order to displace the air before sealing. The exact amount required varies with head space volume and the product. The vacuum purging method requires a more complicated crimping machine and, unless this is a multi-head machine, it may limit the speed of production. It is more difficult to ensure that a satisfactory vacuum has been drawn, although the more sophisticated types have an automatic device for rejecting unsatisfactory cans.

The next machine makes a seal around the boss of the valve, or with certain valves around the valve stem, and then under high pressures (between 400 and 1000 lb/in^2) forces the correct amount of propellant into the container.

The amount of propellant shown in the space between the seals is lost when the head is removed, and it is important to keep this loss as low as possible. By suitably designing the head, it can be so arranged that the button of the valve is actually in place during the gassing operation.

Where the internal dimensions of the valve are restrictive to the flow of propellant, or the amount of propellant to be filled is very large, an "under the cap" filling method may be used advantageously. In this method, the container with the valve in place but not sealed is presented to the filling head. The valve is lifted out of the way, and a vacuum drawn on the container. The correct amount of propellant is then metered in, the valve pushed back into place, and the crimping operation performed. The amount of propellant lost is rather larger than with the conventional pressure-filling method, but the extra cost may well be offset by the increased speed obtainable, or the advantage of filling with the button in place. The filling head is, of course, much more complicated. There are, for instance, three seals to be made: air/propellant, propellant/vacuum, vacuum/air, and the chances of failure are correspondingly greater. However, these machines have now been in use for a number of years, and the system of filling is well proven.

A variation of this system is to place the valve in position, and on one machine fill the product and propellant consecutively around the valve into the container, and then seal the valve to the container.

The specialized filling having been completed, it is now necessary for the containers to be tested to ensure the safety of the consumer. This is usually done in a hot water-bath, which should be large enough to ensure that not only are the containers totally immersed, but that they are there long enough to raise their internal pressure to a suitable level. BS 3914: Part I: 1974 for aerosol dispensers specifies this, but for smaller sizes of metal cans it is common practice to use hot water at 55°C and immerse for three minutes. The time must be increased for large containers, or containers which have low heat conductivities, such as plastics coated glass or products which have low heat-transfer coefficients. The test not only exposes the containers to severe conditions and reveals any mechanical weaknesses, but it also detects gross over-filling and enables leaking cans to be rejected.

The cans are then dried, using hot-air blasts or sometimes infra-red drying tunnels. Buttoning, spray testing, capping, labelling, and packing then follow, together with any special operations required for presentation of the final package. These operations are similar to those carried out on other types of package and will not be described in detail here.

CHAPTER 5

Metal Drums and Kegs

R. M. C. Logan

The steel drum took its name from the musical instrument when in the 1840s a parallel-sided iron cylinder of riveted construction was first made as a substitute for the traditional bilged (or bulged) wooden cask. Over the next 60 years various developments were made, such as welding and double seaming instead of riveting; and the terminology and the definitions of the trade became a combination of those used by blacksmiths and those used by coopers. The First World War and the expanding petroleum industry no doubt created the demand which turned the trade into a mass-production industry.

Raw materials

Steel drums are normally made of low-carbon mild-steel sheet, traditionally called *black iron* because of the blue-black scale that covered it in the days when it was produced by hand-rolling. Today drum steel is rolled either hot or cold on continuous strip mills. In the United Kingdom, most is produced by cold-rolling steel strip which has been cleaned by pickling before rolling; it is subsequently annealed in an inert atmosphere to prevent oxidation, and consequently has a bright silvery appearance. For some purposes the steel is coated in various ways. The most common of these are tinplate, terneplate and galvanized steel. Terneplate employs a coating of a mixture of tin and lead, while galvanized steel is produced by coating the steel with zinc, either by hot dipping or electroplating. It should also be remembered that some of the drums with tin, lead or zinc coatings may have had the coatings applied not to the sheet steel but to the partly finished drum.

When drums are made from stainless steel, the alloy selected is usually an 18/8 austenitic variety, probably with a stabilizer against weld decay. Aluminium drums are made from commercially pure aluminium or a magnesium alloy.

By tradition the thickness of metal was described in terms of *gauge*, mild steel being given in Birmingham Gauge (B.G.) and stainless steel or aluminium in Standard Wire Gauge (S.W.G.). These gauges were defined in terms of weight per unit area, which after metrication, meant grams per square metre. For a given specific gravity this can be converted to thickness. So gradually the term *gauges* is being dropped and replaced by millimetres.

B.S. 1449 covers the kind of steel used in drum making and mentions thickness tolerance. The relationship between the gauge of the sheet metal employed in its construction and the strength of the finished drum is not a simple one, because the drum can become damaged in different ways. We can relate the difficulty of piercing a hole in the metal to its tensile strength, but resistance to deformation is a different matter—not merely stiffness of metal but also design and mode of construction must be taken into account. In fact, selection of suitable gauge is usually done on a trial-and-error basis. A trend over the years has been to use lighter gauges.

Manufacturing processes

The exact method of manufacture varies from manufacturer to manufacturer, but in principle centres round two machines—one to produce a tube by joining the two end edges of a sheet and forming a side seam, and the second to fix the ends to the tubular bodies (figure 5.1). Drums with a folded side seam have been produced for molten solids such as rosin or bitumen, but usually side seams are welded by one of two mass-production methods. Both are electrical, one being *resistance lap welding* and the other *flash butt welding*.

The principle of flash butt welding is to strike an electric arc between the two edges and then butt them together hot. In the case of small tinplate drums, soldering can be used. Whether or not the cylinder should now have corrugations is a matter of opinion, but it will certainly have beads or rolling hoops, either by pressing them out of the cylinder or by fitting separate bands. If the drum is to have a square cross-section, the circular cylinder is squared, and then the reinforcing corrugations are pressed in.

The ends of the drum are usually fixed to the bodies by a process known as *seaming*. The double-seaming process consists of rolling the edge of the flange of the ends round the flange of the body, and then

METAL-BASED PACKAGING

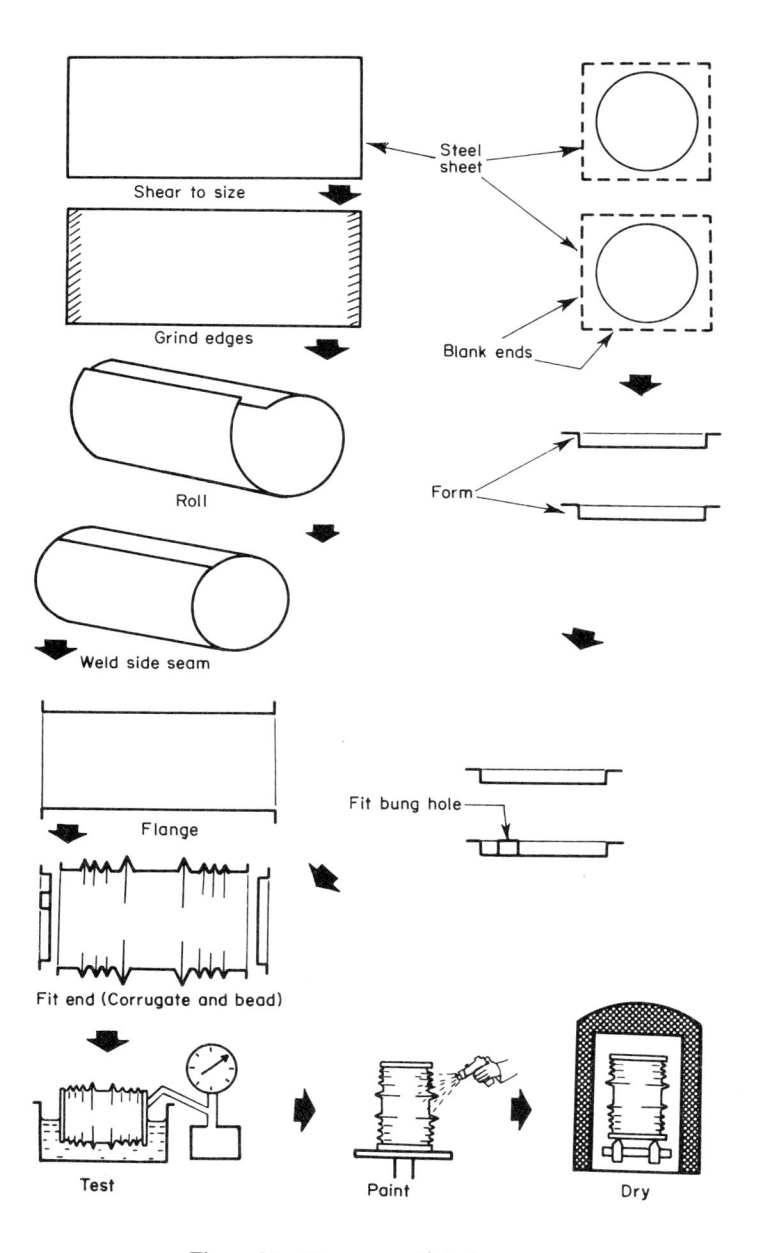

Figure 5.1 Drum manufacturing process

METAL DRUMS AND KEGS

folding this seam flat against the body (figure 5.2a). It is usual to put a layer of seaming compound between the two flanges to ensure a liquid-tight seam. This seam may be reinforced by inclusion of another band of metal, or it may be made more liquid-tight by resistance welding the two single thicknesses on the drum side of the seam (figure 5.2b). More recently a new design of end seam was patented, whereby the metal of the end flange and that of the body flange were rolled together in the form of a

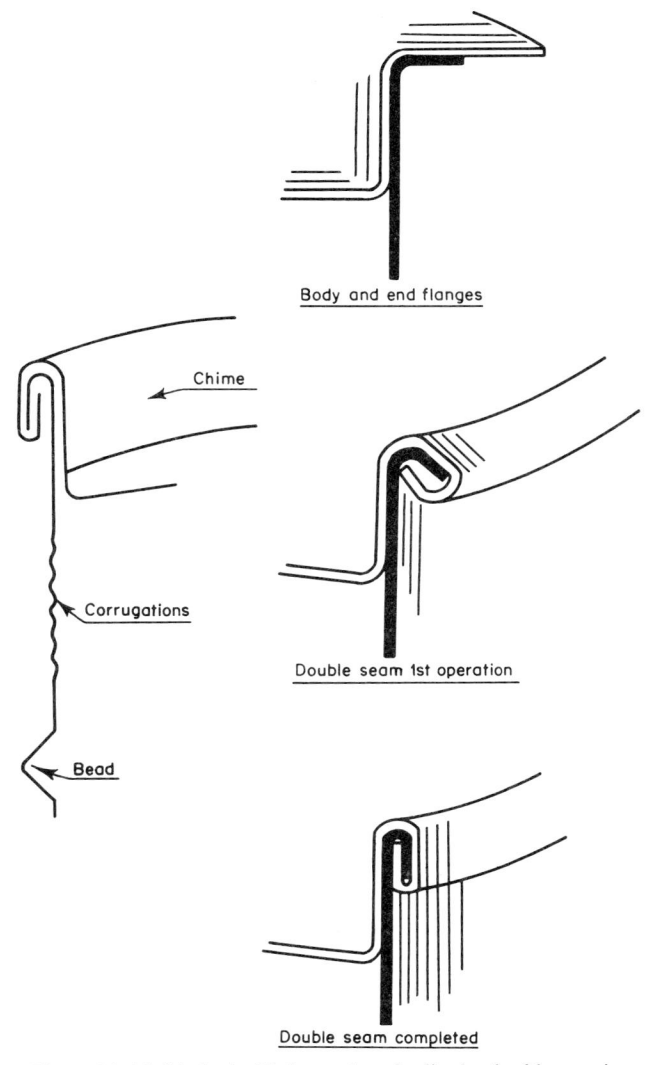

Figure 5.2 (a) Method of fixing ends to bodies by double seaming

METAL-BASED PACKAGING

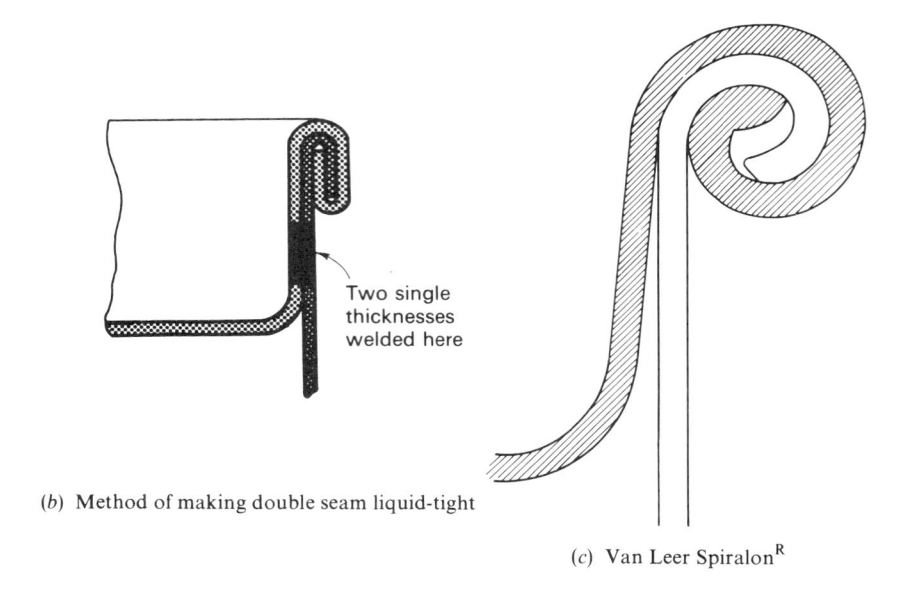

(b) Method of making double seam liquid-tight

(c) Van Leer Spiralon[R]

spiral (figure 5.2c). This is claimed to give much greater resistance to damage when dropped. Heavy-gauge drums (say 1·9 mm or thicker) do not have rolled seams but have the ends welded to bodies, usually by the electric arc process.

Drum types and standardization

Drum manufacturing grew up with certain arbitrary dividing lines, e.g. small drums and large drums, divided at about 15 gallon (70 litre) size; large heavy-gauge drums and large light-gauge drums divided at about the 16 B.G. (1·5 mm) thickness. The heavy-gauge drums were once further divided into bilged and straight-sided drums, but the bilged drum is now no longer made. They are always of an all-welded construction, whereas the light-gauge drum can be double-seamed or welded. These classifications are no longer quite so clear and, for example, large and small now tend to be judged by diameter. Both large and small drums are available, with either a tight head or a fully-open top, and in the small-drum range open-top drums are subdivided into kegs (figure 5.3) and pails (figure 5.4). In the United Kingdom, a pail is usually assumed to mean one where the lugs are an integral part of the lid, whereas a keg lid is usually held on, either by lugs on the body or by a band or closing ring. If the tight-head drum is in the small light-gauge group, it may have a normal chimb at each end, or it may be fitted with one of a variety of interrupted chimb heads to give a top which has no water-collecting

METAL DRUMS AND KEGS

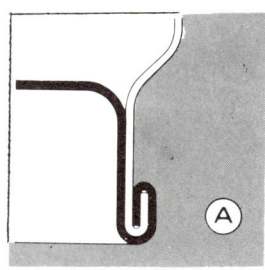

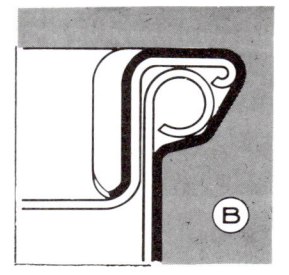

Figure 5.3 General-purpose steel kegs

    A reduced bottom   B tight-fitting lid
        ideal for stacking   suitable for powders, crystals, etc.

Side seam
    electrically welded
Bottom seam
    solutioned and double seamed
    reduced diameter to permit stacking (A)
Aperture
    body die-curled (B)
Strengthening
    one pressed-out bead near top of body
Head
    formed to fit curl and retained by bend over clips attached to body (B) or by bend-over clips attached to head (p.t.l.) or latch and eyelet closing ring
Finish
    interior plain steel, exterior painted
    interior lacquered, exterior painted
    body decorated by lithography or screening process

2.89

METAL-BASED PACKAGING

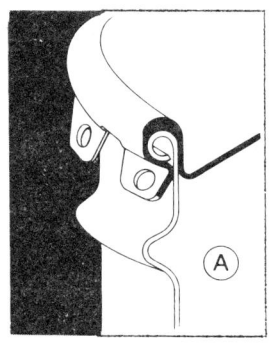

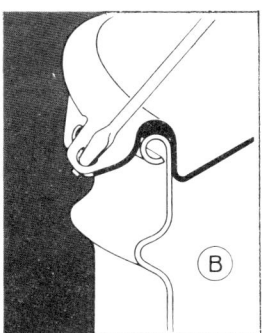

Figure 5.4 Steel pails

Body seam
  soudronic or conventional electrically welded
Bottom seam
  solutioned and double-seamed bottom
  reduced in diameter to permit stacking
Aperture
  body, die curled
Handle
  pail handle with grip attached to body if required
Head
  a gasket is incorporated in the multi-lug cover (A) which is easily opened (B)
Closures
  various neck fittings possible in lid
Strengthening
  one pressed-out bead near top of body

This package for powders, semi-solids, greases, etc., becomes a liquid pack by the addition of conventional fitting in the cover

2.90

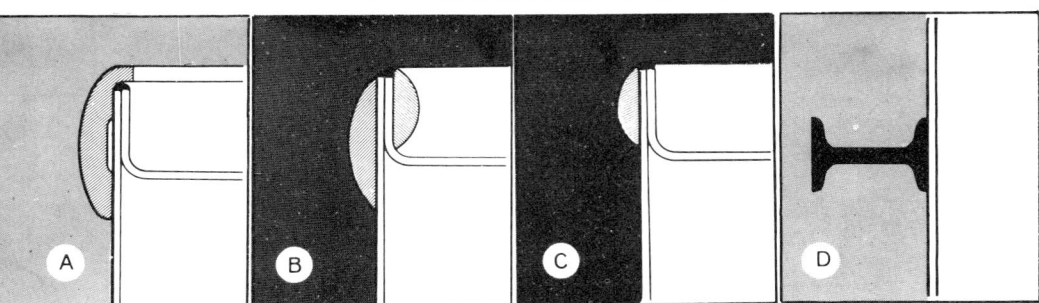

Figure 5.5 Heavy-duty steel drums

Body seam
  electrically welded
Top and bottom seam
  submerged arc welded
  and reinforced with lip section (A)
  or inner and outer convex bands (B)
  or convex outer band (C)
Rolling hoops
  two I-section rolling hoops shrunk on to the body (D)

pockets. Large tight-head drums may have pressed out rolling hoops, with or without reinforcement; but if they are of heavy gauge, it is usual for them to be fitted with I-section rolling hoops (figure 5.5).

In the range of small drums, taper-top drums with conical top sections are handy for pouring, but uneconomic of space, so they have almost disappeared from the market. There are other small drums which have specialized uses and are available in a very restricted size range, examples in the rectangular shape being the Robbican and the Tandrum. Another development applicable to small pails and kegs is a slight taper to the body which permits empty drums to be nested one inside another, and thus to reduce the space occupied in the storage and transport of empty drums. "Nesting" should not be confused with "stacking". Whether the sides are tapered or parallel, if the outside diameter of the base fits within the inside diameter of the top when closed, more stability in stacking is achieved.

As the ends for drums are produced by presswork involving the use of expensive dies, so each manufacturer would operate a limited range of diameters, and achieve a range of capacities by varying the height of the drum. This resulted in a very wide variety of drum sizes. Since the 1930s rationalization has been going on and, with the assistance of the British Standards Institution, three sets of specifications have been issued. In the latest editions classification is now by *duty*, so in the specification for light-duty drums with fixed ends (BS 814) both large and small drums are referred to, but in a restricted range of diameters. Similarly BS 2003 covers the range from small kegs to full-aperture 205-litre drums. BS 1702 covers heavy-duty drums with fixed ends. With the adoption of metrication in the United Kingdom, many drum capacities have been changed to litres. Initially this led to extra variety in the range, but eventually the "gallon" capacities will disappear. Incidentally, the capacity of a drum usually describes the volume filled. The total volume of the container will be greater than the volume filled, and the difference between the two figures is known as *ullage*. Ullage provides the safety margin to cope with any expansion of the contents due to heat.[1]

Standardization should take into account international trade, and a good example is the acceptance in Europe of the light-gauge single-trip container made to meet American Department of Trade specification 17E (formerly Interstate Commerce Commission 17E). The standardization in materials and method of construction does not necessarily meet users' problems if the package will be used for dangerous goods. Various bodies have produced lists of hazards and described performance tests on packages to carry them. For European Railways these are known as the R.I.D. tests. At the time of publication work is still going on sponsored by the United Nations to try to arrive at truly international tests of performance of packages for the carriage of dangerous goods.

METAL DRUMS AND KEGS

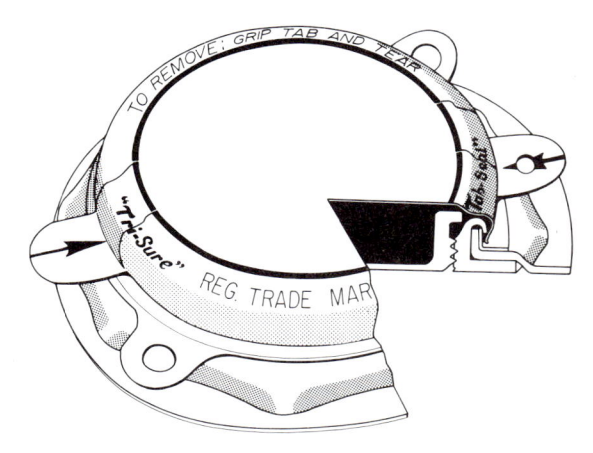

Figure 5.6 The Trisure closure with lacquer-coated steel or zinc plug, or polyurethene plug

Closures

Large tight-head drums for volatile liquids are normally fitted with two small screw-threaded openings, one with a 2″ and the other with a ¾″ pipe thread (now called G2 and G¾ in BS 2779). Into these are screwed a steel, zinc or plastics plug with external thread to match.

A few all-welded drums may still have a welded-in flange, but the majority of drums have a pressed-in flange with plug and cap to match, the most widely known set being the "Tri-sure" closure (figure 5.6).

Small-size liquid-tight drums often have tinplate closures such as a threaded screw cap or a presscap, or a friction plug with spun-on cap. The most popular friction plug is in the 3″ diameter (76 mm) size (figure 5.7). There are also closures which combine with pouring devices such as the "Flexspout". Drums for powders can be quite satisfactorily closed by large diameter (7″ or 9″) push-in friction lids. If the flange on the head is downwards into the drums these friction lids can be expanded in position, but, of course, once removed cannot be resealed (figure 5.8).

There is also a range of drums with fully removable heads. In the case of

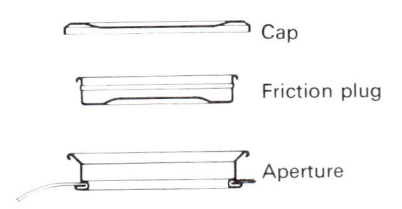

Figure 5.7 Friction plug for small liquid-tight drum

2.93

METAL-BASED PACKAGING

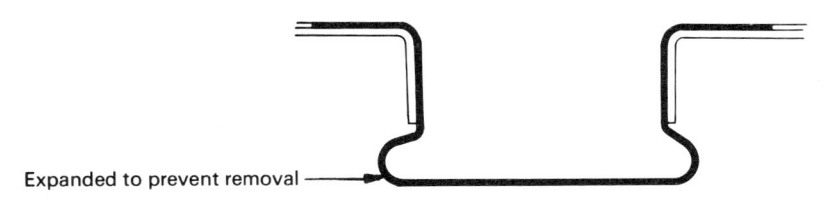

Expanded to prevent removal

Figure 5.8 Standard expanded lid

the larger drums, these tops are held on by a closing ring, consisting of a channel-section steel band held together by a bolt, latch or lever. Usually a gasket made of expanded or foam rubber acts as a seal between the lid and the curled-over top of the drum. In the case of small drums, an alternative to the cover held on by a closing ring is the lug cover sometimes known as the "American pail" cover (figure 5.9). In this type the washer or gasket is usually cast and cured in situ.

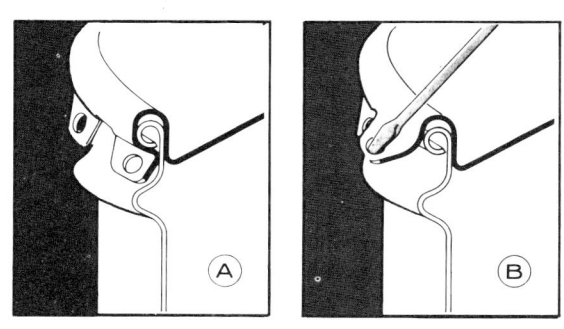

Figure 5.9 American pail cover
HEAD formed to take sealing gasket and to fit on to body curl with multilugs (A)—easily removable (B)

Protective and decorative finishes

At one time a thin coat of cheap bitumen paint was considered adequate for a drum, but today both interior and exterior surfaces are the subject of much specialized attention. The subject is best considered under three heads: painting, decorating and internal coating.

(a) *Painting*

Wherever practicable, drums are sprayed with stoving paints of a quality somewhere between that of a toy enamel and that of a refrigerator finish, according to circumstances. It is not usual to phosphate the outside of drums in Britain (although it is in the United States because of an inside treatment frequently given there). Air-drying paint is still used on some smaller production lines. The advent of roller coating to provide a base on which to print has led to many drums being painted entirely by roller coating the sheets before the drums are formed.

(b) *Decorating*

For many years it has been possible to print by lithography body sheets for small drums on tinplate printing machines, but it is only since about 1950 that litho machines robust enough to deal with sheets for larger drums have become available. Even so, there are disadvantages in the lithographic process for small orders, and the silk screen process is still used by many drum makers. Sometimes this silk screening is done on the finished drum, particularly if only one colour is involved.

(c) *Interior linings*

Probably the oldest and most expensive interior coating for mild-steel drums is pure tin (tin-lined drums should not be confused with tinplate drums). The coating thickness of tinplate is often a mere $0.25\,\mu m$, whereas in tin-lined drums the coating is nearer $25\,\mu m$. Unfortunately tin is cathodic to iron, so that any pinholes in the coating will have a tendency to turn brown when the iron base rusts. Lead-lined drums and drums made from terneplate (95% lead, 5% tin) were used in the past, but are now less common. Another popular metallic lining was zinc; galvanized drums are still used where long life is demanded and the zinc itself does not create new problems. More recently, however, zinc has been replaced by organic lacquer linings. First of all the drum trade used the oleoresinous-type lacquer used by tinbox makers to line food cans. These were known as *sanitary lacquers*. Then in the 1930s these were followed by phenolic resin coatings with good acid resistance but poor flexibility, vinyl resins with good flexibility but poor solvent resistance and, after the war, by epoxy-phenolic resins aimed at the best all-round properties of flexibility and chemical resistance. These were followed by lacquers based on polyurethane.

There is, of course, no universal coating and, even within each group, there can be a wide divergence of properties. Part of the success of a lining lies in the correct surface preparation of the underlying metal, which may consist of zinc or iron phosphating or some form of roughening of the surface such as grit blasting.

Some small drums are, like tin cans, made from pre-lacquered sheet, but the majority of drums are pretreated and lacquered in the prefabricated state, i.e. body and two ends separately. All conventional lacquer linings are heat-cured, and so the components are stoved before final assembly. This restricts the possibility of having all-welded lacquered drums, as the lacquer will get burned in the region of the weld. The Chemical Industries Association[2] once discussed lacquer lining pros and cons, and have published the opinions expressed. Conventional lacquers consist of synthetic resin in solution or dispersion, and one that had a limited use was a PVC plastisol. This led to the idea of making a lined drum from PVC-coated steel. In the United Kingdom the idea was abandoned, again because of the high cost and the limited chemical resistance of plasticized PVC. Recently the idea has been developed again, this time using unplasticized PVC and applying it to the components, as is done with spray lacquer instead of to sheet steel.

Another process involves the application of a dry powder to the steel, then sintering or fusing it. This is especially suited to a thermoplastics resin like polyethylene, though epoxy resin powder coatings are now available. Loose plastic liners can also be fitted to drums to give a composite having the strength of steel with the chemical resistance of plastics.

1   *J. of Inst. of Packaging*, Sept. 1965.
2   ABCM Packaging Convention, Sept. 1959.

CHAPTER 6

Collapsible Tubes

R. F. D'Lemos

Collapsible tubes are flexible composite containers for the storage and dispensing of product formulations which usually have a pasty consistency. They are made either in metal or of thermoplastics material, but the greater proportion of the demand for this type of pack is met by metal tubes.

Collapsible tubes are used in large numbers for cosmetics, pharmaceuticals, toiletries and adhesives. These containers also find some acceptance in the food industry in the United Kingdom for a variety of savoury pastes. A collapsible tube will store its contents for long periods without loss of product quality; as the techniques of processing and the protective coatings for both internal and external surface of the tube improve, an even larger variety of products will become packable.

The collapsible metal tube has been developed to a high degree of perfection; in it we have a comparatively cheap, very efficient and easily handled container. There may not be great changes in shape of the conventional tube, but there will be progressive improvements in the quality of lacquers, enamels, inks and varnishes. Of these, internal lacquers are the most important, as they can extend the use of collapsible tubes to cover a greater variety of products, some of which are particularly corrosive.

Plastics tubes have some advantages over metal from the corrosion aspect, but many problems have to be solved before they will replace the present metal. Nevertheless their use extends the range of products packed.

General description

A collapsible tube without contents is essentially a cylindrical container with a shoulder, nozzle and closure at one end; the other end is open to allow the filling of the product before this end is finally crimped to provide a completely sealed hygienic dispensing pack.

The size of the tube selected for any particular product will depend on the volume of product it is intended to hold. Charts giving capacity against dimensions are available from tube manufacturers. The nozzle and orifice sizes must be chosen to suit the dispensing properties of the product, and the approximate quantity required to be dispensed at a time. Dimensions of tube range from a minimum of 9 mm ($\frac{3}{8}$ in) diameter up to 76 mm (3 in); lengths corresponding to these diameters are dependent on the ultimate capacity of the closed tube and range from 40 mm ($1\frac{1}{2}$ in) to about 250 mm (10 in), but they can be longer if required. There is also a range of standard nozzle sizes and orifice diameters. The availability of various sizes in this large range, especially when combinations with nozzle sizes are considered, depends on the particular capabilities of the processing plant; sometimes one manufacturer of collapsible tubes cannot supply all sizes in the range with all varieties of nozzle.

Collapsible tubes are made in tin, lead, tin-lead alloys, lead coated with tin, aluminium, and certain thermoplastics such as polythene and polyvinyl chloride. Aluminium is now the most commonly used material for metal tubes and polythene for plastics tubes. There are a number of problems to be solved before plastics tubes can be used as widely as metal tubes. Briefly, some of the problems still to be overcome are the "collapsibility" of plastics tubes, their permeability with regard to moisture and essential oils, and the migration of plasticizers, especially where polyvinyl chloride materials are concerned.

Methods of production

As most products packed in collapsible tubes require to be hygienically stored, a brief description of the method of production will serve to illustrate that the risk of contamination of the product by the container is negligible, provided always that the necessary features of the tube to provide maximum protection of the product have been thoroughly investigated prior to filling and marketing.

Collapsible tube production may be divided into two main processes:

1 Mechanical fabrication of the tube.
2 Treatment of the tube surfaces.

Mechanical fabrication

Mechanical fabrication of aluminium tubes involves the following stages of processing. Most of these are carried out automatically on a mass-produced flowline basis, with manual handling reduced to a minimum.

Aluminium of 99·7% purity is cast, rolled to a predetermined thickness, and blanked to provide cylindrical pieces of metal of a defined diameter. These pieces of metal, generally referred to in the industry as *slugs*, are heat-treated to bring them to the desired metallurgical state for fabricating. The slugs are then lubricated and, by a process of impact extrusion, the lubricated slugs are converted in a specially designed press to the tube shape, with a shoulder and nozzle.

The tubes are transferred from the press to an automatic machining lathe, where they are trimmed to the correct length, a thread rolled or cut on the nozzle portion, and the external surface of the tube shoulder machined to give a decorative pattern if desired.

These five stages of processing provide tubes to the desired dimensions. The severe impact-extrusion forces involved in the transformation of slug to tube do, however, leave the tube in a work-hardened state. To provide the flexibility or "collapsibility" associated with these containers, they are annealed at temperatures not far removed from the melting-point of the metal itself. This stage in the processing removes all traces of lubricant used in the extrusion process; it anneals the tube and so provides "collapsibility". Moreover the high temperatures involved sterilize the tube completely.

Treatment of the surfaces

Surface treatment of the tubes follows and includes the coating of the internal surface of the tube with a protective lacquer (if this feature is desired), followed by a high-temperature stoving of the lacquer. The tube is then coated over the whole of the outside wall with an enamel coating, followed by a moderately-high-temperature stoving period. This external enamelled surface can then be printed in from one to four colours, as desired, on offset printing machines. A further stoving period dries the inks and sets the enamel coating. Tubes coming off the processing line at this stage require only a cap or closure to be fitted to the nozzle before packing into suitable boxes or cartons for dispatch. Throughout the processing, inspection procedures ensure that a consistently clean and good-quality container is produced.

Types and styles of container

As explained in the introduction, collapsible tubes are essentially cylindrical containers. Types of tube are generally distinguished by either:

(a) the material used for fabrication (i.e. tin, lead, tin-lead alloy, tin-coated lead, or

METAL-BASED PACKAGING

aluminium where metal tubes are concerned, and polythene or polyvinyl chloride where plastics are involved) or

(b) the type of nozzle on the tube.

The first of these distinguishing characteristics is self-explanatory. The second is rather more involved and some further comment is required. Several types of nozzle may be used.

(1) Conventional nozzles with more or less standard orifice sizes through which the product is dispensed. Typical of this type are the tubes used commonly for toothpaste, hand creams, and similar products.

(2) Nozzles where the threaded portion has an extruded rigid "canula" of small diameter. These are used where the areas of application of the product are specifically defined. Eye ointment and veterinary cerate tubes are typical examples.

(3) Nozzles where the orifice is covered by a thin membrane of metal which must be pierced before the product can be extruded. Such tubes provide a hermetically sealed package and are referred to as "Membraseal" tubes.

(4) "Taper" or "torpedo" nozzle tubes have no threaded portion at the nozzle area and require to be pierced with a pin before the product can be squeezed out. Certain adhesive tubes are typical examples.

(5) Nozzle, canula, and orifice features may be varied to suit particular requirements, provided these are within the practicabilities of the impact extrusion process on a mass-production basis.

(6) Plastics nozzles attached to metal tubes. These are special innovations which have found increasing use (especially where abrasion between the tube closure (i.e. cap) and the conventional metal nozzle produces a blackening of the product). In other instances a plastics elongated nozzle affixed to the tube is often used where the application of the pharmaceutical or veterinary product requires the tube canula to be brought almost into contact with the area to be treated and there is some danger or likelihood of damaging the affected area by metal contact. Certain veterinary cerate tubes with canulas are used as one-shot dispensers where the cerate is introduced into, for example, the udder or uterus of animals for treatment of mastitis or intra-uterine treatment.

These nozzles are attached to the tube by mechanically crimping it to the shoulder, or by spinning metal round the moulded plastics nozzle, or by injection-moulding it directly over a metal nozzle of specific conformation.

(7) Although tubes embodying various ideas for a captive closure were tried over a period, none completely solved the problem. First ideas were based on attaching the cap by some mechanical means to the metal nozzle of the tube. Although this solved the problem of cap loss, it did not remove the problem of abrasion between the metal nozzle and the cap, which

results in discoloration of the product. It is possible to apply a plastics nozzle portion to a tube and to fix a captive closure to it without the necessity of having the nozzle threaded in any way. Such closures have still not gained general acceptance.

Methods of assembly

Transformation of metal from *slug* shape to *tube* shape by the impact extrusion process does not involve any assembly of parts in the fabrication of collapsible tubes. Assembly as such takes place only after the tube is processed, i.e. fitting a closure to the nozzle of the tube.

There are, however, instances where the threaded nozzle portion is a plastics moulding made separately and fixed to the shoulder of the tube by a mechanical swaging or rolling operation, or where the nozzle portion is injection-moulded into a specially designed tube shoulder. In such instances there is another assembly stage prior to closure fitting and packing into cartons.

Types of closure and methods of closing

Closures for collapsible tubes are generally made of plastics materials today, and both thermosetting and thermoplastics types are available with a variety of external shapes and internal threads for screwing tightly on to tube nozzles. Push-on varieties were also available, but are not frequently used.

With certain shapes of nozzle or canula, elongated caps are made to cover both canula and threaded portion, and seal the canula at its top. Membraseal-type tubes are usually pierced open with a "spike" cap of polythene or hard plastics. There are also a few novel closures on the market which offer a captive feature, but these are not used in large quantities.

Packing and dispatch

Collapsible tubes are generally packed in the United Kingdom into paperboard cartons or plywood boxes fitted with honeycomb-type divisions, so that individual tubes do not damage easily in transit, nor have their highly decorative artwork scratched.

Tubes for the home market are delivered by the manufacturer to the filler's premises in plywood or fibre cases, fitted with a flanged fibre lid to exclude dust. Palletized loads are delivered and the cases, which are returnable against a nominal charge, are collected for re-use at the tube works. Tubes are also exported packed in divisional paper-board cartons inside stout wooden overcases.

CHAPTER 7

Metal Foil Packaging

J. R. Green

Aluminium foil is widely used in packaging. Its properties, strengths and weaknesses are described in the following pages and, from a study of these, rules for selecting the best foil combinations become obvious.

Aluminium foil is used for its protective value and its decorative appearance. It may be used either unsupported or laminated to paper and/or films.

Aluminium foil thickness

Aluminium is rolled to foil in thicknesses between 5 μm and 200 μm; the thinnest gauge of 5 μm, however, is not used in packaging, but in foil/paper electrical capacitors. The thinner gauges of foil, normally below 40 μm, are double-rolled, i.e. two webs rolled simultaneously, producing a foil which is bright on one side and matt on the other. Foil of 40 μm and over is usually single-rolled, producing a foil which is bright on both sides.

The thinnest aluminium foil used for packaging is normally 8 μm or 9 μm thick. These very-thin foils contain minute perforations (visible if the foil is placed over a light in a darkened room) which will allow traces of moisture and gases to pass through.

The major proportion of the perforations arise from the casting of the rolling slab and the hot rolling conditions, and there have been improvements over the years, both in methods of casting and in the production of foil. Today 12-μm foil will rarely exceed 200 perforations per square metre when tested by the standard method,[1] whereas 20 to 25 years ago 800 perforations per square metre would be common.

Where foil is to be used as a virtually complete barrier to water vapour and/or gases, consideration must be given to the best thickness to avoid excessive perforations or, alternatively, how to combine the foil with a suitable material which will block the perforations. Even very thin foil is, however, a very effective water vapour barrier. At 9 µm thickness, few samples of aluminium foil have a water vapour transmission rate (W.V.T.R.) in excess of 1 g/m^2 per day at 38°C and 90% relative humidity (R.H.).[2] As the foil thickness increases, the number of perforations drops rapidly, and foil of 30 µm and more has no perforations.

Where the maximum protection is required, e.g. many pharmaceuticals, dried foods, accelerated freeze-dried (AFD) foods, etc., the gauge of foils used ranges from 18 to 40 µm. A recent investigation into the effects of perforations in foil on its barrier properties carried out by Pira* has shown that, when the foil is combined with a film such as polythene, even the thinnest foil has no measurable water vapour or gas transmission rate.[3]

Decoration of foil

For decorative packaging, aluminium foil offers a bright surface and, in the thinner gauges, the alternative of a bright or matt surface which can be decorated by all-over colouring with transparent and opaque colours by printing using flexographic, gravure or lithographic methods, and by embossing.

The possible combination of both transparent and opaque colours gives a wide scope for attractive decoration. With transparent colours, the reflectivity of the foil varies the brilliance of the colour according to the incident angle of the light, causing changes which catch the eye of the purchaser in a supermarket, allowing the foil pack to stand out where it might otherwise be lost among other packages.

Because the reflectivity of bright foil may vary from black to brilliant silver, according to the incident angle of the light, printed information is better produced in opaque colours, or surrounded by an opaque colour, to throw the information into prominence under all lighting conditions.

Foil laminates

Thin aluminium foil is not particularly strong, and it is frequently laminated to stronger material to reinforce it. The cheapest reinforcing material is paper, and laminates of foil and paper are widely used in packaging. For foil gauges below about 18 µm, the strength of the laminate

---

*The Research Association for the Paper & Board, Printing & Packaging Industries.

is almost entirely due to the reinforcing paper or film. However, the barrier properties are dependent largely on the foil. The films normally used for reinforcing foil are regenerated cellulose, cellulose acetate, polyester, polyamide, polyethylene and polypropylene. These films, in addition to providing strength, protect the foil against mechanical damage. They also block any perforations in the foil.

Lamination of foil to paper is usually done by *wet bonding*, where the two webs are combined with an adhesive in the wet state, and drying takes place through the paper web.

Lamination of film to foil is normally done by *dry bonding*, where the wet adhesive is applied to one web, dried out, and then the other web is brought into contact with the adhesive-coated surface, and the combination passed through nip rolls which may be heated.

A third method of laminating is by *hot melts*, where the adhesive is applied in a heated molten state to one web, which is then combined with the other web, followed by chilling to set the adhesive. No drying is necessary, since there is no volatile liquid to remove. *Extrusion laminating*, which in principle is a variant of hot-melt lamination, is done by extruding a film of a molten plastics such as polyethylene through a flat die, bringing the foil and paper or film together with the extruded polyethylene film sandwiched between, and nipping the combination between rolls chilled to solidify the molten plastics.

Foil as a water vapour and gas barrier

Where foil is used as a water vapour and gass barrier, too much emphasis has been given, in the past, to the presence of minute perforations. The recent report by Pira[3] on the effects of perforations suggests that they are of little or no importance where a heat-sealing film is used in combination with the foil. However, large and therefore more serious holes which can be produced in a foil pack by fractures have to be considered. These fractures may be due to mechanical damage, or to stretching of the foil during folding, or to both. They are a more important cause of water vapour or gas transmission than the normal foil perforations produced during rolling, because they are in general much larger.

Water vapour permeability through aluminium foil is broadly proportional to the diameter of the pinholes, the transmission rate through a 1-mm hole in 10-μm foil being 10 mg per 24 hours.[4] If the diameter of a pinhole is only 10 μm, 300 holes per square metre would give a transmission of 0·3 mg/m^2 per day. However, a fracture in aluminium foil due to damage may easily be of 1 mm diameter.

Fractures occur most commonly at the corners of packs, especially at the corners of rectangular packs, and at folds where the foil is bent sharply, causing the material to stretch. Fractures are also more likely to occur when the foil is the outside layer of a thick laminate. To avoid fracturing the foil, it is best to eliminate sharp corners or, alternatively, to reinforce the foil on the outside with a heavy lacquer coating, or laminate to paper or film. Fractures can be virtually eliminated in a triple laminate of film/foil/film, where the foil is in the middle of the laminate and the two films are of approximately the same thickness.

Sealing the package

However good aluminium foil is as a barrier material, unless the foil package is properly sealed, a great deal of moisture or gas can pass through any overlap. Aqueous gums or solvent-based adhesives are generally unsuitable for sealing foil packs, since the foil is impervious, and the water or solvents will dry out only very slowly. Closure of foil packs is normally done by heat sealing using:

(a) Thermoplastic resin coatings on the foil or a paper reinforcement of the foil.
(b) Thermoplastic films on the foil or on the paper of a paper-reinforced foil.
(c) Hot-melt coatings such as wax on the foil or on a paper reinforcing the foil.

Thermoplastic resin coating weights of 5–8 g/m^2 dry weight are commonly applied to foil. Foil of 9 μm or 12 μm thickness, coated with a vinyl copolymer resin, is widely used for wrapping and sealing bars of chocolate as a protection against loss of moisture, contamination by outside odours and access of the coconut moth. Similar foils are also used for packing small portions of processed cheese. The cheese is pasteurized at the time of packaging, and hermetically sealed inside the coated foil. Such packs will keep for up to one year, as long as the foil barrier is unbroken.

Foil of 25 μm to 40 μm, coated with vinyl resin, is used in the pharmaceutical industry for unit packs of tablets. With such products it is important that the coating weight is adequate because, during heat sealing of thin coatings, minute capillaries could remain unsealed between the two layers of foil, allowing moisture to enter. Sufficient weight of coating must be present to flow under the heat and pressure during sealing to fill up these capillaries. Many pharmaceutical products are packed for long storage in the tropics in foil/polyethylene laminates. The polyethylene layer is two or three times the thickness of a normal heat-seal coating; not only does it give a stronger pack, but the film allows more complete flow under the heat and pressure of sealing, which ensures that any capillaries are adequately sealed.

METAL-BASED PACKAGING

Table 7.1 gives figures for various barrier materials[5] and illustrates:

(a) The very good barrier properties of even 9-μm foil (0·00035″).
(b) The lower barrier properties (laminates marked *) when a layer of paper is between the foil and heat-sealable film.

Table 7.1.

Material	W.V.T.R.[1]	O.T.R.[2]
28 lb GIP paper/0·00035 in foil/0·0008 in Pliofilm	0·006	2
0·00035 in foil/28 lb GIP/0·0015 in Pliofilm*	0·02*	67*
28 lb GIP/0·00035 in foil/0·0015 in polyethylene	0·02	1
28 lb GIP/0·0015 in polyethylene/0·00035 in foil/0·0017 in polyethylene	0·003	2
0·00035 in foil/28 lb GIP/0·002 in polyethylene*	0·74*	275*
0·001 in acetate/0·001 in foil/0·0012 in polyethylene	0·005	1
0·001 in acetate/0·001 in foil/0·0015 in Pliofilm	0·005	4

[1] Water vapour transmission rate (W.V.T.R.): g H$_2$O/100 in^2 per day at 100% R.H. and 25°C.
[2] Oxygen transmission rate (O.T.R.): cc oxygen/m^2 per day at 1 atm partial pressure difference at 75% R.H. and 20°C.

Table 7.2[6] gives additional examples of the effect of a layer of paper between the foil and the heat-sealing film. Water vapour transmission rate measured at 25°C and 75% R.H.

Table 7.2.

Laminate	W.V.T.R. uncreased	W.V.T.R. creased
Cellulose acetate/foil/coating	0·1	0·1
Cellulose acetate/foil/Pliofilm	0·1	0·1
Foil/paper/Pliofilm	0·4	1·1

Wax coatings which lack the strength of films or resin coatings can be impregnated into papers so that passage of moisture along the paper fibres is impeded; a laminate of foil/paper/wax will often give better barrier protection than a foil/paper/resin coating.

Heuss[7] discussing coated foil to coated foil seals and the effect of wax impregnation on paper to paper seals (basic weight 80–120 grams/square metre) (g/m^2) gives transmission rates of 0·03–0·06 mg/cm of seal per day for heat-seal coated foil to foil seals, 1·2–1·4 mg/cm of seal per day for non-impregnated paper, and for a wax-impregnated foil/wax/paper laminate of 0·1 mg/cm per day. In addition, leakage of water vapour through perforations or fractures in the foil is less when the paper is impregnated with wax than when the paper is not impregnated. A very efficient overwrap for biscuits and other confectionery is 9-μm foil laminated to 40-g/m² paper, the paper being coated with 25–30 g/m² of a paraffin wax/Elvax blend.

Other research workers[8,9] have demonstrated the principles of selecting the right foil laminates for protection, both by putting paper outside the pack to protect the foil from damage, and by using a flexible polythene extrusion of 12 g/m² as the adhesive between foil and paper to increase the resistance to damage.

Aluminium foil as a barrier to insect infestation

Because aluminium foil is a hard material, it can prove a bigger obstacle than plastics films or paper to penetration by insects, and in gauges above 30 µm it is quite a good but not an insuperable obstacle.

Insects may be divided into *penetrating insects*, those that enter packages by piercing or gnawing their way through the packaging material, and *invaders*, those that gain entry by way of an existing structural weakness or opening, e.g. by following a fold of an overwrap, through the damaged corner of a carton, or through a hole or puncture in the packaging material.[10,11] Penetrating insects are capable of boring through almost every known packaging material, except glass and thick metal. The invading type of insect, however, is an important source of insect damage, and the invasion of the saw-toothed grain beetle *Oryzaephylus Surinamensis*, the confused flour beetle *Tribolium Confusum*, and the flat grain beetle *Laemophloeus Pusillus*, has been examined for heat-sealable foil laminates which gave better protection for certain foods than did a number of alternative materials, including heat-sealable regenerated cellulose film, glassine and waxed paper. Collins[12] makes further comparisons with foil and foil-laminate test packages, cellulose film packages and glassine packages using these three beetle species.

Foil gives good protection, but the results emphasize the need for perfect sealing of all possible channels of entry to insect invaders, and show that with aluminium foil the cross-serrated type of sealing has a tendency to cause punctures which provide entrance for these insects.

Sterilization of foods and other products

The use of aluminium foil combinations, in the production of both pouches and small-unit-size containers suitable for sterilization processes, is becoming more common, both pouches and containers being closed by heat-sealing.

In the pouch field three and four-ply laminates are being used, the laminates being polyester/foil/polyolefin or polyester/foil polyester/polyolefin. Foil gauges in the range of 12–18 µm are normal.

In the small-unit container field, heat-seal coated foil or foil/polypropylene laminates are the base materials from which containers are formed. Foil gauges used are in the heavier range around 100 μm. The lidding material is generally similar to that used for the container.

The filled and sealed pouches or containers are sterilized in autoclaves but, because these types of packages do not have the strength of the normal tin can, there has to be an over-pressure in the autoclave during the processing cycle to balance the internal pressure developed inside the pack. Packaging of ready meals, food pastes and pet foods is already carried out in these types of container.

Sterilization by irradiation is used for surgical instruments packed in sealed-foil packages. The aluminium permits the passage of gamma radiation more easily than most metals. Care must be taken in the selection of the heat-seal coating material. Many of the normal heat-seal coating resins contain halogens and, under irradiation, these coatings break down, producing corrosive by-products which can attack the surgical instruments. Suitable coatings can be formulated, and polyethylene/foil laminates will withstand heavy doses of irradiation without much change, except perhaps an increase in brittleness of the polyethylene at low temperatures. This, however, is not a major disadvantage.

Containers—general

Because of the relative inertness of aluminium to many foods and its dead fold characteristics, gauges of foil from 40 μm to 150 μm are used in the production of semi-rigid containers intended for packaging of many foods, particularly confectionery, frozen foods, convenience foods and various jams. These containers are produced both in the wrinkle-wall form and the smooth-wall form, closures being effected with foil/paper laminates, paper board or plastics film.

The high thermal conductivity of aluminium assists in the rapid freezing of frozen foods, while the high reflectivity reduces the rise of temperature in freezer display cabinets in shops.

Selection of foil for protective packaging

From the above information it is possible to formulate rules for selecting a foil specification for a packaging purpose. Where the foil is to be used solely for decoration, the thickness of the foil is unimportant and, for economy reasons, it is usual to select the thinnest foil which will work successfully on the packaging machine.

In the United Kingdom much unsupported foil is used for decorative purposes, but laminates of foil and paper are also widely used. The grade of paper selected is the cheapest to give the laminate the strength and appearance required. It must, however, be remembered that thin foil easily takes up the surface "grain" of the paper and, because of the foil's high reflectivity, this becomes very apparent if a rough paper is used. The smoother or more highly glazed the paper surface, the better the appearance of the laminate.

Where the foil is to be used as a barrier against water vapour and gases, the heavier foils of 30 μm and upwards are selected, or thinner foils reinforced on the outside of the pack with a film, such as regenerated cellulose, polyester, polyamide or polypropylene films. These films not only protect the foil against mechanical damage but, by blocking the perforations, improve the barrier properties, and materially reduce the thickness of foil necessary. This, however, is not possible when a porous material like paper is used as the reinforcing material. The inside of the foil pack will be coated with a thermoplastic coating or laminated to a thermoplastic film such as polyethylene. High-clarity films used for external reinforcement, such as polyester or cellulose acetate, can, if reverse-printed, give a brilliant finish to the pack.

Aluminium foil is also a light barrier and its high reflectivity reflects heat rays, reducing the rate of temperature rise of a pack. Light promotes the development of rancidity in fats and can also induce undesirable flavours in other foods, such as orange juice. It is destructive to some vitamins in foods. Milk has been shown to undergo losses of vitamins A, B, C and D when exposed to lighting conditions varying from normal well-lit stores with fluorescent lighting to sunlight. The protection of edible fats afforded by the use of laminated aluminium foil is well known, foil laminated to parchment paper or grease-proof paper being widely used for these types of product.

Examples of the use of aluminium foil in packaging

Strip packaging

Strip packaging with foil is extremely valuable, because each item is individually protected against moisture, light and air, which can cause colour changes and deterioration of the active constituents of pharmaceutical tablets.

Two webs of heat-seal coated foil are fed between two electrically heated sealing rollers, the product to be packed being dropped between the two webs just prior to sealing. The thickness of foil used depends on the degree of protection desired, but normally gauges between 25 μm and 30 μm, coated with a heat-seal lacquer or laminated to polythene, are used. The coatings normally seal at about 120°C, but adjustments to this can be made. The heavier the coating weight, the better the degree of protection under high humidity, but coatings of about 6–8 g/m^2 are satisfactory. Polyethylene laminates with greater thickness of sealing layer give the best protection under extreme conditions. Large-diameter tablets or unusual shapes may cause fracture of the metal foil on high-speed strip packaging machines and, to overcome this, pre-forming devices, which shape the web before the tablet is inserted, are employed to reduce the probability of fracture during sealing. For products like hypodermic needles, a heavier gauge of foil is used (90 μm) laminated to polyethylene, the web being preformed to receive the needle before being closed with a foil laminate of polyethylene.

Pouch packaging

Two basic types of pouch exist: the *fin-sealed pouch*, where the pack is sealed on all four sides, and the *pillow pouch*, where the pack is sealed at the top and bottom with one longitudinal seal on one face of the pouch. As already indicated, when protection of the foil from mechanical damage is necessary to preserve the contents, reinforcement of the foil on the outside is necessary. In addition, consideration has to be given to the possibility of some products, such as granular material with sharp edges, puncturing the pack. This may be important in vacuum packaging where the flexible pack is drawn down closely on to the product. Typical constructions for pouches are:

(a) Paper/foil/heat-seal coating or film
(b) Foil/paper/heat-seal coating or film
(c) Polyester/foil/polyolefin
(d) Foil (30 μm)/polyethylene
(e) Foil/paper/heat-seal coating

Two-ply laminates are generally used for small pouches or sachets where the packs are subjected to less handling.

Attention must be paid to the product being packed to ensure that no reaction will occur between the product and the inner layer of heat seal. For example, low-density polyethylene is affected by oils, fats and shampoos and, in such instances, an alternative heat-seal material may have to be used which has the required product resistance. In the above-quoted situations PVDC copolymers are suitable alternatives.

Table 7.3 gives examples of typical specifications and applications.

Table 7.3.

Type of foil or laminate	Use	Remarks
9-μm foil laminated to paper with 20–25 g/m² Hot Melt coating on paper	Over-wraps for confectionery	Gravure-printed on foil surface
9-μm foil reinforced with paper coated with 20–30 g/m² tacky wax covered with porous tissue.	Over-wraps for confectionery	Gravure-printed on foil. Wax penetrates porous tissue when heated to form heat-seal
9-μm foil or 12-μm foil coated with a vinyl copolymer heat-seal	Protection of bar and block chocolate	
12-μm foil coated on both sides with protective and heat-sealable coatings	Protection of processed cheese	
25–40-μm foil coated on inside with a heat-sealable resin coating or 20–40 g/m² polyethylene	Pharmaceutical wrapping	Often gravure-printed
50-g paper/25-μm foil with 25-g/m² polyethylene or Saran or 12-μm polyester/18-μm foil with polyethylene or Saran film	Accelerated freeze-dried food	Heat-seal film depends on packed food. Polyethylene for non-fatty and Saran for fatty foods
12-μm polyester/18-μm foil/25-g polyethylene or 12-μm polypropylene/18-μm foil/25-g polyethylene	Vacuum-packed coffee	
12-μm polyester/12–15-μm foil/50-μm polypropylene, or 12-μm polyester/12–15-μm foil/12-μm polyester/50-μm polypropylene	Sterilizable pouches for foods	
5-g Stoving Lacquer/100-μm foil/10–12 g heat-seal coating. 5-g Stoving Lacquer/100-μm foil/50-μm polypropylene	Container material for sterilizable containers	
60–100-μm foil/6–8 g heat-seal coating	Container material for containers only requiring normal storage	

2.111

Bibliography

1. BS 1683: 1967.
2. BS 1133: Section 21: 1964, 26.
3. Pira Report for BAFRA * 1974.
4. Heuss, *Verpakking*, March 1957, 485.
5. *Accelerated Freeze-Drying (AFD) Method of Food Preservation*, H.M.S.O., 56.
6. *Packaging*, March 1961, 50.
7. Heuss, *Modern Packaging*, August 1958, Vol. 31, No. 12, 122.
8. Brown, L. R., *Modern Packaging*, October 1964, Vol. 38, No. 2, 187.
9. Miller, *Gravure*, January 1961, Vol. 7, No. 1, 24.
10. Collins, *Pest Control*, October 1963, 26.
11. Collins, *Modern Packaging*, October 1963.
12. Collins, *Candy Industry and Confectionery Journal*, August 1961, Vol. 117 (4, 23, 24, 26, 31).

PART THREE
GLASS AND PLASTICS—BASED PACKAGING

CHAPTER 1

Glass Containers

C. Weeden

Glass container manufacturing is a continuous high-speed automatic process. Each year glass container factories in the United Kingdom sell over 7000 million bottles and jars, for food (35%), beverages (45%), pharmaceuticals (9%), and toilet and perfumery (4%). In value these sales constitute about 10% of all types of packaging, some competitive and others not competitive with glass.

Although there is no precise definition, a broad classification for glass containers is that narrow-necked containers are known as *bottles* and wide-necked as *jars*. As with other industries, the glass container industry over the years has acquired a number of terms appropriate only to its own processes and products. Exact definitions are given in BS 3447: 1962, Glossary of Terms used in the Glass Industry.

The nature of glass, the fact that it is processed when ductile, and the manufacturing methods employed, allow considerable flexibility in design. Capacities vary from 2 ml to 10 litres.

Compared with many others the glass container industry was late in developing mechanical processes. Once machines had been introduced, however, their spread was rapid. Nevertheless, the next decade or so is likely to see striking innovations in various aspects of the process, such as the melting of raw materials, the inspection of finished products, and the forming process itself. Current research indicates that there will be a trend toward methods of utilizing the theoretical tensile strength of glass to a far greater extent than hitherto.

Specialized surface coatings, better distribution of glass, and attention to design means that glass containers can now be made with as good or better resistance to breakage but lighter in weight than hitherto.

GLASS AND PLASTICS-BASED PACKAGING

The nature of glass

As a packaging material, glass has advantages in that it will not affect the product, nor does it need additional treatment to make it impervious. It is a rigid inflexible material that does not alter its characteristics with the passage of time. Odour, which can be of considerable concern to food packers, does not arise as a form of contamination.

There are, however, a limited number of products, blood transfusion fluids and certain drugs, which are extremely alkali-sensitive and for which special glasses or treatment is required. A process known as "sulphating" is used to remove most of the sodium from the surface of the glass and alternatively, when the quantities are small in number, containers are made by semi-automatic methods or from tubing in a glass of low alkali content.

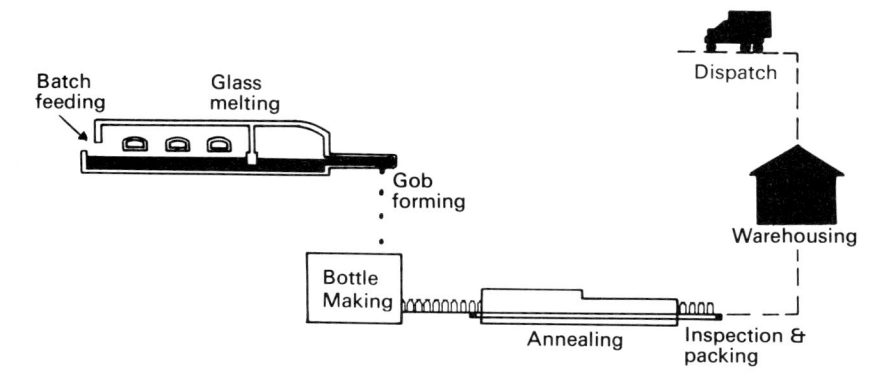

Figure 1.1 Glass container manufacturing is a continuous process from the time that the raw materials are weighed and mixed, to delivery to the customer or the warehouse

The production of glass containers

The process is continuous in the sense that the raw materials enter at one end and emerge as finished products at the other. It is also continuous in the sense that it does not stop. The controlling factor is the life of the glass melting furnace, which is up to about 7 years, at which point it is dismantled and rebuilt. The length of service it gives is known as a "campaign". The sequence of operations is shown in figure 1.1.

Selection and mixing of raw materials

A typical formula for clear glass is:

Silica	SiO_2	73%
Lime	CaO	11%
Soda	Na_2O	14%
Alumina	Al_2O_3	1%

with small proportions of other raw materials introduced for specific purposes, or accepted as minor constituents of the main materials, e.g. magnesium oxide (MgO) or potash (K_2O). The silica (sand) should have an iron oxide (Fe_2O_3) content not exceeding 0·04% and decolorizing agents such as selenium (Se) and cobalt oxide (CoO) are added, to give maximum clarity.

Most container glass is colourless, and is known as "colourless", "white flint" or "clear" glass.

A wide range of coloured glasses is commercially available, and some products are traditionally associated with containers of a particular shade:

Amber—a brown-coloured glass obtained by combining moderate amounts of iron oxide, carbon and sulphur.
Green—different shades of green are obtained by varying the iron, manganese and chromium oxide content of the glass.
Blue—obtained by adding cobalt oxide.
Opal—an opaque glass in which minute crystals, obtained by adding a fluorine compound, act as light scatterers.

To assist melting, "cullet"—broken glass of similar formula—is added to the batch. The proportion used is on average about 25% by weight.

It is important to ensure that the raw material specification and the formula are strictly adhered to, since no further correction to the manufacturing process can rectify a faulty glass composition or poor colour.

Melting

After mixing, the raw materials are mechanically conveyed and charged into the melting chamber of the furnace. The furnace is rectangular in shape, holding from 4 to 400 tons according to the number and type of machines it is feeding. The raw materials fed in at one end react, fuse and circulate until they become a homogeneous molten liquid, with a viscosity similar to that of treacle. The glass passes through a throat into a zone known as the *working chamber* before it is fed to the forming machines. Most furnaces are heated by oil or gas flames sweeping across the surface of the glass, up to temperatures of 1500°C. A typical tank furnace is illustrated in figure 1.2.

Melting by electricity, using the glass as a conductor, is technically possible but, although there are many experiments in this field, it is at

GLASS AND PLASTICS-BASED PACKAGING

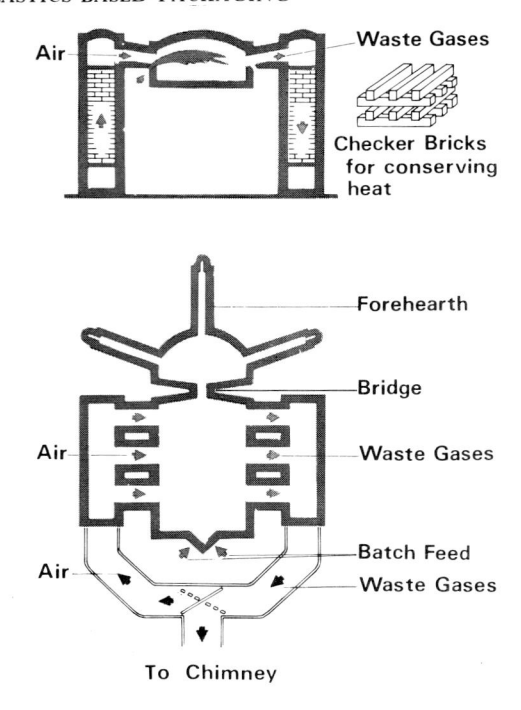

Figure 1.2 Elevation and plan view of a tank furnace

present more expensive than the conventional oil or gas-fired method. Electricity is, however, used for boosting oil and gas-fired furnaces.

The aim at this stage of the process is to produce a homogeneous glass of good colour, and free from blemish. Temperatures should remain steady, and the pull on the glass should be such that it is given sufficient time to fuse and throw off the gases caused by the reaction of the materials. Otherwise the glass will contain the imperfections of insufficiently melted raw materials (known as *striae*), small bubbles of unreleased gases (known as *seeds*), or crystalline inclusions in the glass (called *stones*).

Feeding to the forming machines

When the glass has been refined, it flows from the working end of the furnace along a channel known as the *forehearth* which is situated above the forming machine. At the end of the channel, a mechanism known as a *feeder* extrudes the glass in the form of *gobs*, and these gravitate by chutes to the *moulds* on the *forming machine* below. This operation is shown in figure 1.3.

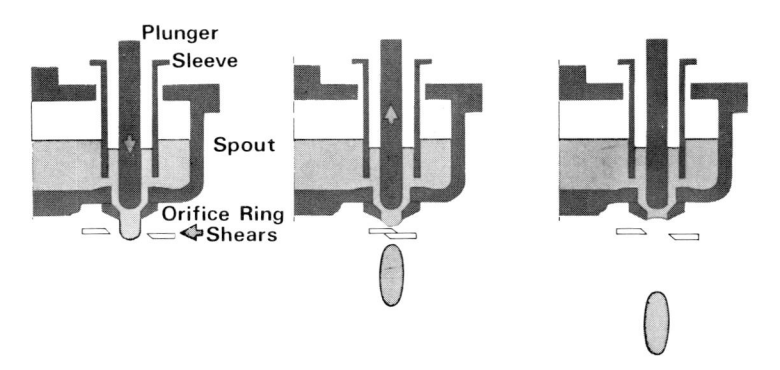

Figure 1.3 Gob forming. The feeder consists of a plunger which extrudes the glass that flows along the forehearth. A pre-arranged weight of glass is cut off to form the gob.

Forming the container

All glass containers are made in two stages, and therefore two sets of moulds are required. First, an initial blank, or *parison*, shape is formed and then transferred to the *blow mould* (figure 1.4) where the final shape is blown. The parison can be blown or pressed. These processes are shown in figures 1.5; 1.6; 1.7 and 1.8. Illustrations of the two sets of moulds can be seen in figures 1.9 and 1.10.

The accuracy of the final mould will determine the ultimate shape of

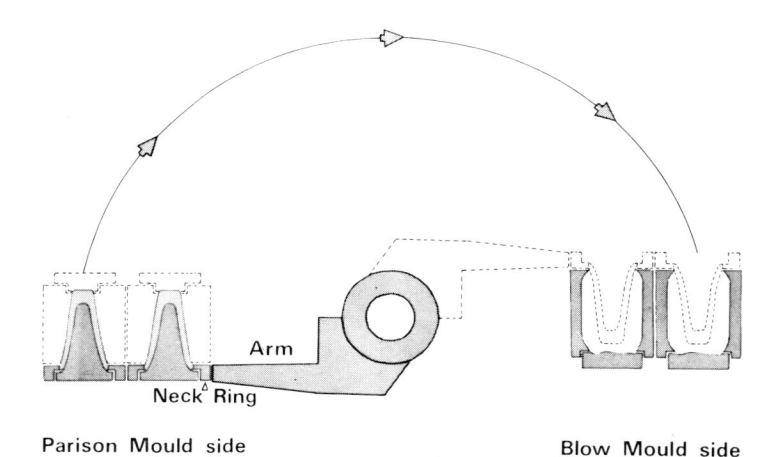

Figure 1.4 Transfer mechanism. The gob, having dropped into a parison mould has been pressed into its initial, or parison shape. It is then transferred to the blow mould for completion.

GLASS AND PLASTICS-BASED PACKAGING

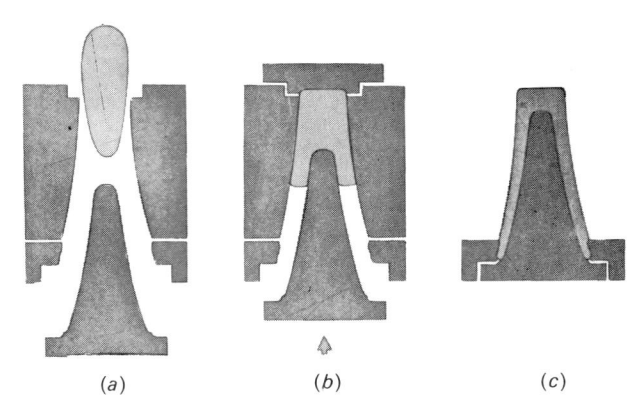

Figure 1.5 Parison pressing. The first stage of the manufacturing of a glass container in which the parison is pressed.
(a) Gob drops into parison mould (b) Plunger starts to press parison
(c) Parison completed

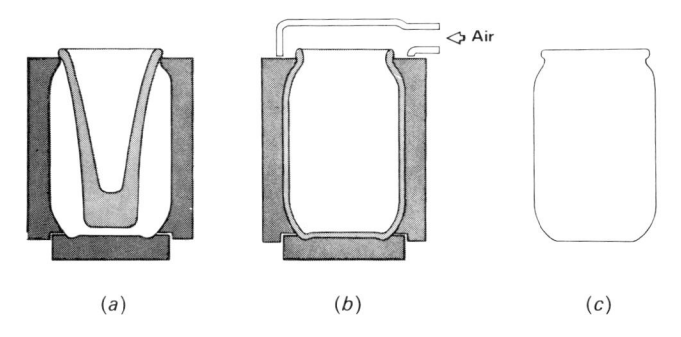

Figure 1.6 Final blowing. The second stage of the manufacturing of a glass container in which the parison is pressed.
(a) Parison in blow mould (b) Jar blown to shape (c) Finished jar

the glass container, but of equal importance is the distribution of glass and wall thickness. It is in determining this that the parison plays an important part. The shape of the parison, and the temperature of both the glass and the mould, will affect the distribution of the glass when the final shape is blown. The initial temperature of the glass is about 1000°C and of the mould about 500°C, and the aim is to keep these as

GLASS CONTAINERS

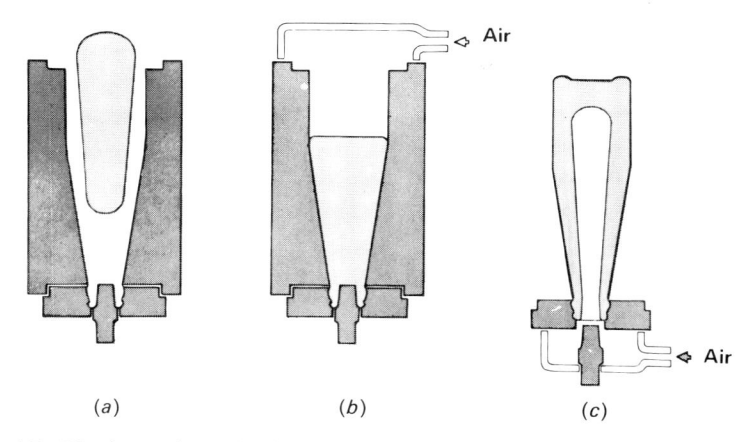

Figure 1.7 Blowing parison. The first stage of the manufacturing of a glass container in which the parison is blown.
(a) Gob drops into parison mould (b) Settle blow to form the finish
(c) Counter blow to complete parison

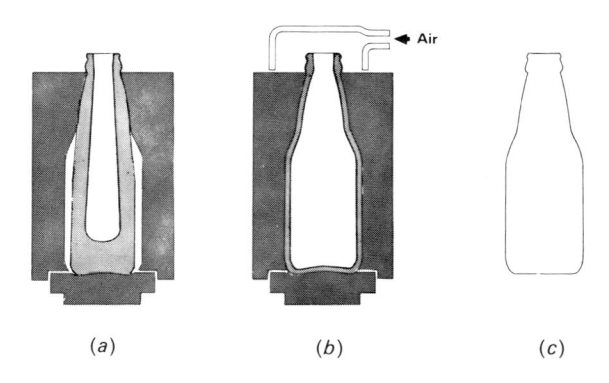

Figure 1.8 Final blowing. The second stage of the manufacturing of a glass container in which the parison is blown.
(a) Parison in blow mould (b) Bottle blown to shape (c) Finished bottle

constant as possible. If the mould cools, the glass will not flow easily; if it is overheated, the glass will stick to the surface.

The moulds are made from fine-grained cast iron, with the working surfaces highly polished. The subsequent application of lubricants containing graphite forms a very fine carbon film. The heat of forming causes oxidation, which has to be cleaned off. This, together with fair

GLASS AND PLASTICS-BASED PACKAGING

Figure 1.9 A parison mould. At this stage the neck only is completed.
Glass Manufacturers' Federation

wear and tear, means that the mould cavity will gradually enlarge. Since tolerances on the body diameter and the finish are important, it is customary to machine a new mould to slightly below the mean specified dimensions of the container. The life of a set of moulds will depend ultimately upon the agreed tolerances.

Annealing

When the finished bottle leaves the final mould (figure 1.11) its temperature is in the region of 450°C. If it were left to cool on its own,

GLASS CONTAINERS

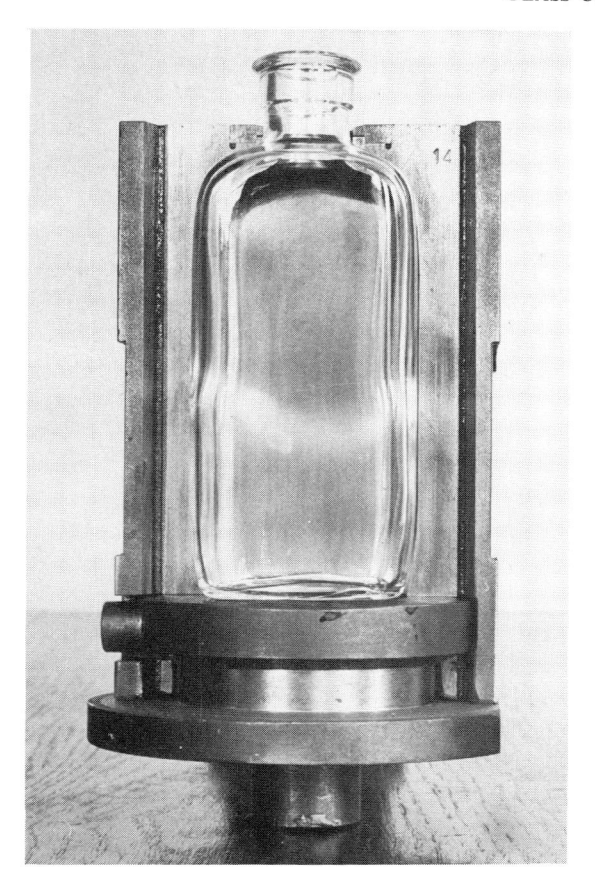

Figure 1.10 The blow mould in which the shape of the bottle is completed
Glass Manufacturers' Federation

a differential rate of contraction across the glass would set up sufficient strain to make the bottle unstable. This is because glass is a poor conductor of heat. Because of the heat retained within the bottle, the internal surface of the bottle wall will cool less rapidly than the outer surface, and the body of the glass between the surfaces even more slowly. The aim, therefore, is to ensure that the rate of cooling, and hence any contraction, is as even as possible throughout the body of the glass. To do this, bottles are fed from the forming machines to an annealing *lehr*, which is a long tunnel, where they are heated to a temperature of about 600°C and gradually cooled (figure 1.12).

GLASS AND PLASTICS-BASED PACKAGING

Figure 1.11 The bottles emerge from the blow mould and are conveyed to the lehr
Glass Manufacturers' Federation

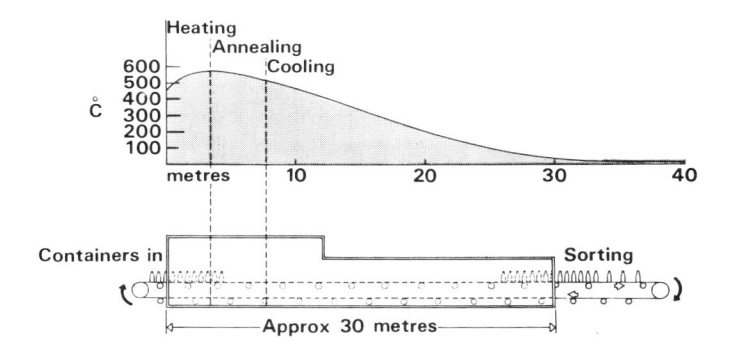

Figure 1.12 The annealing process

Inspection

When glass containers emerge from the annealing lehr (figure 1.13) each one is inspected for faults such as *striae*, *seeds*, or *stones; checks*, *smears* or *crizzles*, which are cracks on the surface of the glass; or *chokes*, constrictions in the bore of the neck and other forming defects. The bottles and jars are then packed in outer containers, or bulk-palletized and dispatched to the warehouse.

3.12

Figure 1.13 As the glass containers emerge from the annealing lehr, each one is inspected for faults

Glass Manufacturers' Federation

Semi-automatic process

The process that has been described is automatic, and almost all glass containers manufactured in the United Kingdom are made in this way. But automatic high-speed processes in any industry, if they are to be economic, depend on long runs. The automatic glass container process, with mould costs in the region of £1000, is no exception to this rule.

In the semi-automatic process, the glass is gathered and the moulds are operated by hand, with the result that, although production speeds are lower, so too are the costs of the moulds. The semi-automatic process can cater, therefore, for special orders, trial runs, and those markets where small quantities are required. Figure 1.14 shows the sequence of operations.

GLASS AND PLASTICS-BASED PACKAGING

Figure 1.14 A semi-automatic blowing system. The gob is gathered by hand on an iron, and the correct amount of glass is dropped into a preliminary mould. Compressed air is introduced to form the neck of the article. The embryo shape (parison) is then transferred to the finishing mould in which the final shape is blown.

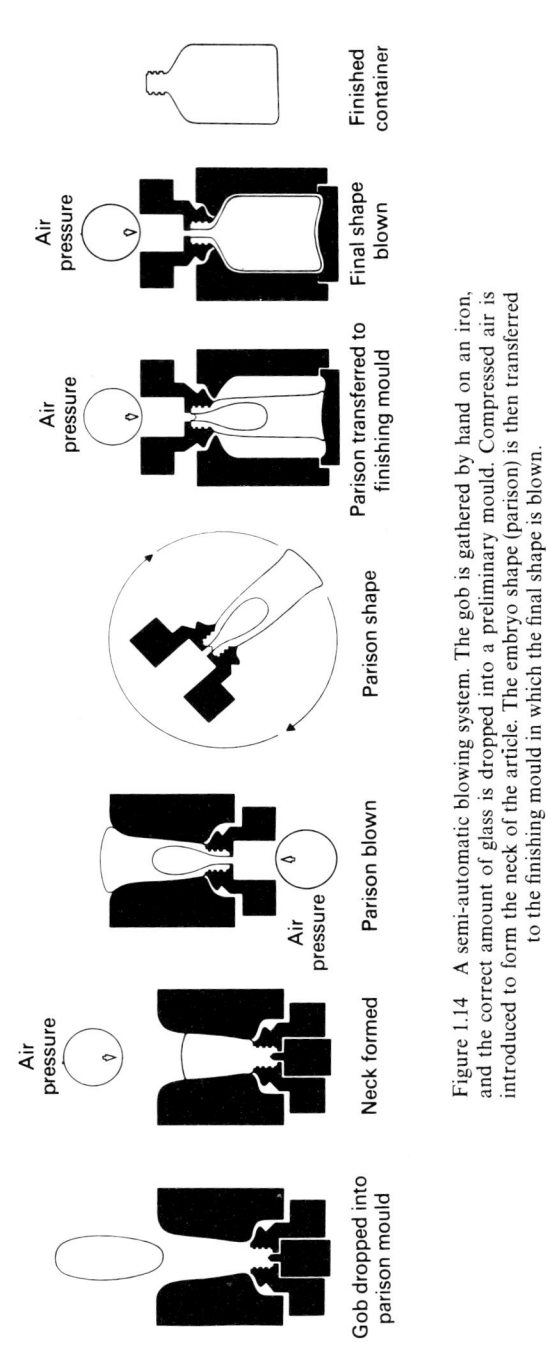

Glass Manufacturers' Federation

GLASS CONTAINERS

Tubular containers

Amongst the smallest glass containers today are the ampoules and vials used in the pharmaceutical industry. The *ampoule* is a single-dose container manufactured fully automatically from glass tubing and used to contain liquid or powder injection medicines. The top of the stem is sealed after filling and then snapped off when the contents are required. *Friable ampoules*, which can be easily crushed and are used mainly for inhalants, are manufactured by hand from thin-walled tubing. *Vials* can be single or multi-dose containers, and are manufactured either from tubing or on conventional bottle-making machines. They are often sealed with closures that have rubber inserts, which can easily be pierced by the needle of a hypodermic syringe, but they are also designed for sprinkler necks, screw caps or cork stoppers.

Quality control

In ensuring that a run of containers will meet the specification, a wide range of factors has to be considered. It is important, therefore, that a constant check be kept on every part of the process where variability is likely to occur. In a highly productive process, where two days may elapse between the feeding of raw materials to the furnace and the emergence of finished containers, a considerable number of faulty containers could be made if there were not a recognized procedure.

Already an indication has been given of the factors likely to affect the quality of a glass container. The manufacturing process is continuous and, from the point of view of control over quality, it can never be put in reverse, i.e. an error at any one stage of the process cannot be corrected later. Apart from the obvious checks on the quality of raw materials, and the glass, mould condition, temperatures and rate of annealing, sample bottles are taken immediately they leave the finishing mould, and are checked for weight and dimensions. A general impression of these controls is shown in the summary given in figure 1.15. Many of these are routine, necessary for the efficiency of the process as a whole.

At the inspection stage a more rigorous check is made to ensure that the containers meet the specification. The number involved prevents a complete check on every bottle and jar. Not only would such a procedure add considerably to the cost, but it is unnecessary. However, for some specific characteristics, e.g. height, diameter, neck bore, cracks and crizzles, each container may be subjected to an electronically operated inspection. Detailed inspection for other factors is done on a random sample selected at regular intervals from the production line.

In drawing up a quality-control procedure, defects are classified into

GLASS AND PLASTICS-BASED PACKAGING

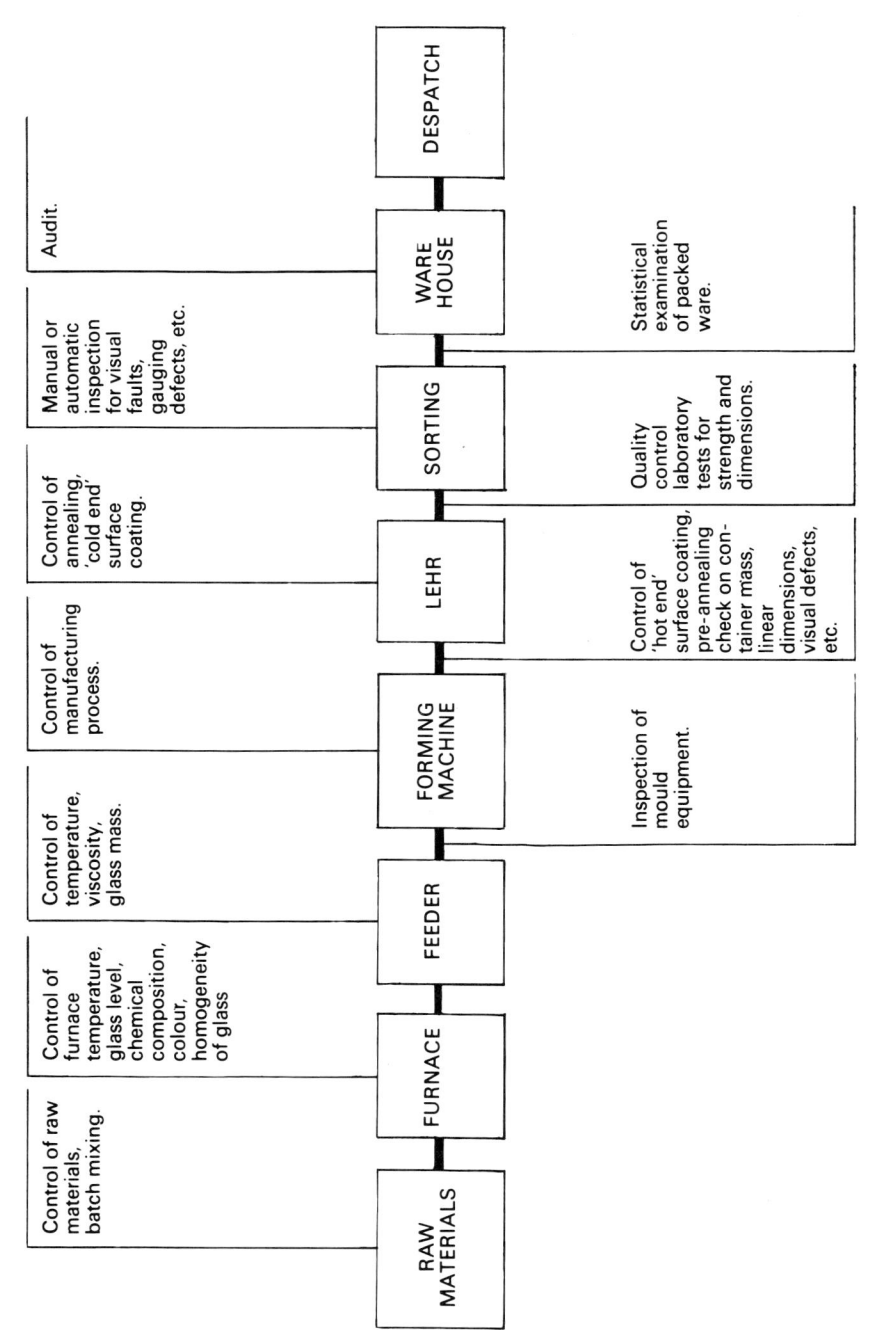

Figure 1.15 The zones of quality control

3.16

GLASS CONTAINERS

three categories: minor, major and critical. Minor defects are those which will not affect the utility of the container, but which may look unsightly. In this category come seeds, striae, prominent mould marks, and the like. Major defects are those which could, but will not necessarily, cause the container to give an unsatisfactory performance. Fine cracks or checks could give trouble if the container were to meet considerable tensile stress, but such stress may not be encountered during the life of the container. Critical defects are those which render the container unserviceable. These are rare and, given normal control over production, such containers should not get to the annealing lehr.

The aims of quality inspection are to maintain the reputation of the glass manufacturer, and to ensure that the customer receives containers of the standard required. For minor and major defects, an acceptable limit for the percentage of defectives is set, and this determines whether or not the batch of containers from which the sample has been taken is acceptable. For major defects this level is more severe than for minor defects and, in the unlikely case of critical defects, the limit is zero. The over-riding aim throughout is that the levels should be set from a knowledge of what will give a satisfactory performance.

In order that its customers may study in more detail the quality-control procedures in current use, the glass container industry has published guides.[1]

Acceptance sampling

Not all customers will want the same *quality* of bottle. The product, how it is filled and capped, the method by which it is to be sold, will each in its own way determine what is, and is not, important about the glass container.

When a customer for one reason or another wishes to carry out additional checks on quality, an acceptance sampling scheme is agreed. This clearly is a matter for negotiation and may well affect the cost of the bottles.

Tests and test procedures

The most critical of the dimensions on a glass container are those relating to the height, the body diameter (or relevant dimensions if the container is not cylindrical) and the finish. These are important to the packer, especially on high-speed lines where the filling heads, the star wheels and the capping machinery can cause damage if there is too great a variation in these dimensions between one container and the next.

3.17

Reference has already been made to the variation in the dimensions of a set of moulds during its working life, and the tolerances that can be reliably maintained by glass manufacturers have been set out in detail.[2]

The cavity of the moulds will not, of course, wear evenly over its surfaces during the working life. This differential wear on the surface causes *ovality*, which is usually expressed as the difference between the axes. In practice, ovality is kept to a minimum, particularly in those instances where bottles are conveyed on the sides or rolled for labelling.

Verticality is the relationship of the centre of the finish of a bottle to the centre of the base. This is an important feature, particularly in tall bottles where an excessive eccentricity could cause difficulty in locating the filling head.

To check the contour of the finish, a shadowgraph is used in which the enlarged outline of the finish is projected against the specified contour.

The introduction of legislation calling for an exact definition of the weight or quantity of goods offered for sale has emphasized the need for accurate "capacities". In recent years, improved control over production has resulted in closer control over the capacity of glass containers. Nevertheless, since capacity depends upon the volume of the mould, the volume of glass in the mould, and the accuracy with which the glass reproduces the mould shape, variations are bound to occur. The existence of these has led to the adoption of the *bulk test* as the normal method of control. In this, the capacities of a given sample of bottles, selected at random, are measured, and statistical calculations indicate whether the mean capacity, the sample and the spread of individual measurements satisfy predetermined requirements.[3]

When glass containers are to be filled with hot products, or heated during processing, a test for resistance to *thermal shock* has to be carried out. The procedure is to immerse the bottles in hot water for a given time, and then to plunge them into cold water. This is designed to test the effect of thermal shock, and the important points are the close control of the temperatures, the time the containers are immersed in hot water, and the transfer time. The details of the tests will vary according to the nature of the bottle and the severity of shock it is likely to meet on the packaging line. All glass containers, once they are packed and closed, will be subjected to variations in internal pressure, although these will be minimal in many instances. Variations are likely to be more severe when the containers hold carbonated liquids. Where internal pressure is known to be an important feature, bottles are regularly subjected to pressure tests.

Design

Aesthetic as well as technical considerations play an important part in the development of new designs. The range of designs is very wide. Nevertheless, no matter how exotic the designer's ideas, or how compelling the marketing manager's argument, in the final analysis the container has to be made on a machine, it has to hold a given product, and it must be capable of being filled, closed, transported and used with ease by the consumer. Each design, therefore, is a compromise, the ultimate aim being to achieve the most effective pack.

It is important, therefore, that packers should consult all those concerned with the marketing of the product at a very early stage. Furthermore, the glass container manufacturer, and the manufacturers of closures and filling machinery, must be given precise details of what is required, otherwise something less than optimum efficiency will be achieved.

Since glass containers are blown from a ductile material, the basic shape is a sphere. This is the shape of many early bottles, whose only departure from the spherical shape was the addition of a neck for filling and decanting the product, and a pushed-in base to enable the bottle to stand upright.

For most practical purposes the cylindrical shape is the best. It can be manufactured relatively easily; it bears a close relationship to the sphere, which gives maximum strength; and because it is stable, it lends itself to high-speed filling and automatic handling. A departure from this shape, although adopted in almost every design, brings certain disadvantages. The following is a rough guide to the comparative strength of various shapes.[4]

Cylindrical cross-section	1
Elliptical cross-section	$\frac{1}{2}$
Square cross-section (with well-rounded corners)	$\frac{1}{4}$
Square cross-section (with sharp corners)	$\frac{1}{10}$

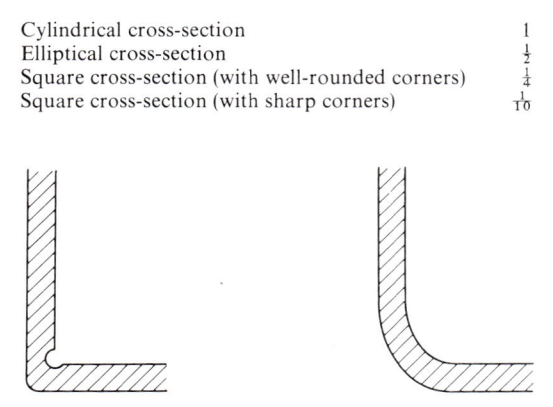

Figure 1.16 The "thin spot" produced by a right-angle junction and the remedy, the "inswept junction"

GLASS AND PLASTICS-BASED PACKAGING

There are, of course, many bottles on the market which have a square cross-section with sharp corners, and their strength is adequate. The guide is for bottles with a comparable distribution of glass. An adjustment can be made by varying the amount of glass in order to increase the strength of the container. This ensures that the distribution of glass over the surface is adequate to resist any shock the container is likely to meet in use.

Since the natural shape of blown glass is spherical, there is a resistance to flow into rectilinear shape. A square corner is thus likely to have less glass in it. This weakness is accentuated by the fact that, during handling, corners are more likely to be knocked. This problem has to be faced in every glass container, since they have to stand vertically, and it is overcome by "insweeping" the side wall to join the base (figure 1.16).

Ideally, any change of shape, from the base to the sidewalls, from the sidewalls through the shoulder to the neck, for example, should flow smoothly. The glass manufacturer will then achieve a good distribution of glass, which is of the utmost importance. It is important, for instance, that the walls of the container should be slightly thicker at likely points of impact. If a lightweight container is required, then the design should

Figure 1.17 Example of lightweighted bottle. On the right is the lightweight version — compare the wall thicknesses and the distribution of glass.

Glass Manufacturers' Federation

GLASS CONTAINERS

be such that the distribution of glass is well under control and approaches that of the ideal. Handling methods and high speeds in packing plants impose their own discipline on the designer. Simpler shapes can be handled at lower costs, and care in matters such as the position of the centre of gravity can enable the filling line to run faster. The label is not a matter to be left to the last moment, since the bottle designer must know the area needed, and whether a recessed panel is required to protect the label from scuffing.

The subsequent usage of the bottle, and how it is to be marketed, are also of importance; the designer should be provided with these facts. Most product markets use non-returnable bottles but, where the delivery of filled and the collection of empty bottles is under the effective control of the packer, as in the case of the brewing and dairy industries, a multi-trip bottle is an economical pack. Where distribution is such that collection is difficult or costly, the non-returnable bottle is more effective and can be lighter in weight (figure 1.17).

Marketing through self-service stores suggests the need to show the housewife what she is buying, and this in turn imposes conditions on the shape of the container and the label. The ability to stack jars is another factor which calls for stable containers and attention to the closure.

Specifications

The specification should set down in precise terms the details of the container the packer wants and which the glass manufacturer can supply. Ideally, it specifies those requirements that are vital to the efficient handling of the container. In practice, there is always a danger that unimportant, or relatively unimportant, details will be included.

The specification should, therefore, deal only with essentials: the height, diameter and capacity, for example, and the weight and colour of glass; and if resistance to pressure or thermal shock is important, then details of the tests should be set down. Dimensions that are critical should carry tolerances, and these should be realistic; otherwise the container will carry unnecessary additional cost.

In short, the specification is the basis of the contract to which all parties will work, and to which dispute, should it arise, can be referred. It must, therefore, be both aceptable and practicable.

The terms used in a specification are set out in fig. 1.18.

Developments

One of the most interesting features in glass-container manufacturing is the varied nature of the developments that are occupying the time and

GLASS AND PLASTICS-BASED PACKAGING

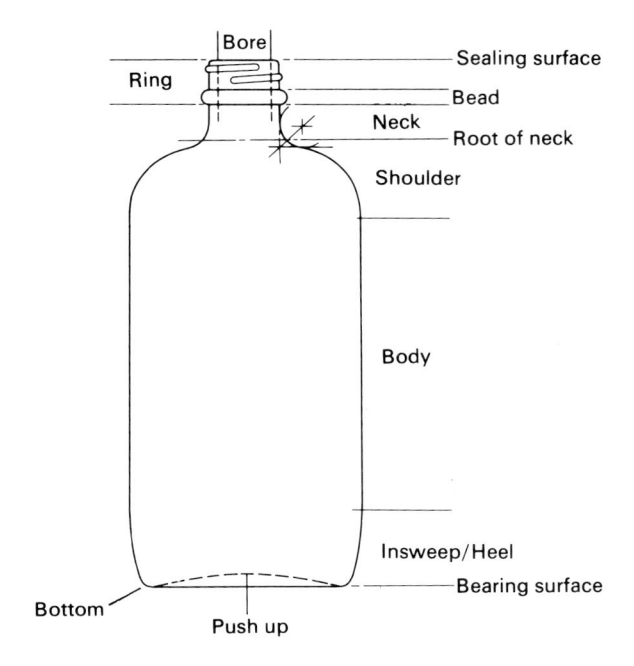

Figure 1.18 The glass bottle showing glass container terms. The terms *ring* and *bottom* have, in general, been used in this terminology. The term *finish* is sometimes used to denote the *ring*, and the terms *base* and *punt* to denote the *bottom*.

energy of engineers and glass technologists.

Within recent years, the heating of furnaces has changed from gas, produced on the premises from coal or piped from an outside source, to furnaces fired by natural gas or fuel oil, boosted on some occasions by electricity. The all-electric furnace has advantages, particularly in that it cuts down extraneous heat, but at present running costs are against its widespread use in the United Kingdom.

At the forming stage, productivity has been increased by blowing two containers in a single set of moulds. Since, at the same moment, two identical gobs of glass have to be cut off to feed the machine, this is known as *double gobbing*. The use of *triple gobbing* for small containers is now well past the experimental stage.

The inspection of glass containers as they leave the annealing lehr is receiving attention from electronic engineers. Until recently, inspection at this stage was entirely manual and visual, but the maintenance of the present high standard as manufacturing speeds increase imposes additional strain on the inspectors and increases the cost. Automatic equipment is already used, and mechanized inspection will increase considerably over the next few years.

In recent years much research has been brought to bear on the strength of glass. Success in this field of research is, of course, of importance to the manufacturers of many glass products other than containers. The fact is that in use the tensile strength of glass is far below its theoretical value. For example, the theoretical tensile strength of glass is estimated to be in the region of 1 000 000 lb/in^2 (70 000 kg/cm^2), and in laboratory experiments values of 200 000 lb/in^2 (14 000 kg/cm^2) have been achieved. In practice, however, a bottle will have a tensile strength ranging from 10 000 to 20 000 lb/in^2 (700–1400 kg/cm^2). For normal usage this is sufficient and, although a bottle breaks occasionally, this is sufficiently infrequent to cause little bother. Nevertheless, with each increase in filling speeds a hold-up on the filling line becomes more costly, and each reduction in breakage makes the glass container more economic.

There is already one commercial method of increasing the strength of glass products, which is employed for motor-car windscreens. This uses the high compressive strength of glass, and is achieved by rapidly cooling the outer surfaces of the glass so that they contract at a greater rate than the volume of glass that lies between them. The basic requirement for this process is that the whole of the outer surface area must be cooled at the same rate, otherwise the compressive stress will be unbalanced. This is clearly impracticable in the case of a narrow-necked bottle, for example, since the surface inside the bottle would have to be cooled at the same rate as the outer surface. An alternative method is under development in a number of countries, where the potassium ions on the surface are exchanged for sodium ions, and the subsequent contraction creates the necessary compressive stress. A variation of this process is to induce crystallization in the outer layer after ion exchange. Each of these processes is commercially successful in its respective field of application, but both are far too costly for packaging.

It has been known for many years that the loss in tensile strength is due to *microcracks* in the surface of glass which arise as the result of, or develop after, the manufacturing process. In the container field, therefore, research has tended to concentrate on the less costly method of treating the surface in such a way that microcracks are less likely to develop or, alternatively, to create a more stable surface layer. Silicones and other chemicals have been used commercially to give greater lubricity, but in some instances these have led to difficulty in handling. A more recent development in the United Kingdom has been to spray a liquid metal organic compound on to the surface of the glass whilst it is still hot. The main effects of this are to reduce the development of micro-cracks, and at the same time to increase the lubricity of the surface. A reaction between the compound and the glass takes place, such that some penetration of the surface occurs, inhibiting the development of micro-flaws. The addition of an organic lubricant when the bottles are cold increases the resistance to damage during handling.

From this brief statement it can be seen that glass container manufacturers are acutely interested in improving the performance of their products. But experiments in one field are never divorced from experiments in another, and further success for the container industry could well come from experiments elsewhere.

An increase in tensile strength will also hasten the development of lighter-weight containers. Thinner-walled glass has a greater resistance to thermal shock but less resistance to mechanical shock. A greater resistance to impact will create the opportunity of reducing the amount of glass, and increase at the same time speeds of production.

Additional reading and references

A textbook which sets out to cover the entire field of packaging of necessity limits the extent to which any one subject can be discussed in detail. Those who wish to following the subject of glass for packaging more closely are recommended to read *Packaging in Glass* by B. E. Moody (Hutchinson, 1963). The glass container industry publishes leaflets on specific aspects of glass containers, to some of which reference has already been made and details of all its publications can be obtained from the Glass Manufacturers' Federation, 19 Portland Place, London W1N 4BH.

REFERENCES

1 *Quality Control in the Glass Container Industry* (leaflet); *Quality Control of Glass Containers* (Lecture Aid Kit), both available from the Glass Manufacturers' Federation.
2 *Glass Container Tolerances*, available from the Glass Manufacturers' Federation.
3 *Accurate Determination of Glass Container Capacity*, available from the Glass Manufacturers' Federation.
4 BS 1133, section 18.

CHAPTER 2

Moulded Plastics Containers

J. H. Briston

Definitions

Moulded, in the context used here, may be defined as *shaped, under the influence of heat and pressure, either in or around an appropriate form*.

Plastics are not so easily defined, but the BSI *Glossary of Terms used in the Plastics Industry* (BS 1755:1951) defines them as *a wide group of solid composite materials which are largely organic, usually based on synthetic resins or upon modified polymers of natural origin and possessing appreciable mechanical strength. At a suitable stage in their manufacture, most plastics can be cast, moulded or polymerized directly to shape. Some plastics are rubber-like, while some chemically modified forms of rubber are considered to be plastics.*

Plastics drums are dealt with in another chapter, so that the emphasis here will be on the smaller containers associated with retail use.

Methods of production

Blow moulding

The basic techniques of plastics blow moulding were derived from those used in the production of glass bottles. In each instance air is forced under pressure into a sealed molten mass of material which is surrounded at the right distance by a cooled mould of the required shape. The pressure of the air causes the molten mass to move out to the mould walls where it cools on contact. Finally, the mould is opened and the moulded article ejected. In the moulding of plastics bottles, the blowing processes in commercial

GLASS AND PLASTICS-BASED PACKAGING

use can be divided into two main classes, namely, *injection blowing* and *extrusion blowing*.

(a) *Injection blowing*

Injection blowing is the process that most closely resembles the blowing of glass bottles. The plastics material is moulded round a blowing stick, inside a more or less conventional injection moulding machine. This gives a thick-walled tube called a *parison*. The parison and blowing stick are then transferred to another (blowing) mould where compressed air is passed down the blowing stick. The injection moulded parison is still molten at this stage, and is thus blown out to the shape of the second mould. The sequence is shown in figure 2.1.

Injection

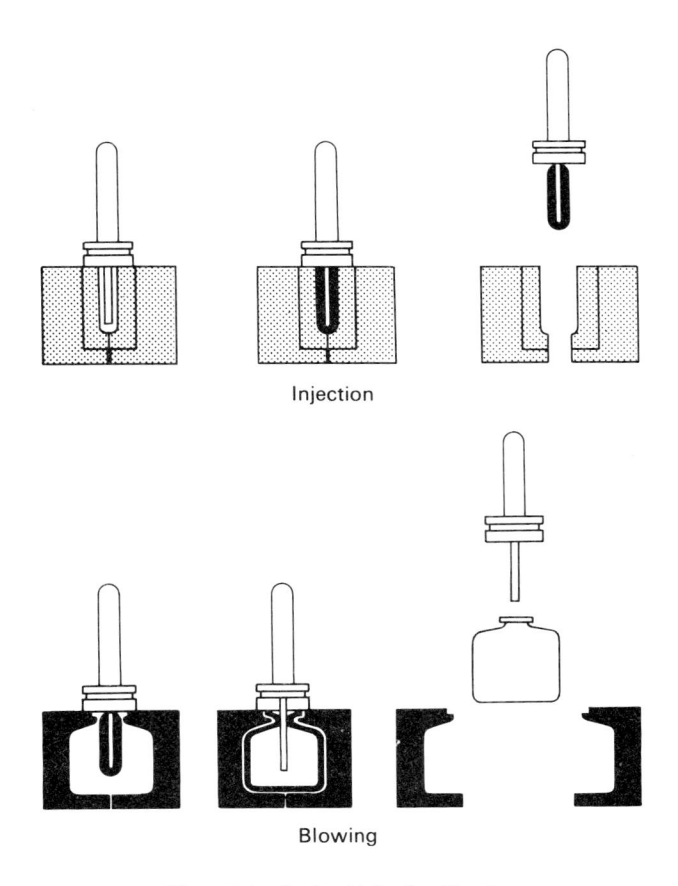

Blowing

Figure 2.1 Basis of injection blowing

MOULDED PLASTICS CONTAINERS

In the early days of the injection blow-moulding process it was hand-operated, i.e. the blowing stick and parison were removed from the injection moulding machine and transferred to the blowing mould by hand. Later developments include the provision of blowing moulds on either side of each injection mould, the parisons being transferred automatically into one or other of the blowing moulds alternately. Another development has been the use of multi-cavity injection and blowing moulds.

(b) *Extrusion blowing*

Extrusion blowing was a later development, but accounts for the greater percentage of blow mouldings produced today. Instead of an injection-moulded parison, we have a continuously extruded tube. At intervals a predetermined length of tube is trapped between the two halves of a split mould. Both ends of the tube are sealed as the mould closes, and the trapped portion is inflated by compressed air introduced via a blowing pin. Once again, the inflated plastic is cooled by the cold mould walls, the mould is opened and the bottle ejected. The sequence of operations is shown in figure 2.2.

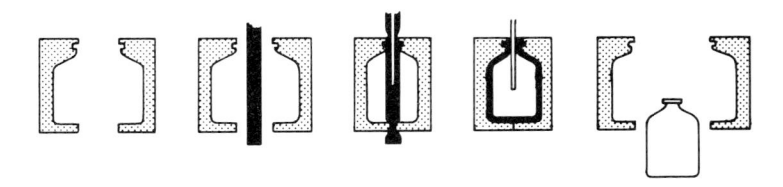

Figure 2.2 Basis of extrusion blowing

A variation on this is to install the blowing pin at the base of the mould. The molten plastics tube is then extruded vertically downwards so as to fall over the pin. The two halves of the mould then close around it as before, sealing it at both ends. This enables moulding to be speeded up, since the mould plus blowing pin can be moved away from under the extruder, allowing a second mould to be put in its place. Extrusion of a tube for a second bottle then takes place while the first bottle is being inflated, cooled and ejected.

This technique can be extended by the use of rotary moulds, or by having a series of matched mould halves mounted on two endless belts, so that they first close on the extruded tube, travel with it during blowing and cooling, then diverge, thus ejecting the bottle. Another way to speed up production is to fit the extruder with multiple die-heads, and feed the various extruded tubes to a multi-cavity mould.

One disadvantage of extrusion blow moulding is that when a length of plastics tubing of even wall thickness is blown into a bottle shape, the material is thinned more at the extremities of the mould than elsewhere. This is important because it is at these points (the base corners and the shoulders) that extra strength is needed in the bottle. Machines are now available that will extrude a tube of variable wall thickness, such that extra material is available where it is most needed.

One method is known under the descriptive name of *dancing mandrel*. The mandrel forms the inner part of a circular die gap, and in the dancing mandrel extruder it is conical, and so produces a variable die gap as it is moved up and down.

A more recent development is the extension of the form-fill-seal principle to blown bottles. There are quite a number of ways in which this can be done, but the general principle for all is much the same. The blowing tube is integral with the extrusion die, i.e. air is blown from the top. Concentric with the blowing tube is a filling tube. The bottle is blown as usual, then the liquid to be packed is metered into the bottle. The liquid helps to cool the bottle and so reduces the cooling cycle. A further concentric tube is also fitted to allow the air displaced during the filling operation to escape. When the bottle is full, the neck is heat-sealed and the filled bottle ejected from the mould. The process often involves a series of split moulds, the matching halves of which are mounted on two endless chains, positioned vertically beneath the filling/blowing head. The bottles are heat-sealed while still inside the moulds, and are ejected when the mould halves diverge again at the base of the machine. One advantage of the form-fill-seal bottle-blowing approach is that aseptic filling is possible: another is the elimination of empty bottle storage and transport.

Injection moulding

In the injection-moulding process the material is softened in a heated cylinder, then forced under high pressure into a closed mould. After the material has hardened (by cooling) the mould is opened and the moulding is removed by ejector pins or compressed air.

The essential elements of an injection-moulding machine are shown diagramatically in figure 2.3.

The complete cycle of operations is as follows:

(i) The mould is closed and a locking force is applied. This force must be large enough to prevent the two halves of the mould being pushed apart when molten plastic is forced in.

(ii) The plunger moves forward, taking a fresh charge of granules with it into the heating zone of the cylinder. At the same time it displaces already molten material left there during the previous cycle. This molten material flows out of the nozzle, through the "sprue" opening in the die, and into "runners" which terminate in a "gate", leading to the mould cavity itself.
(iii) The pressure on the plunger is maintained during the period while the material in the mould cools and contracts. If no pressure were applied during this period, the contraction during cooling would cause depressions or "sink marks" to occur on the surface.
(iv) The plunger returns to the fully retracted position.
(v) The mould opens and the moulding is removed.

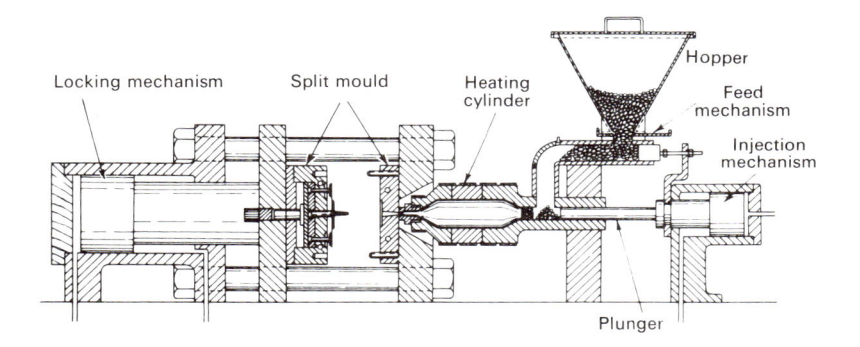

Figure 2.3 Diagrammatic representation of an injection-moulding machine

The cycle is then repeated. Continuity is maintained by coupling the feed mechanism to the plunger. Fresh material is then deposited during the return stroke of the plunger, ready to be pushed forward during the next cycle.

Many modern injection-moulding machines have a screw feed instead of a plunger. The screw aids plasticization and so reduces the cycle time.

Some definitions may be helpful at this stage:

A *gate* is a point of entry into the mould cavity. Gates can be of various sizes and shapes, depending on the viscosity of the molten plastics, and the size and shape of the mould cavity.

Runners are the channels that feed the molten material to the mould, when there is more than one gate, and they connect with the "sprue".

The *sprue* is the path from the external entry of the molten material into the die, to the gate (if there is only one) or to the runners.

One reason for having more than one gate is that the mould is a multi-cavity one. This is often the case when small items are moulded. Another reason is the necessity to have more than one point of entry into a large mould cavity, to give a more even flow of material into the cavity.

The following points about injection moulding should be noted.

3.29

GLASS AND PLASTICS-BASED PACKAGING

(a) High production rates can be achieved, particularly with multi-cavity moulds.
(b) Reverse tapers and pronounced undercuts cannot be moulded by this method, because the finished article could not be removed from the mould.
(c) Mould costs are high, because of their massive construction (necessary to withstand the high pressures involved in moulding). Short runs are, therefore, uneconomic.

Thermoforming from sheet

In this process a plastics sheet is softened by heat and then forced either into or over a mould. There are three main methods of thermoforming, namely, *vacuum forming, pressure forming* and *forming between matched moulds*.

(a) *Vacuum forming*

The simplest vacuum forming has the plastics sheet clamped over a box containing the mould. The sheet is heated by electric panel heaters. The air in the box is withdrawn through holes in the mould, thus creating a vacuum between the sheet and the mould. The atmospheric pressure of the air on the sheet forces it on to the mould, where it cools sufficiently to retain its shape when removed from the mould (figure 2.4).

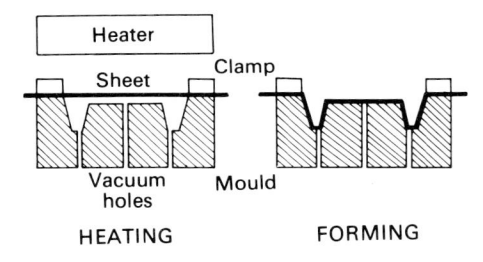

Figure 2.4 Vacuum forming into male mould

(b) *Pressure forming*

Pressure forming is very similar to the vacuum-forming method, but air pressure is applied from above to push the softened sheet on to the mould. One important difference is that the pressure that can be applied during forming can be greater than atmospheric pressure, so that better mould definition can be obtained.

(c) *Forming with matched moulds*

This method is particularly valuable when complex shapes are required. As the name implies, the heated sheet is pressed into shape by trapping it between matched male and female moulds.

Mould costs for vacuum and pressure forming can be low, although they obviously vary with the complexity of the mould, the length of run, and the degree of automation. For short runs and prototype moulding, wood and plaster moulds are often used, while for longer runs aluminium or filled epoxy resin moulds are popular. Matched mould forming is more expensive, because two moulds are required and dimensions are more critical.

Moulding of expanded polystyrene bead

The moulding of expandable polystyrene beads is different from other processes and deserves special mention. It is carried out in three stages, the first of which is a pre-expansion using steam. This softens the polystyrene and increases the internal pressure in the beads (due to the liquefied gas, pentane, contained therein). The degree of expansion, and consequently the density, is controlled by the time of heating and the temperature. The pre-expanded beads are then allowed to cool. At this stage there is a partial vacuum inside each bead and they are, therefore, weak and easily collapsible. The beads are left for about 12 to 24 hours to mature, during which period air permeates into the beads, which are then ready for the moulding process.

In the moulding operation, the mould is completely filled with pre-expanded matured beads, then closed and heated by injecting with steam. The steam softens the polystyrene and causes any remaining pentane, and the air that entered the beads during maturing, to expand. Since the beads are confined in the mould, they simply merge together and fill all voids in the mould. Cooling then hardens the moulded block.

Rotational casting

Rotational casting is particularly suited to the production of small numbers of large mouldings, although it can also be used for smaller articles. A powdered plastics substance, such as polyethylene, is placed in a heated mould, mounted in such a way that it can be rotated in two directions at right angles to each other. The powdered plastics melts and spreads over the inside of the mould. The mould is then cooled and, when the plastics material has hardened sufficiently, opened and the moulding removed.

Types and styles of container

Moulded plastics containers have no special terminology and are usually named after traditional containers of similar shape. Thus, the plastics containers are referred to as bottles, jars, pots, tubes, jerricans, punnets, etc., but within these broad classifications there are ranges of size, colour and style.

There is also much overlap between the types of containers and the methods of manufacture. For example, bottles can be made by blow moulding and by injection moulding (in two parts subsequently joined), while thin-walled tubs can be made by injection moulding, vacuum forming and blow moulding.

Blow-moulded containers

The blow-moulding process is principally used to produce bottles, jerricans, carboys and other containers where the neck diameter is small compared with the overall diameter of the container. Wide-mouthed containers can be produced by blow moulding by the "lost neck" process, or by blowing two containers together, head-to-head. These methods are shown diagramatically in figures 2.5 and 2.6.

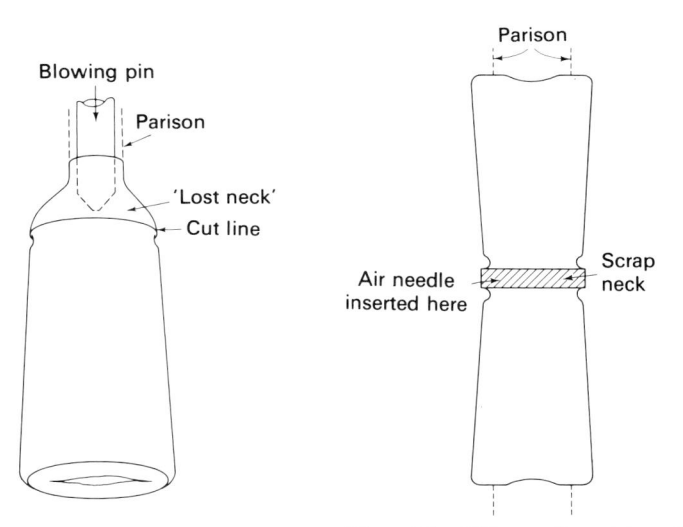

Figure 2.5 Wide-mouth container blown by "lost neck" process

Figure 2.6 Wide-mouth containers blown head to head

The materials used for blow-moulded containers include polyethylene (both low and high density), PVC, polystyrene, polypropylene, polyacetals, polycarbonate and cellulose acetate. By far the most common of these is polyethylene. Up to about 1961 low-density polyethylene dominated the

3.32

bottle-blowing market in the United Kingdom, but large numbers of high-density polyethylene bottles are now produced, and the production of PVC and polystyrene bottles is increasing. The choice of material is governed by cost, ease of processing, chemical resistance, permeability to water vapour and gases, and whether or not a transparent or opaque, flexible or rigid container is required.

Injection-moulded containers

Injection-moulded containers include jars, tubs, bottles (moulded in two halves subsequently welded together) phials and boxes. The main polymer used in the field of injection-moulded containers is polystyrene, but the use of polypropylene is growing rapidly. Both polystyrene and polypropylene are available in special grades where the containers are to be used in the packaging of foodstuffs.

Polystyrene containers can be produced both with a very wide range of colour effects and in clear grades to allow full visibility of the contents (figure 2.7a and b). Polypropylene containers can also be obtained in a range of colour effects, but the clarity of polystyrene cannot be matched in the clear grades.

High-impact grades of polystyrene are also used, particularly for those thin-walled pots where general-purpose polystyrene would be too brittle. High-impact polystyrenes are also non-transparent, and the choice between polypropylene and high-impact polystyrene is usually resolved by consideration of the required chemical resistance, permeability and impact strength. In the absence of any technical advantage, the issue will probably depend upon the cost. The prices of the raw materials on a volume basis are usually fairly close together, and vary from time to time. Hence, it is impossible to give a once-for-all answer to the relative economics of these two materials.

Bottles, or jars with necks narrower than the maximum internal wall-to-wall diameter of the containers, cannot be injection-moulded in one piece, because of the difficulty of removing the male half of the mould. On the other hand, injection moulding gives perfect control of wall thickness, which is important when the container has to withstand internal pressure. The problem has been solved by injection-moulding the bottle in two parts, which are subsequently joined together, either by an adhesive, by solvent welding (particularly useful in the case of polystyrene, for which simple solvents such as ketones are available) or by spin welding. In the last process, one part of the bottle is held stationary in a clamp, while the other is rotated at high speed in a magnetic clutch. The frictional heat which is generated causes the polymer at the two surfaces in contact to melt. At this point the rotation is rapidly halted by the

GLASS AND PLASTICS-BASED PACKAGING

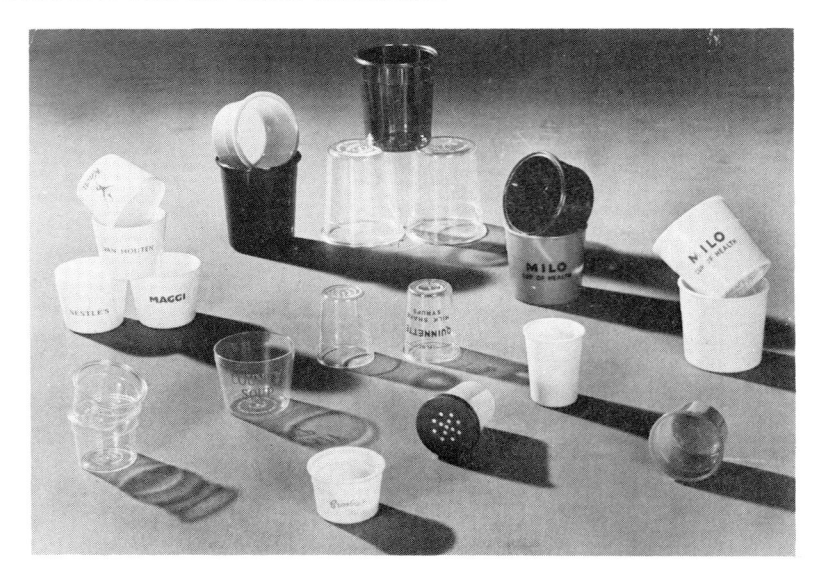

Figure 2.7a   A selection of food packs and sample tasting cups made in a special heat resistant high molecular weight grade polystyrene
Shell Chemical Co. Ltd.

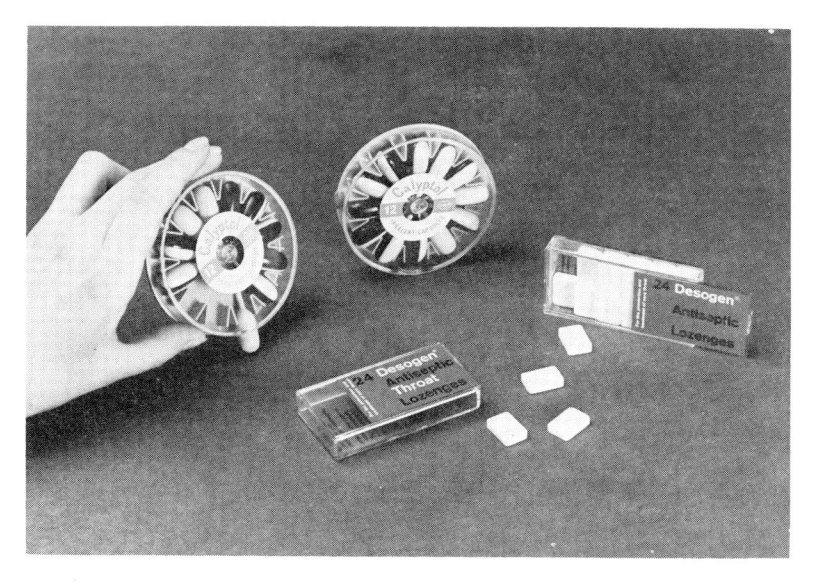

Figure 2.7b   Dispenser packs for pharmaceuticals moulded in polystyrene
Shell Chemical Co. Ltd.

MOULDED PLASTICS CONTAINERS

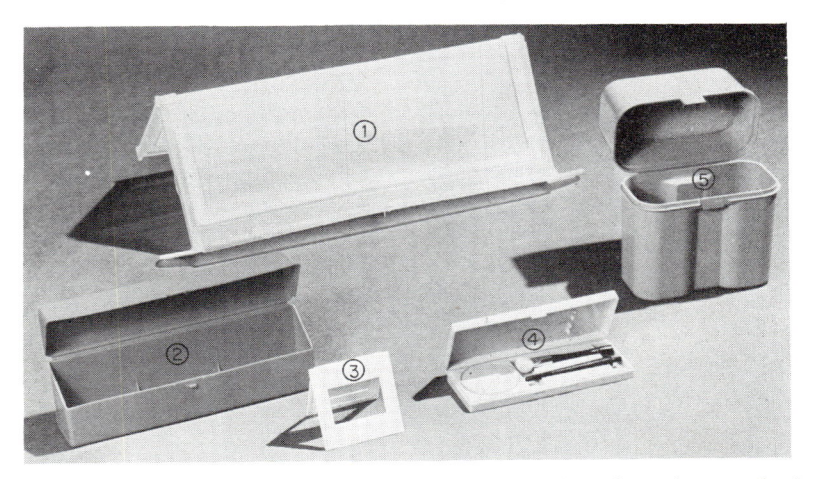

Figure 2.7c Hinged components in polypropylene. The excellent flex resistance of polypropylene gives wide scope for the production of articles with integral hinges, and introduces many interesting design possibilities for anything which folds, opens or swings. A polypropylene hinge will not stick, corrode or squeak and can be flexed over a million times without failure; multi-part assemblies and expensive finishing processes are eliminated. The items shown are: 1. filter unit for dishwashing machine, 2. box for photographic slides, 3. photographic slide holder, 4. cosmetic case, 5. water softener test kit case.

Shell Chemical Co. Ltd.

magnetic clutch, and the polymer allowed to cool. The seam produced by this method is said to be as strong or stronger than the rest of the bottle.

One of the most important developments in the field of injection moulding is the production of a container and lid in one piece, using polypropylene. This polymer has the property of forming a practically indestructible hinge when once a thin section has been flexed. Repeated flexing of such a hinge fails to cause breakage, even after as many as a million cycles. The moulding of a container with an integral hinge cuts assembly costs, and of course it cannot corrode or stick, as may sometimes happen with a metal hinge (figure 2.7c).

Thermoformed containers

Tubs, trays and box inserts are the commonest containers formed by this method, particularly where very thin walls are required, such that it would be difficult for a polymer to flow between the mould walls in an injection moulding.

Polystyrene, cellulose acetate, PVC, polypropylene and high-density polyethylene have all been used for thermoformed containers. The latest material in this field is foamed polystyrene sheet, which can be thermoformed to give trays, and containers with built-in cushioning properties.

GLASS AND PLASTICS-BASED PACKAGING

Expanded polystyrene containers

Expanded polystyrene mouldings are widely used as contour-spreading packages for items such as typewriters, cameras, microscopes, hi-fi equipment and similar complex-shaped and shock-damageable products. More recently, higher-density material has been used for moulding boxes for the carriage of 5 to 6 kg of fresh fruits and vegetables, such as cherries, grapes, and tomatoes.

Fittings for corrugated outer cases are also moulded, to be used for the packaging of television sets and similar items.

Rotationally cast containers

As mentioned earlier, rotational casting is particularly useful for the production of large containers. Plastics drums may be made by this process, as may storage tanks and intermediate bulk containers. However, the process can equally well be used for smaller containers by cycling a larger number of moulds at one time. With smaller sizes, however, the economics compared with blow moulding or thermoforming are not so favourable.

Low-density polyethylene is by far the most common material used, but both high-density polyethylene and polypropylene have been successfully employed.

Types of closure

Screw caps

Screw caps are injection-moulded from polyethylene, polystyrene or polypropylene, or compression-moulded from phenol-formaldehyde and urea-formaldehyde. They are commonly used on plastics bottles or jars, and conventional capping equipment is usually suitable for applying them.

Polypropylene caps are particularly valuable as closures for cosmetic preparations, because of their design possibilities. Polypropylene has good resilience, so that mouldings having slight undercuts can be "jumped off" the mould core without damage to the moulding. Decorative inserts, such as imitation jewels, can then be pressed into these moulded-in undercuts to give caps with highly effective sales appeal. The resiliency of polypropylene also makes possible the design of linerless closures.

Plug fittings

Plug fittings are normally injection-moulded from low-density polyethylene, since its softness and flexibility enables it to give a good seal, even against hard smooth surfaces, such as the walls of a polystyrene tube. The plug itself is often ribbed to give even better sealing.

An interesting example of the design possibilities inherent in the use of plastics is the plug closure which incorporates flexible prongs on the underside of the plug. The use of this closure for tablet tubes eliminates the necessity for a wad of cotton wool on top of the tablets to prevent their movement, with consequent risk of breakage during transport. Again no special equipment is necessary for closing.

Push-on covers

Push-on covers are normally used for injection-moulded plastics pots or jars, and for some types of vacuum-formed containers, e.g. beaker-shaped containers with a curled rim. As well as plastics push-on covers, paperboard ones are still used in some instances. Before applying the push-on cover, a foil diaphragm is often crimped over the top of the pot or jar. This serves to give extra protection or a tamper-proof seal. In thicker gauges, aluminium foil is sometimes used as the only closure, e.g. in the case of yoghurt containers. Flexible push-on covers (in low-density polyethylene) can also be used for bottles.

Heat-sealed covers

Heat-sealed covers are often used for closing vacuum-formed containers of the tray type, or the deep-drawn pyramid type used for fruit drinks. The cover may be flat or recessed to give a shallow plug-type fitting to the container, with a consequent increase in rigidity and strength. Sealing is carried out with a heated jig. With certain types of equipment it is possible to vacuum-form plastics sheet continuously from the reel, fill the depressions so formed, then cover them with another plastics sheet fed from a separate reel. After heat-sealing the cover on, the containers are cut out and trimmed.

Miscellaneous closures

The possibilities for design inherent in plastics moulding have led to many special types of closures, such as combined plug and snap-on covers, and plug or screw-caps which incorporate means of dispensing the contents in droplet form or as a jet or spray. Such closures are usually fitted to "squeeze" bottle designs. Another interesting design feature which is often incorporated in plastics closures is the integral moulding of a nozzle and cap to give a captive closure. The assembly is fitted as a plug in the bottle neck. The same result has also been achieved by the use of a snap-on-action retaining ring.

Uses of moulded plastics containers

Detergents

The first large market for plastics bottles was for washing-up liquids. The great majority of liquid household detergents are now packed in these bottles, the most common material being low-density polyethylene. This is used because it combines chemical inertness, low weight and resistance to breakage, with a flexibility that allows the production of "squeeze" bottles, thus giving the convenience of easy dispensing to the housewife. The bottles can be attractively printed to give them "sales appeal". In the United States, liquid detergents are commonly packed in high-density polyethylene. These bottles are more rigid than those in low-density polyethylene and were brought on to the US market as direct replacements for cans. If required, high-density polyethylene can be used for the manufacture of "squeeze" bottles by reducing the wall thickness and designing the bottle to improve "snap-back". One such bottle used for car shampoo has an elliptical cross-sectional shape to give the required "snap-back".

More recently, PVC bottles have been used in large numbers for the packaging of detergents, particularly when the detergent has an attractive colour. The clarity and sparkle of a PVC is often effective in such instances in increasing the sales of the product.

Foodstuffs

A problem of importance in the use of plastics for the packaging of foodstuffs is that of tainting, but special grades of several plastics have long been used for packaging a wide range of food products. Polystyrene, for example, is used in the form of pots or small jars in the packaging of butter, jam, and cheese, while in bottle form it has been successfully used to give a "new look" to vinegar retailing (figure 2.8). The vinegar bottle, incidentally, is a good example of an injection-moulded two-piece container.

Polystyrene containers are both light in weight and chemically inert. Polystyrene can also be used in crystal-clear grades if visibility of the contents is required. Vacuum-formed trays or punnets with heat-sealed covers are made from polystyrene and PVC, and used for a variety of foodstuffs. Both materials have good rigidity, so that very thin-walled one-use containers can be produced.

A toughened grade of polystyrene (containing a small percentage of a synthetic rubber) is used to produce deep-vacuum-formed containers for fruit juices, while special high-flow grades are used to give thin-wall injection-moulded tubs for such products as cream and yoghurt. The

MOULDED PLASTICS CONTAINERS

Figure 2.8 Vinegar bottles: injection-moulded two-piece containers in polystyrene
Shell Chemical Co. Ltd.

squeeze-action dispensing possibilities of polyethylene have also been exploited in the foodstuffs field, with the production of containers shaped and coloured like lemons and containing pure lemon juice, plastics "hot dogs" (containing mustard), and many variations on the same theme.

One of the most important developments of recent years in foodstuffs packaging is the use of rigid PVC blow-moulded bottles for cooking oils, wine and fruit-juice concentrates. PVC is an ideal material for many foods (especially oils) because of its chemical resistance, clarity and cheapness, but it had always been difficult to blow into bottles because thermal degradation occurs at temperatures very little above that needed to melt the material and give adequate flow properties. Advances both in blowing equipment and in compounding PVC have since made it possible to produce bottles of extremely good quality.

Cosmetics

The wide range of colour effects possible with polystyrene has led to its extended use in the packaging of cosmetics where "sales appeal" is extremely important. In addition to being chemically inert to many cosmetic products, it has a pleasant touch and is light in weight. Typical uses are for face-powder boxes, cream jars, holders for sticks of deodorant and shaving soap, compacts for face-powder, and holders for lipsticks. Later developments are blow-moulded containers for talcum powders using polystyrene.

Polypropylene is also finding uses in cosmetics containers because of its good gloss, coupled with the possibilities for moulding containers with hinged lids in one piece. Powder compacts and eye-shadow cases are typical examples. One use of polypropylene in cosmetics packaging is a blow-moulded polypropylene aerosol container for perfume. It is extremely difficult to break and, even if breakage does occur, the container only splits to allow the gradual release of its contents.

Both low- and high-density polyethylene are used for packaging shampoos, hand-creams, talcum powders, sun-tan preparations and deodorants. The absence of rusting when kept in a moist bathroom atmosphere is an advantage when compared with tins, while resistance to breakage gives an advantage over glass containers.

Pharmaceuticals

Polystyrene is used successfully for the packaging of tablets in injection-moulded phials. Unbreakable, but with the visibility of glass, they have the added advantage over glass of being easier to print.

Where greater protection against moisture vapour is required, high-density polyethylene tubes can replace those of polystyrene. Another advantage of high-density polyethylene is its higher softening point, which enables it to stand up to sterilization temperature.

Low-density polyethylene is also used for a wide variety of pharmaceutical packs, ranging from hydrocortisone to foot powder. A squeeze action makes it particularly suitable for dispensing powders, and for eye-drops and nasal sprays. Toughened polystyrene has been used for prescription containers, for pills, tablets and ointments.

Household products

Resistance to corrosion and to breakage have been the most important factors leading to the use of plastics containers for packaging liquid household products. Examples include rust removers, window cleaners, oven cleaners, writing inks and fly sprays. One oven cleaner pack makes full use of the potential of plastics to give a novel dispenser and applicator. High-density polyethylene is used for lavatory cleaners in both liquid and powder form. Liquid bleaches are also packaged in high-density polyethylene bottles, thus giving an unbreakable container for a potentially hazardous product.

Plastics containers are also being used for scouring powders in place of the spirally-wound composite containers. They are made either by vacuum-forming high-impact polystyrene in two halves, subsequently spin-welded together, or by blow-moulding high-density polyethylene.

Miscellaneous

An interesting example of the packaging of a corrosive product is the use of polyethylene bottles for hydrofluoric acid. This acid attacks glass and was formerly packed in either guttapercha or wax bottles. Polyethylene bottles are now used, giving a safer pack and one in which the liquid level is visible. Warning notices and identification can be printed directly on the bottle, so that there is no possibility of losing the label. Both low- and high-density polyethylene are used for the production of blow-moulded jerricans. These are used as water containers for camping and many aqueous solutions, while those made in high-density polyethylene can also be used for the short-term storage of paraffin. The normal design of a jerrican is not suitable for petrol, but specially designed containers have been approved for such use in Germany. The main difference lies in an increased wall thickness of the jerrican designed for petrol.

Use of newer materials for containers

Polyacetals are more expensive than the materials already mentioned, and are used mainly as alternatives to light metals, such as aluminium, in engineering applications. The high strength of the polyacetals has led to their use for aerosol containers, and usage can be expected to increase.

Polycarbonate is again a rather expensive material, but is of interest in packaging because of its combination of high impact strength, high softening-point, and good clarity. It can be blown into bottles and has been used in Japan for the manufacture of soda-water siphons.

Polymethylpentene is more widely known under the name **TPX** (trade name of Imperial Chemical Industries Ltd.). It has extremely good clarity and a high-softening-point, but is a poor barrier to gases and moisture vapour. Its packaging uses would appear to be mainly for coating and laminating, rather than in blow-moulded or injection-moulded containers.

Ionomer is basically a modified low-density polyethylene with ionic bonds between the polymer chains. These ionic links give it a high melt strength (very suitable for extrusion coating) and also modify its crystallinity, consequently improving clarity. Bottles can be produced, therefore, with the high impact resistance of low-density polyethylene, but with significantly improved clarity.

Modified acrylics. Straight acrylics such as polymethyl methacrylate (Perspex, widely used in World War II for aircraft windshields and radomes) have many excellent properties, including extremely high clarity, but have rather high melt viscosities which make it difficult to blow them into bottles or extrude them into film. Modified acrylics have been developed to overcome these processing defects, without losing the valuable end-use performance characteristics of unmodified materials. One of the earliest modified acrylics was developed in the United States under the name XT polymer. It has been used to produce bottles having a high resistance to oils and greases, together with good contact clarity (although the bottles are somewhat hazy when empty). XT polymer has also been extruded into sheet, which can subsequently be thermoformed into trays and tubs. Products packaged in this material include medicinal paraffin oil and peanut butter.

More recent examples are Lopac (trade name of a material manufactured by Monsanto Chemical Company) and Barex (trade name of a material manufactured by Vistron Division of Standard Oil of Ohio). Both were developed in the United States. The main monomer from which Lopac is produced is meth-acrylo-nitrile, with small percentages of styrene and methyl styrene. Lopac is a hard, rather brittle polymer, but it has excellent barrier properties, good resistance to cold flow or creep, and is very clear. Barex consists mainly of acrylonitrile, copolymerized with methyl acrylate, together with a small percentage of a butadiene/acrylonitrile rubber. This is also a clear polymer with good creep resistance and excellent barrier properties. It has, in addition, good impact strength. Both Lopac and Barex are being developed, mainly as bottle-blowing materials for the packaging of carbonated soft drinks.

Developments in container manufacture

One of the disadvantages of normal thermoforming techniques is that two separate heating processes are involved, one when the granules are heated during extrusion of the sheet and the second when the sheet is heated prior to forming.

Much work has been carried out on the cold forming of sheet, and some success has been achieved. Present cold-forming techniques, however, have not yet reached the high degree of sophistication attained in thermoforming methods, and the cost of equivalent-sized containers is still normally greater by cold forming. A saving is shown, though, when the comparison is made for printed containers. Cold forming then becomes potentially cheaper, because printing of the flat sheet, prior to forming, is possible. This is quicker and cheaper than rotary printing of formed containers. Printing in the flat can be carried out, because cold forming subjects the sheet to less movement by stretching than does thermoforming. Most of the work on cold forming has been carried out on acrylonitrile/butadiene/styrene copolymer (ABS).

A process which does involve some heating of the sheet is *stretch forming*. A heated disc of the plastics sheet is placed over a split mould, the disc fitting into a recess at the top of the mould. The disc is then clamped into position and a forming punch descends, stretch forming the disc into the shape of the mould. The method is suitable for polypropylene, which is at a disadvantage in normal thermoforming processes, because of the difficulty of controlling it at the high temperatures involved and also because of its longer cooling time. Stretch forming has another advantage, since the resultant containers have good transparency, which cannot be achieved by most normal methods of processing polypropylene.

Stretch forming involves the use of special equipment, but it is also possible to convert polypropylene into transparent containers by more or less conventional pressure-forming equipment, using sheet heated to a temperature below its crystalline melting-point. Containers with diameter to depth ratios of 2:1 have been produced in this manner.

Developments such as these could increase the use of polypropylene in areas where its other properties give it an advantage over polystyrene and PVC, which are the commonly-used transparent plastics. These other properties of polypropylene include a higher softening-point, resistance to oils and greases, and toughness.

CHAPTER 3

Packaging with Flexible Barriers

C. R. Oswin

A package is essentially a substitute for a human hand. It frees the hand for other work by "holding" a handful of possessions—isolating them from other matter and securing against loss. The modern package serves also to identify its contents and often to embellish them. Almost all wares tend to deteriorate under the influence of some extraneous factor; properly selected wrappings will delay this process by excluding that factor.

Primitive wrappings were improvised from available materials such as leaves and skins. Deliberate wrapping in a flexible material only became feasible a century ago, when paper-making machines and new raw materials reduced the cost of paper sufficiently.

The earlier flexible wrappings, apart from metallic foils, are derived from vegetable crops, purified and fluidized by chemical processes. The sheet is deposited from a suspension (paper) or a solution (regenerated cellulose, cellulose esters) and dried. The later wrapping materials have been produced from fossilized vegetable remains (coal and oil) purified and reacted into a solid polymer which is fluidized simply by melting, and extruded into a film by forcing the liquid through a slit. Wrappings are usually produced and supplied in the form of reels, but cut sheets can be obtained where necessary. Melt-extruded films are often produced in tubular form: when cut and sealed simultaneously they then provide bags.

There is thus a wide choice of commercial wrapping materials, each of which has its particular value for wrapping certain wares in particular ways. No one flexible wrapping is suitable for all wares or all types of package. Even the size of the pack will help to determine which wrapping

PACKAGING WITH FLEXIBLE BARRIERS

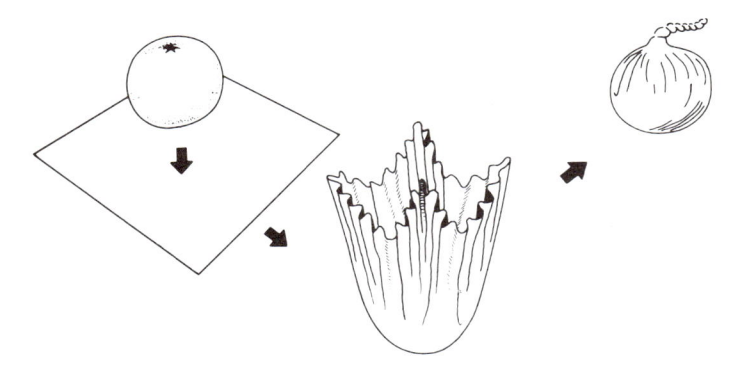

Figure 3.1　Hand-wrapping (twisted ends)

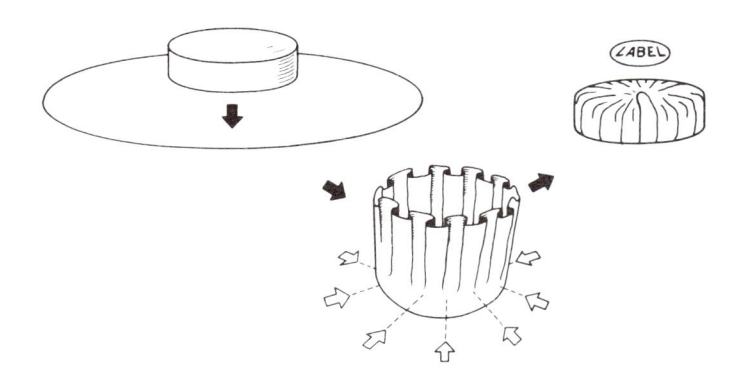

Figure 3.2　Hand-wrapping (pleated ends)

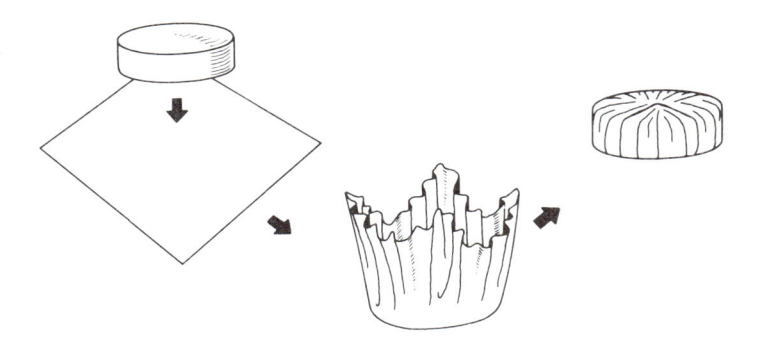

Figure 3.3　Hand-wrapping (folded ends)

3.45

GLASS AND PLASTICS-BASED PACKAGING

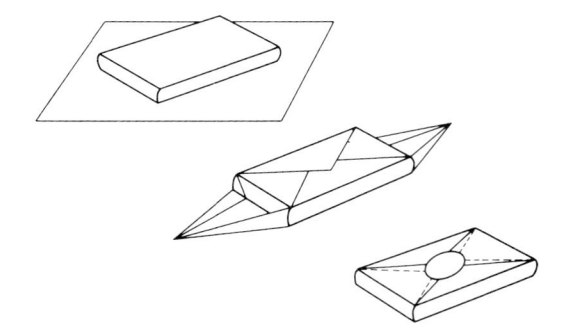

Figure 3.4　Hand-wrapping—the bias wrap

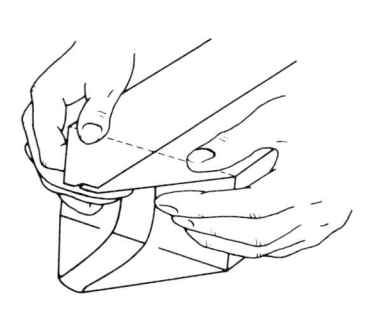

Figure 3.5a　Parcel wrap

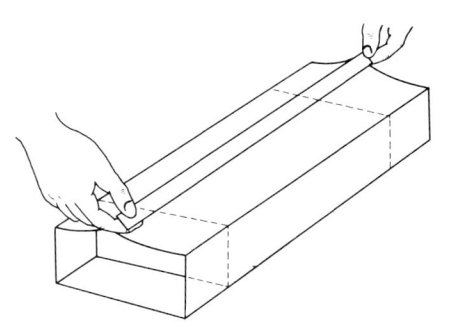

Figure 3.5b　Parcel wrap with "grocer's fold"

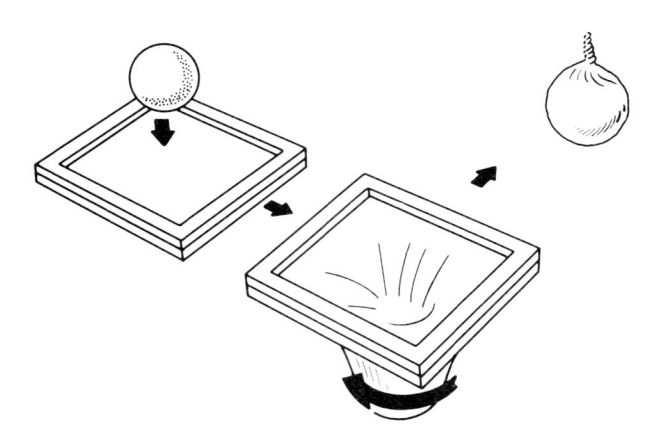

Figure 3.6　Wrapping in distensible film. Temporary distensibility can be produced by heating the wrapping

PACKAGING WITH FLEXIBLE BARRIERS

is best. In general, no absolute rules can be laid down for selection but, in the pages which follow, there will be an indication of the sort of wrappings which can be used, and the way in which they can be applied, for packing different wares. This indication is in broad principle only, and no responsibility is accepted for the results of trials; but it should enable the user to narrow his choice to a few wrappings. He must make the final selection on questions of price, practical performance and personal preferences.

The styles of wrapping which are possible will first be discussed, then the properties of particular wares which demand certain functions of protectiveness in the wrappings, and finally the wrappings with their individual advantages and limitations. All three of these must be considered together. A final section considers some incidental processes which can be applied to improve the pack, shrinking, gas-packing, etc.

Styles of wrapping

Hand-wrapping

The simplest styles of wrapping involve applying the sheet to an object and folding or gathering the loose ends as nearly as possible. The loose ends may be twisted (figure 3.1), pleated (figure 3.2) or folded (figure 3.3).

A simple fold, which uses the minimum amount of film for a slab-shaped object, brings the corners together on one face of the slab (figure 3.4, bias wrap).

More elaborate folds are used for parcels (figure 3.5a), with or without a "grocer's fold" (figure 3.5b) at the overlap, to assist in pulling the wrapping taut.

With suitable dead-folding or tacky wrapping materials, the folds or twists stay in position and no further closure is necessary. This is chiefly with foils, laminates, thin clinging films and heavily-waxed papers. Other wrappings have to be held in place by an adhesive patch, or by fusing the over-laps together. The adhesive patch can range from a simple piece of pressure-sensitive adhesive tape to a more elaborate printed label with delayed tack.

Heat-sealing may be done by touching the over-laps briefly with a hot iron, or by sliding the pack over a table which has a heated bar let into the surface. The temperature of the bar must be regulated to suit the wrapping used. Distensible films can be stretched tightly over the wrapped object when wrapping by hand. This improves the appearance of the pack, although it may reduce the protection offered by causing thin spots. Temporary distensibility can be produced by heating the wrapping (figure 3.6).

3.47

GLASS AND PLASTICS-BASED PACKAGING

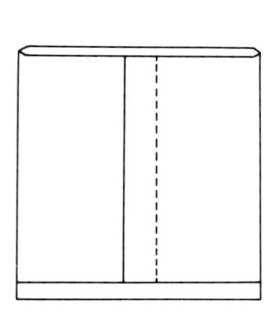

Figure 3.7 Bagging—the simple pouch

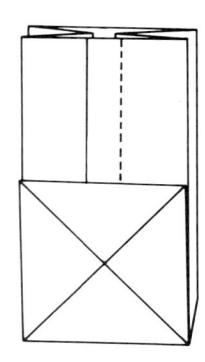

Figure 3.8 Bagging—hand-made "block-bottom" bag with square cross-section

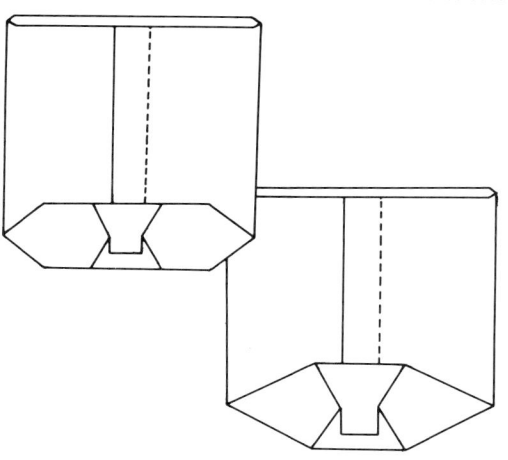

Figure 3.9 Block-bottom machine-made bags of rectangular section

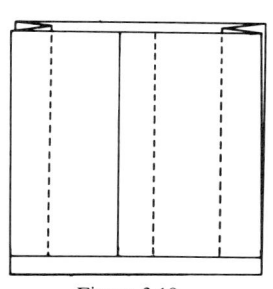

Figure 3.10
Pouch made from gusseted tube

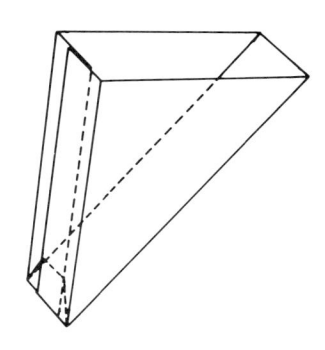

Figure 3.11 Triangular bag

3.48

Tight packs can also be obtained if a film is used which will shrink when the pack is subsequently passed through a warm air tunnel. Shrink wraps are particularly appropriate to fragile or irregularly-shaped objects (see page 3.73).

Assisted hand-wrapping

When a number of similar packs have to be wrapped, the slow process of hand-wrapping can be assisted by simple devices which eliminate repetitive or slow movements. In particular, the end folds can be completed (after a longitudinal parcel-fold has been made by hand—figure 3.5) by pushing the pack through a series of folding-ploughs which form the folds mechanically. If heat-sealing plates are placed beyond the ploughs, the packs can be sealed after being pushed through the complete aid as they are made. This is particularly helpful when bulk-overwrapping blocks of small packets. Such aids put the rate of wrapping up from about 4 per minute to about 16, but they depend on a constant supply of similarly shaped wares being ready to wrap.

Bagging

Where the objects to be wrapped together in quantities are not easily stacked, or are of varying shape, it is convenient to use bags. These are virtually prefabricated wraps, with only one end remaining to be closed. Various forms of bag can be produced, each being particularly convenient for its own range of shapes. The simple pouch (figure 3.7) suits flat wares (stationery, handkerchiefs, stockings, shirts, etc.) but bulky objects are more easily inserted into bags which can open to a rectangular cross-section. Hand-made "block-bottom" bags can be made with a truly square cross-section (figure 3.8) but, if a rectangle is acceptable, machine-made bags are available (figure 3.9). Several variations are known: with thermoplastic wrapping materials a similar effect can be obtained by making a pouch from gusseted tube (figure 3.10). Block-bottom bags can also be made by modification of a finished pouch by sealing down the "ears" which distinguish a distorted flat pouch. Shaped bags may sometimes wrap a specific object economically so that, for example, a triangular bag (figure 3.11) can be used for sandwiches.

Filling the bag is usually assisted by mechanical means, a funnel attached to weighing scales, or a guide directing the wares into a bag which has already been opened by a jet of air directed at its mouth. For repetitive bagging, the supply of preformed bags can be held in a magazine, with the top one opened by an air-jet, or they can be fed from a continuous roll or zig-zag. Wares to be bagged can be pre-weighed, counted or otherwise pre-measured.

GLASS AND PLASTICS-BASED PACKAGING

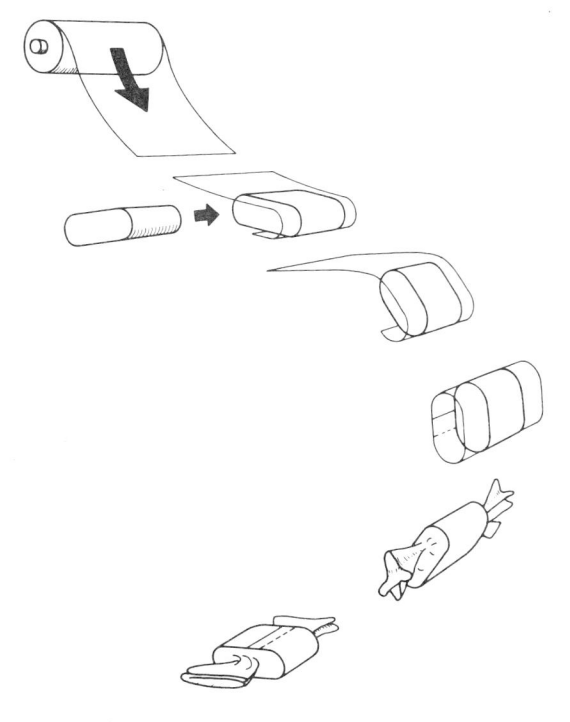

Figure 3.12 Twist wrapping by machine

Bags cannot usually be closed securely by twisting the ends. For security in transport, the top of the bag can be folded over and secured by an adhesive tape or a label, or a printed cardboard "header" can be stapled over the mouth of the bag. More simply, the mouth of the bag can be gathered together and held in place by a twist of adhesive tape or of wire. Moulded thermoplastic clips perform the same function and are conveniently re-usable, but they are usually more expensive than simple tapes. Bags made from a thermoplastic material, or from a material having a thermoplastic coating, can be heat-sealed by heated jaws or between two endless belts of heated metal. The latter process lends itself to continuous operation on a conveyor belt, which relieves the operator of part of the procedure.

Mechanical wrapping

Mechanical wrapping produces far more packages than hand-wrapping —and consumes correspondingly more wrapping material. The previous section is therefore out of proportion to its commercial importance: but it has been given in some detail because the principles of mechanical wrapping are the same.

For wrapping mass-produced articles in a constant flow, automatic wrapping machines replace the manual operator. The speed of packaging is greatly increased, and in the case of small objects which are convenient to feed and wrap, speeds of up to 600 pieces per minute may be achieved by cutting a piece of film, forming it into a tube around the object and twisting the ends of the tube (figure 3.12).

Rectangular objects lend themselves to mechanized versions of the parcel wrap (figure 3.5). There are several versions which can be selected according to the size and shape of the wares to be wrapped. In each case the principle is broadly similar: a length of wrapping material is drawn or fed from a reel, and cut off. The object to be wrapped is pushed into this length and the ends folded around the object to form a "tube", with an overlap of ¼ to 1 in. The open ends of the tube are tucked in appropriately, and the overlap and tucks are sealed in place.

Neatly rectangular packs, such as packets of cereals or cigarettes, can be given a fold similar to that of figure 3.5. One end is tucked in by grippers, and the package is then pushed through pairs of ploughs which fold the other ears in order (figure 3.13).

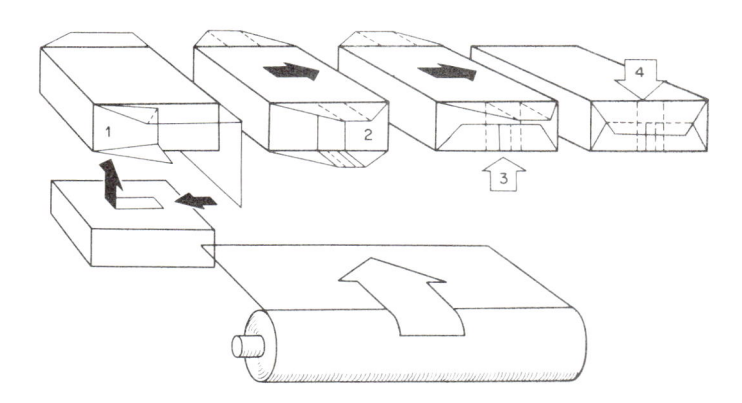

Figure 3.13 Mechanized "parcel" wrapping

When the objects to be wrapped are not constant in size, or are soft, the ends are more accurately folded in by grippers as in figure 3.14. In order to secure each fold as it is made the ends are then folded down in continuing order, a style more suitable for large loaves of bread.

3.51

GLASS AND PLASTICS-BASED PACKAGING

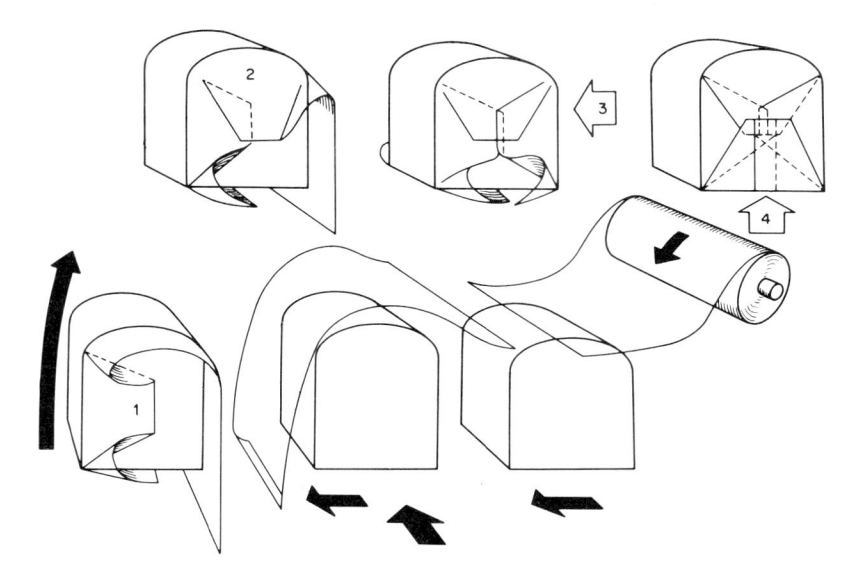

Figure 3.14   Bread wrapping

Modifications of these folds are used for regular but comparatively small objects. If the last "ear" is folded outwards and underneath, as in figure 3.15, all heat-seals can be made in one face. The appearance of this style is tidier, and it is also used for dead-folding wraps such as foil laminates, as well as for sugar confectionery—caramels, etc.

The sealing of the overlap on the bottom face of the parcel in the style of figures 3.13 and 3.14 can damage the barrier properties of coated wrappings, and with small packets such as cigarettes the overlap is made on one of the narrow edges. The mechanics of this is shown in figure 3.16, the "overlap" in this version being made during the later stages.

In all these versions of parcel wrapping it is necessary to feed the wrapping materials from a reel and to cut off lengths for folding over the wares. With very thin flexible wrappings this can be difficult, and careful adjustment of the cut-off knives is essential to prevent misfeeding. Pull-feed of the material by grippers is also preferable to the simpler method of push-feeding by intermittent rollers.[1] When the end-seals have been formed by ploughs, they can be secured by adhesive labels or by heat-sealing if suitable wrappings are used. Pressure-sensitive labels are not easy to dispense at the highest speeds, and it is usual to apply roll-fed printed labels which are coated with a thermoplastic adhesive. The adhesive is formulated to be super-coolable, so that it softens on being heated but does not "set" for a few seconds.

PACKAGING WITH FLEXIBLE BARRIERS

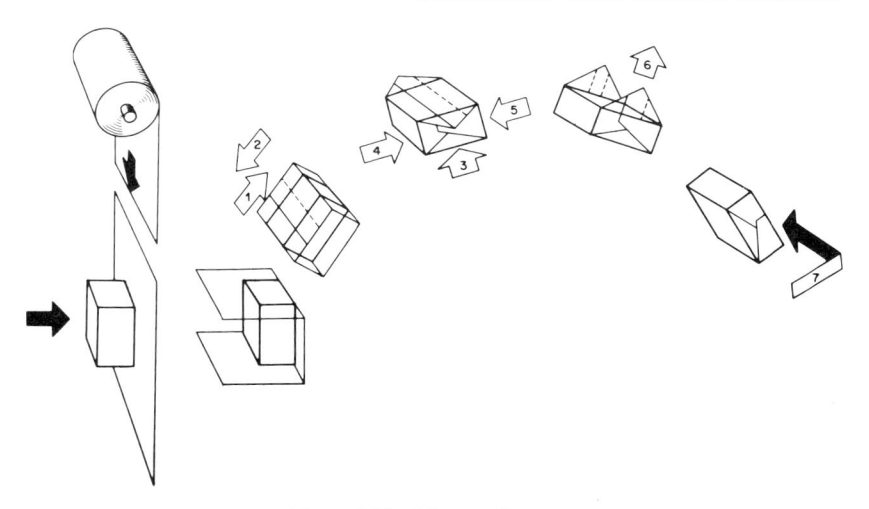

Figure 3.15 "Caramel" wrap

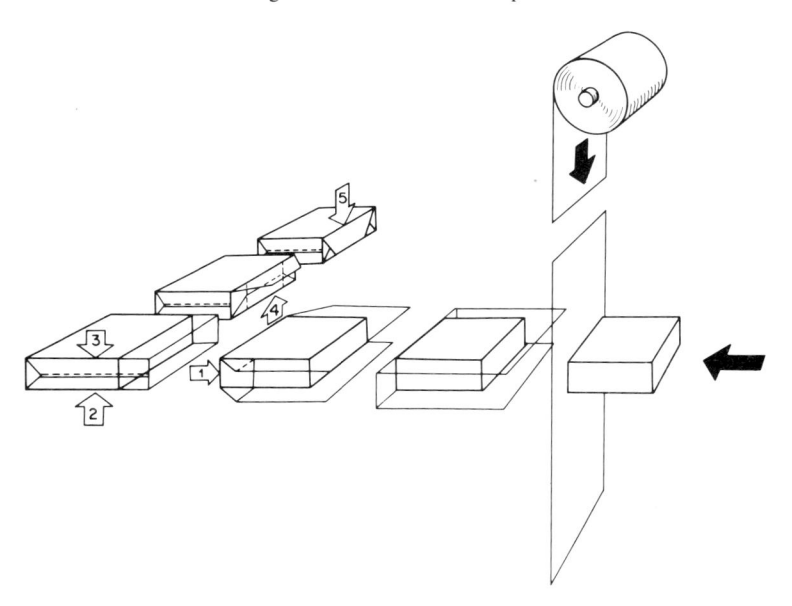

Figure 3.16 Small packet wrap with overlap on edge

If the assembly of objects to be wrapped is cylindrical, e.g. a pack of round biscuits, it is neater to replace the four folds at the end by a number of pleats, which can be done by rolling the pack between two rows of pleating teeth (figure 3.17).

3.53

GLASS AND PLASTICS-BASED PACKAGING

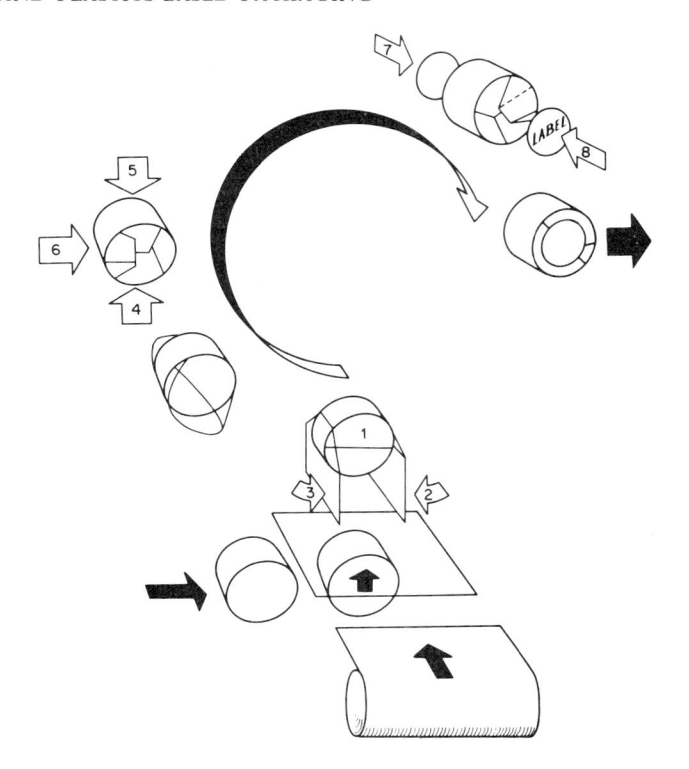

Figure 3.17 Wrapping with pleated ends

The "parcel" style of pack is not convenient for thin, irregular or numerous objects, and machine wrapping of these is more conveniently done by strip-packing. In one version of this (figure 3.18) the web of wrapping material is formed into a tube with a heat-sealed horizontal seam. The wares to be packed are fed into the tube, which is sealed across at regular intervals. This method can be used for powdered or granular materials and, if the direction of the transverse seal is alternated at right angles, a tetrahedral pack suitable for liquids is obtained (figure 3.19). A modification (figure 3.20) will handle wrappings which have a fusible layer on one side only. It is also suitable for larger packs.

Wrapping materials which are not rigid, or which require special care in heat-sealing, are most easily used in the form of two webs which are sealed together round the edges of the wares. This gives a flat pack (figure 3.21) when both webs are thin and flexible.

If one of the webs is rigid, a "skin" pack is obtained (figure 3.22) and the appearance can be improved by shrinking the thin web so that

3.54

PACKAGING WITH FLEXIBLE BARRIERS

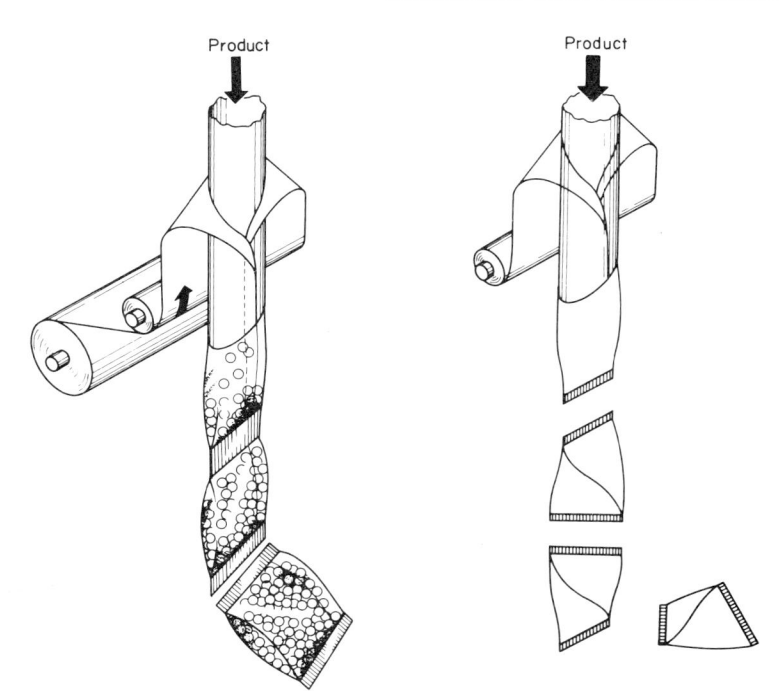

Figure 3.18 Pouch pack

Figure 3.19 Tetrahedral pack for liquids

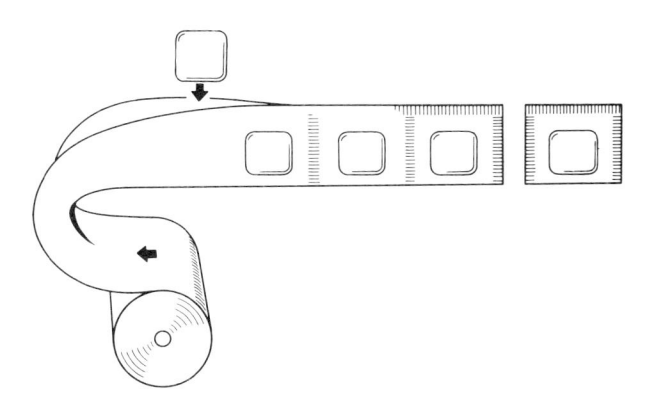

Figure 3.20 Strip pack with fusible layer on one side only (suitable for larger packs)

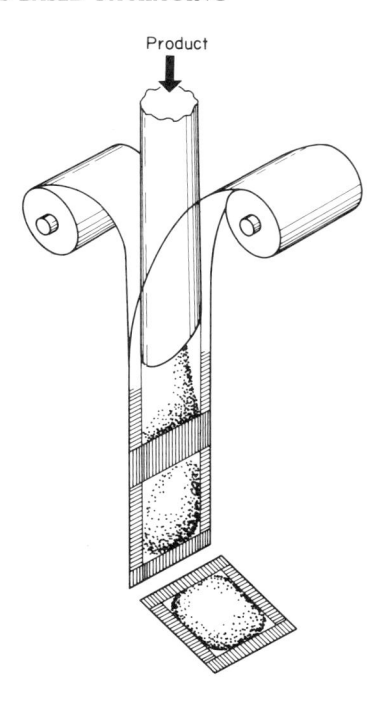

Figure 3.21 Flat pack—both webs thin and flexible

the wrinkles disappear and the web comes into taut contact with the wares. If both webs are relatively rigid one or both must be preformed to provide receptacles for the wares (figures 3.23a and b).

Variations of figure 3.23b will give shaped packs which can be used as containers for liquids, wine, oil, etc., as substitutes for small bottles. A preformed web with thin "lid" sealed over is a derivative of figure 3.23a used for pastes such as jam or yoghurt.

The wares to be packed

General

The style of wrapping selected from among those illustrated in the previous section is largely determined by the size, shape and quantity of objects to be packed. The nature of the wares themselves has less influence on the style of wrapping, although fragile wares may be unsuitable for machines which have coarse feeding mechanisms.

The type of wrapping material to be used is much more critically

PACKAGING WITH FLEXIBLE BARRIERS

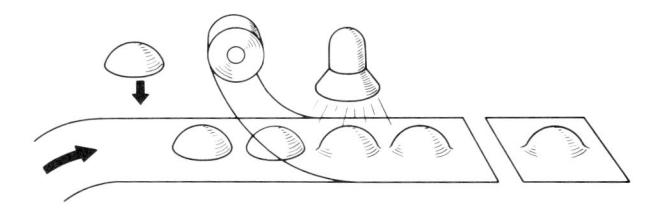

Figure 3.22   Skin pack—one web rigid

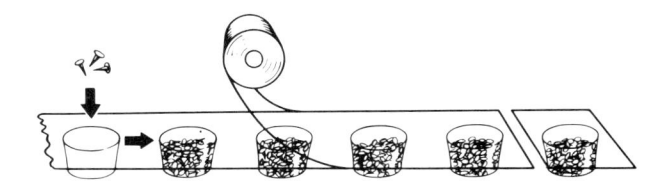

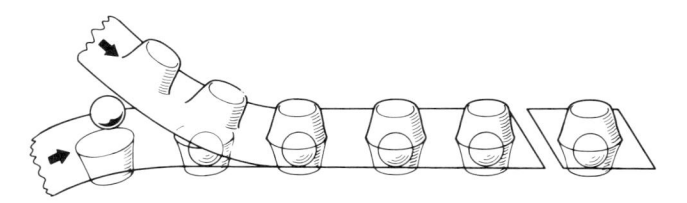

Figure 3.23a

Figure 3.23b   When both webs are relatively rigid, one or both must be preformed

determined by the nature of the wares to be packed. If the wares are themselves attractive, a glossy transparent wrap may enhance their appearance; if not, an opaque wrap may be preferred. If they are sensitive to light, an opaque or coloured wrap may give the best protection, even at the sacrifice of appearance. There have always to be compromises between display and protection, between preservation and ease of use, and above all between durability and cost.

Some wares are not very susceptible to damage; they need wrapping only for convenience, and almost any decorative wrap can be used. Many wares, however, are wrapped because they might otherwise deteriorate, and a wrapping must be selected to fend off deterioration as long as the wares remain unused. Wares can deteriorate in various ways and, although the mechanisms of deterioration are fairly well understood, the quantitative theory[2] has not yet been developed far enough to select the wraps without final trial under practical conditions.

3.57

Dry wares

Many foods sold in a dry form are resistant to microbiological deterioration as long as they remain dry. Biscuits, breakfast foods, boiled sweets, cereals, potato crisps, soup powders and dehydrated foods are all examples of wares which "keep" so long as they are not allowed to absorb excessive amounts of water from the surrounding air. They would soon do this if they were not wrapped in a barrier material which resists the passage of water vapour. Dry foods which are ready to eat are particularly sensitive to spoilage, because their texture as well as their flavour depends on their dryness.

These dry foods must be wrapped in good water-vapour barriers which are themselves efficiently sealed. Coated paper and cellulose film and, more recently, coated polypropylene film are commonly used for biscuits. The coating is a highly-resistant barrier to water vapour, such as a vinylidene chloride copolymer. For wares which are very sensitive to moisture, such as some accelerated freeze dried (AFD) meat or fish products, or where very long shelf life is needed, laminates of two or more barrier materials (often including aluminium foil) are used.

If sufficient data are available about the moisture content of the wares in equilibrium with atmospheres of different relative humidities, and the initial and critical moisture contents, calculation will show what order of water-vapour resistance will be required for their expected shelf life. Suitable wrappings can then be tested in the field.

Air-dried foods, such as dried fruit, pulses, pasta, starch, are not so sensitive to small changes in water content, but they can usefully be packed in a wrapping of moderate resistance to water vapour, which will protect them from damage by excessive moisture when they happen to encounter extremes of atmospheric humidity.

Moist wares

These goods lose moisture when exposed to the average atmosphere, and the barrier wrap is needed to prevent a loss of moisture (instead of a gain as was the case with dry wares). Calculation once again can indicate the type of wrapping needed for any required shelf-life. This can be done with cigarettes, tobacco, and certain kinds of flour confectionery.

The problem is more complicated, however, when the wares contain nutrients and water sufficient to support microbial growth. If the wares are moist enough, any mould spores or bacteria present will begin to grow. After two or three days at normal temperatures this would spoil the appearance of the wares. Where foods are concerned, microbial growth will also change the flavour, and can often render foods harmful if consumed.

The problem can sometimes be solved for long storage by sterilizing the contents of hermetically-sealed packages by heat or by irradiation. Short-term storage can be achieved if the packages are stored in refrigerators at temperatures just above the freezing-point. Chemical preservatives can also prolong the shelf-life, but their use is generally disliked unless they are surely innocuous.

If none of these methods of preservation is acceptable or practicable, a compromise in packaging must be adopted. A wrapping is selected which will enable the surface of the wares to dry out sufficiently to prevent mould growth, while keeping enough water within the bulk to preserve the texture. Cakes having a low sugar content can be packed in this way. Only a limited life is possible from either cause of deterioration, and the optimal solution to the problem is one which postpones both faults until they develop simultaneously.

More complicated instances occur, for example with bread, which also deteriorates by "staling". This is a chemical process which occurs independently of the method of wrapping, and any inexpensive wrap which will give protection for the duration of the period of "freshness" is good enough for the purpose. The use of permitted preservatives and anti-staling chemicals would lengthen the realizable shelf-life, but the penalty is inevitably a softened crust.

"Breathing" wares

Some wares are sensitive to oxygen, and the action is often increased in conjunction with water vapour.

Iron and steel objects, such as motor spares, tools and surgical instruments, will rust in air if the relative humidity exceeds 40%. The rate of attack is very slow at 40% humidity, but increases rapidly as the relative humidity rises. Such articles could theoretically be protected by packing in a wrapping material which is impervious to oxygen, but so little oxygen is needed to cause noticeable damage that this is impracticable. It is simpler to wrap such hardware in a good water-vapour barrier, and to enclose a small quantity of a desiccant, such as silica gel, inside the package. The desiccant will absorb a considerable amount of water before allowing the humidity inside the pack to rise to the danger-point. Volatile corrosion-inhibitors can be enclosed inside the pack instead of desiccant, and a wrapping must then be chosen which will not allow the volatile vapour to escape too quickly.

Oxygen is also harmful to fatty or oily wares such as butter, fried snacks or nuts, and oxidative rancidity develops more quickly under the influence of light. An opaque pack, printed or foil-laminated, will give good protection for these wares. Where the shelf-life is short and the wares (such as potato crisps) more sensitive to other factors, rancidity can usually be ignored. However, an oil-resistant wrapping must be selected.

Oxygen is consumed from the air inside a package by green vegetables, fruits and similar crops which continue to respire and produce carbon dioxide after they are harvested and packed. If the package were to be hermetically sealed, the oxygen would all be consumed, creating conditions favourable for the growth of certain bacteria, and the texture and flavour of the wares would be spoiled. To prevent this, a regular supply of oxygen is necessary, and arrangements must be made for the removal of carbon dioxide. Unfortunately, most vegetables respire at a fairly rapid rate[3] and only porous or perforated wrappings let oxygen and carbon dioxide through quickly enough to preserve the contents. A compromise is again necessary, because excessive ventilation would also encourage the loss of water, with consequent wilting of the vegetables. The usual solution is to bag or pack the wares in an impervious wrapping having 3 or 4 holes about 5 mm ($\frac{1}{4}$ in) in diameter, or to leave the pack only partially sealed. It should be stressed, however, that cooled storage during the marketing period is very beneficial for prepacked fruits and vegetables.[4]

Meat

Meat presents a difficult problem in packaging, because it can be spoiled by so many factors. Excess of moisture will encourage microbial growth, while excessive dryness spoils the texture and appearance. Excess of oxygen can change the colour of the meat, but so can inadequate oxygen supply.[5] Even too bright illumination will bleach the colour.

Cured meats are less perishable than fresh meat, but at some sacrifice of the original flavour. They are still sensitive to mould growth unless they are dry or contain added preservative. Semi-cured meats such as bacon are even more sensitive, and even in cool storage they need protection against oxygen (for example, by vacuum packing) and bright light to have a reasonable shelf-life. Fresh meat is most critical and will quickly spoil for one reason or another unless the biochemical processes involved are slowed down by storage at reduced temperatures. Storage at 1° to 3°C will give a shelf-life of a few days if the wrapping is permeable to oxygen, and special grades are usually made for this purpose. (See pages 3.66 and 3.67.) Below this temperature there is a "forbidden" region where the texture of the meat is destroyed by ice-crystal growth. Prolonged storage calls for quick freezing, with storage at temperatures below $-15°C$ and, under these conditions, a wrapping of fairly high resistance to water vapour is needed to prevent freezer-burn by surface desiccation.

Liquids

The use of flexible packages for containing liquids has come to be accepted as a practical convenience. The leader in securing this acceptance was the blow-moulded bottle which can be regarded as an extreme case of figure 3.23, fitted with a re-closable cap. These bottles were first used in fairly thick gauges for detergents and bleaches, but thinner bottles are now used for milk, oil, fruit squashes and wine. High-density polyethylene is mostly used, but PVC and nylon are also suitable.

The main requirements in packs for liquids are leakproofness and resistance to attack or penetration by the liquids packed. Some rigidity, coupled with a flat base so that the pack can stand alone, is also necessary unless a rigid container is supplied for simple packs (figure 3.20).

Leakproofness demands a fusible material, or coating, which is thick enough to give an autogenous weld. A layer of low-density polyethylene at least 50 µm thick will provide this, while a second layer of paper, regenerated cellulose or nylon laminated to it, will provide rigidity. The second layer, and the adhesive employed in laminating, covers any pinholes which may be present in the polythene film.

The style of pack can be a tetrahedron (figure 3.19), or a bag (figure 3.7), or pillow pack (figure 3.18) modified by gussets to give a flat bottom. This enables the pack to be stacked for display. The flexible laminate which is used can be printed before the pack is made, so that there is no need to apply a separate label. The packs are intended for relatively small quantities of beverages such as wine or fruit juices. Staple commodities such as milk are packed in simple pillow packs (figure 3.18), with a rigid container provided for permanent use. These packs are not quite as economical as a re-usable glass bottle, but they eliminate the necessity to return the bottle.

Flexible packaging materials

General

Flexible sheet materials depend for their usefulness on the properties of a special kind of molecular structure: long flexible molecules, interlocked into a strong and not-brittle lattice.

The first packages were improvised from whatever sheet materials happened to be available. Animal products such as the skins, bladders and intestines were widely used, but they suffer the disadvantage of being difficult to shape and impossible to seal perfectly. They are also subject to microbial attack by organisms which have evolved to prey on the animal kingdom. Their edibility is still of value in packaging foods such as sausages.

Bulk production of wrappings has come to depend on the vegetable kingdom—living crops which can be harvested or fossilized vegetable deposits from coal mines or oil wells. Mineral products, other than foils of the ductile metals, are usually too inflexible to withstand folding.

This section will describe the salient properties of the commercially available wrappings, and will attempt to draw together the previous sections into a general conspectus of flexible packaging possibilities.

Most of the wrappings are available in several thicknesses. Thick grades will obviously be stronger and more protective, but the thin grades will "go further". The range of properties quoted must be interpreted accordingly.

The following set of arbitrary units has been chosen to simplify comparisons between the wrappings:

Yield. One unit represents a yield of $42\,m^2$ of wrapping per kilogram weight. Two units signifies a higher yield, $84\,m^2/kg$, and so on.

Strength. Each unit represents a breaking strength of 1 kg for a 28 mm wide strip or 350 N/m.

Stretch. One unit indicates that the wrapping can be stretched about 400% (i.e. to 5 times its original length) before it breaks. Half a unit implies that it can be stretched 200% (i.e. 3 times its length).

Water vapour resistance. One unit corresponds to a (tropical) permeability of about $18\,g/m^2/d$, but two units signifies *half* that permeability, since barrier properties and permeability are related reciprocally. The resistance is 800 MNs/mole.

Oxygen resistance. One unit corresponds to an oxygen permeability of about $8000\,cm^3/m^2 day/atm$. Two units signify *half* that permeability. Unit resistance is 23 GN s/mole.

Although these units are arbitrary, they are manifest as the typical properties of a sheet of normal low-density polythene, 25 μm thick. This convenient reference point has been chosen because it is not extreme in any value, and of all the familiar wrappings it probably shows less variation between one manufacturer and another, irrespective of the country of origin.

Paper

Paper is made by depositing matted cellulose fibres from a dilute suspension in water. Almost any vegetable fibre can be used for making paper, but the great bulk is produced from wood. The wood is pulped either by special grindstones (mechanical pulp) or by chemical processes (chemical pulp). The dilute suspension of fibres in water is poured on to a moving wire sieve, where it is formed into a continuous sheet. The dried sheet is porous and quite permeable to gases and vapours. It can be treated, in the process of manufacture, to have improved wet-strength or grease resistance. It does not form a barrier package by itself, but it is a cheap base for barrier coatings, and is still the most widely used wrapping material. Paper which has been impregnated with wax, or coated with a synthetic resin, is a useful barrier material, but an even better barrier can be made by laminating preformed sheets of other barrier materials to the paper (see page 3.71).

The properties of some typical paper-based barrier materials are given in Table 3.1, in the units defined on the page opposite. They are given to only one significant figure, because of the variation between papers (and tests).

Table 3.1. Paper-based barriers

	55/70 gsm waxed paper	35/50 gsm lacquered glassine	Polyethylene coated paper	P.V.D.C. coated paper
Yield	0·3	0·5	0·2	0·2
Strength	4	2	4	3
Stretch	0·01	0·01	0·01	0·01
Water vapour resistance	0·03	0·2	1	2
Oxygen resistance	0·01	0·01	1·0	500
Heat-seal temperature	60°C	115°C	120°C	130°C

Waxed paper is a cheap and moderately protective wrapping, suitable for wrapping loaves of bread on automatic machinery (style of figure 3.14). In much thicker form it produces cartons for quick-frozen foods and for milk (figure 3.19). Coated papers, especially those coated with a thick layer of polyethylene or vinylidene chloride copolymer, are better barriers and can be used for packing dry wares. When the coating is thick enough to bestow heat-sealing abilities, mechanical wrapping is feasible, e.g. biscuits as in figure 3.17 and snacks as in figure 3.18.

For hand-wrapping large mechanical objects, a heavily waxed paper, or textile laminated to a barrier material such as coated cellulose film, has been used. It has suitable conforming and dead-folding characteristics, and it is easy to seal the folds and overlaps with a warm iron.

Regenerated cellulose film

Regenerated cellulose film is produced from a specially pure wood pulp by a chemical process which takes the individual fibres into solution and then regenerates a transparent film by precipitation. The material is chemically similar to paper, except that it is no longer porous. It is a poor barrier to water vapour but, when dry, a good barrier to oxygen. Its water-vapour resistance is greatly improved by coating: it is a more economical base for the application of good coating materials than are porous materials such as paper. It needs to be plasticized with glycerine, or similar humectants, if it is to be used in a dry atmosphere, and the extent of plasticizing can be varied to adjust the flexibility to suit the conditions of use. Some grades are quite stiff in use, while others are flexible enough for twist-wrapping (figure 3.12) around boiled sweets. The degree of resistance to water vapour can be adjusted by suitable choice of coating, and very highly moisture-resistant grades are suitable for packing biscuits (figure 3.17) and snacks (figure 3.18). The properties of some regenerated cellulose films are given in Table 3.2, quoted to one significant figure because of the variation in products and in tests. At high relative humidities the strength falls, and the stretch increases somewhat; oxygen resistance also falls.

Table 3.2. Regenerated cellulose films

	32 g uncoated	Normal coated	Flexible coated	P.V.D.C. coated	High barrier
Yield	0·7	0·6	0·6	0·6	0·5
Strength	7	7	5	8	10
Stretch	0·05	0·05	0·08	0·05	0·05
Water vapour resistance	0·01	1	1	2	7
Oxygen resistance	700	800	500	1000	1200
Heat-seal temperature	–	135°C	135°C	135°C	125°C

The films are available in yields ranging from 0·8 down to 0·3, with corresponding strength properties. The resistances to water vapour and to oxygen of the various grades of coated films do not vary, because it is the base film which is thickened not the coating.

Other grades of coated film may have a water-vapour resistance of 0·01 or 0·02. Normally moisture-proof grades are suitable for cigarettes (figure 3.13), fruit cake (figure 3.12) and many other commodities. The grades of intermediate resistance are made for wrapping moist goods (see page 3.58) including meats, and one-side-coated film is used for fresh meat (see page 3.60).

The coated cellulose films can be sealed together by heat. They are favoured for their ability to run on automatic wrapping machines; all styles of mechanical wrap can be used, except those involving thermo-forming or shrinking. A degree of shrinkage can even be achieved, as when uncoated cellulose film is moistened, applied to a bottle and allowed to dry. Cellulose films are easy to print, and can be given a tear-strip for convenience in opening.

Cellulose acetate

Cellulose acetate is the most widely used of a number of cellulose derivatives which are intended to be somewhat less sensitive to water than are paper or cellulose film. It can be formed from solution or by melt-extrusion. Cellulose acetate film 25 µm thick has a yield of about 0·8 unit, a strength of 5 units and a stretch 0·1. The resistance to water vapour is 0·02 unit and to oxygen 2 units. Thicker grades have strengths and resistances in inverse proportion to their yields. It is a poor barrier to water vapour and oxygen, and can be used for vegetables and fruit, although perforation is usually desirable. It is a constituent of laminates, with paper as a decoration over print, or with thermoplastics to hold liquids (shampoos packed as figure 3.21). In thicker grades it is rigid enough to be formed into boxes for presentation packs or to be thermo-formed (figure 3.23).

Polyvinyl chloride

Polyvinyl chloride is a synthetic polymeric material derived first from coal (via acetylene) but now from oil (via ethylene). It must have plasticizers and stabilizers added if it is to be used for packaging. Non-toxic additives must be selected if the film is to wrap foods which might absorb them and become contaminated.[6] Non-migratory plasticizers are better, and internal plasticizing (by copolymerization) is best of all, though it may be more costly. The degree of softness can be adjusted from quite rigid to limp and clinging. Each type has its virtues in packaging. The properties of some PVC films are given in Table 3.3.

Table 3.3. PVC films

	Oriented rigid	25 µm normal	20 µm flexible
Yield	0·7	0·7	0·9
Strength	7	2	0·6
Stretch	0·05	0·5	0·6
Water vapour resistance	0·5	0·2	0·1
Oxygen resistance	50	30	2
Heat-seal temperature	—	150°C	120°C
Shrink temperature	90°C	90°C	80°C

3.65

GLASS AND PLASTICS-BASED PACKAGING

The rigid grade is easily thermo-formed and can be used as preformed trays or for packing as in figures 3.22–23 for small ironmongery preserves, jelly, toys, dairy-products, etc. It is not a very good barrier to water vapour, but in the thicker thermo-forming grades this is usually unimportant. The more flexible grades can be stretched over breathing wares, cheese and processed meats, to give attractive packs which have adequate shelf-life for the relatively perishable wares. If the sheet has been stretched during manufacture, it can be made to shrink after application by passing the package through a tunnel heated to about 90°C.

The more flexible grades can be heat-sealed at 100–120°C, but they do not run easily on fully automatic wrapping machines. Semi-automatic and assisted hand-wrapping are feasible.

Polystyrene

Polystyrene is also derived from oil. It can be extruded in film form, but is rather brittle unless prepared as an "impure" copolymer. It does not heat-seal easily, tending to run and stick on the sealing jaws, but impulse sealing will give good seals on thermo-formed packs (figures 3.22–23). Oriented polystyrene film with a yield of 0·8 unit has a strength of 5 and stretch of 0·03. Resistance to water vapour is 0·01 unit and to oxygen 3 units. It is relatively cheap, but a poor barrier to water vapour. It can be used to pack small ironmongery, preserves, lettuce, dairy-products, etc., but its chief application is in thermo-formed containers. These have been used for soft drinks, preserves, cigarettes and, as trays, for meats and produce. Although it is available in transparent grades, styrene copolymer sheet is often used pigmented.

Polyethylene

Polyethylene is synthesized from ethylene, a by-product of the petroleum industry. It combines the chemical properties of a wax with the physical properties of a long-chain polymeric compound, and in a quarter of a century it has risen to the second largest weight of output amongst wrappings. Many grades with specialized properties have been produced in the process.

The properties of some polyethylene films are shown in Table 3.4 with 100 gauge (25 μm thick) as the reference for the whole section. The properties of this film are closely proportional to thickness, and so inversely proportional to the yield.

The normal (low-density) polyethylene has fairly good resistance to water vapour, and it can be used for packing dry foods which are not extremely

PACKAGING WITH FLEXIBLE BARRIERS

Table 3.4 Polyethylene films

	LDPE 25 μm	Irradiated 18 μm	HDPE 12 μm
Yield	1	1·3	2
Strength	1	4	1
Stretch	1	0·2	0·3
Resistance to water vapour	1	1	1
Resistance to oxygen	1	1	1
Heat-seal temperature	120°C	—	140°C
Shrink temperature	—	70°C	—

sensitive. It is also less prone to puncturing than most other wraps, and is suitable for textiles, hardware and toys. It should not be used for vegetables or fruit unless it has first been perforated to allow "breathing".

It can be fused to itself for sealing, but special equipment is necessary to prevent complete fusion or sticking to the jaws.[1] The seal is very strong when formed, and tearing points may have to be provided for easy opening. It is too distensible to run well on fast folding machinery, but bread can be wrapped (figure 3.14) with modified sealing means. Side-welded bags (figure 3.20) can also be used. For hand-wrapping, some form of twisting (figures 3.1 and 3.6) is possible if the twist is cut off and fused by a hot wire. Bags are convenient forms to use (figures 3.7 and 3.10) and the filling can be assisted by having the bags supplied in continuous perforated lengths, with a jet of air to open the end bag. Closure is by strip, clip or fusion.

Polyethylene film which has been stretched in manufacture will shrink on heating to 70°C. Greater shrinkability is conferred if the film is irradiated before stretching. Strip packing (figures 3.18 and 3.21) is possible with polyethylene so long as the sealing jaws are set to a carefully controlled temperature, and faced with a non-stick material such as PTFE (polytetra-fluoro-ethylene). Polyethylene film is flexible at sub-zero temperatures and suitable for the packaging of quick-frozen foods.

For blister packing, the high-density grade of polyethylene is preferred. It is more rigid than the normal grade, and the forming and sealing temperatures are higher. It is more resistant to water vapour than the low-density grades, but the transparency is inferior. The crisper "feel" of HDPE, especially when it is stretched during production, enables it to be used in thinner gauges which offset the higher costs. Some traditional uses for paper bags have been displaced by 12-μm HDPE, e.g. in butchers' shops where the wet strength and juice-retention are important advantages.

If polyethylene film is suitably treated beforehand, it can be printed. Normal polyethylene film is slightly hazy; a clear grade is available, but it is not quite so strong. Other grades can be formulated to give better

3.67

GLASS AND PLASTICS-BASED PACKAGING

"slip", or to reduce condensation from wet wares. The thinnest films of polyethylene can usefully be perforated to make them more suitable for packing "breathing" wares, and to reduce the danger from suffocation of small children who may be tempted to pull them over their heads.

It is also used as one ply of a laminate, because it can be readily applied by direct extrusion and confers good durability, some barrier properties and heat sealing. Coatings on paper can be used for milk (figure 3.19). With an impervious ply such as aluminium foil, the complex can be used for AFD foods (figure 3.21), but a reinforcing layer of paper or strong film is advisable.

Polypropylene

Polypropylene is synthesized from another petroleum by-product. It has a higher melting-point than polyethylene. By suitable stretching and annealing, it can be made into very thin and clear, but relatively rigid, film. The low-temperature flexibility is improved by the stretching, and so is the general machine runability. Simple polypropylene film is difficult to seal. It is more convenient to use when coated with PVDC or PP copolymer.

The properties of a simple (chill cast) polypropylene film, and of a stretched and (polyvinylidene chloride) coated film are indicated in Table 3.5.

Table 3.5. Polypropylene films

	25 μm chill cast	10/15 μm PVDC coated	Copolymer coated
Yield	1	1·3	1
Strength	2	4	6
Stretch	2	0·1	0·1
Resistance to water vapour	1	3	1·5
Resistance to oxygen	2	500	3
Sealing temperature	170°C	130°C	130°C
Shrink temperature	—	—	—

Both polyethylene and vinylidene chloride copolymers are used as fusible coatings for oriented polypropylene film; the former is cheaper, but the latter confers far better resistance to water vapour and to oxygen. These coated films can be run, with small modifications, on automatic wrapping machines with suitable feeding means (pull-feed). Coated film is used in England for dry wares such as biscuits (figure 3.17) and snacks (figure 3.18). Polypropylene film is more costly than polyethylene film and is used only where its properties of rigidity, clarity, stiffness or protection prove it superior. It can be printed and made into bags. The very thin grades of coated polypropylene film are comparable in cost and clarity with coated cellulose film, and approach it in machine-running properties.

Unstretched film has a limited use for wrapping bread (figure 3.14) and textiles (figures 3.7 and 3.20). Thicker films of polypropylene can be thermo-formed (figure 3.23), while the thin films which have not been annealed after stretching can be used for shrink-wrapping. Thin film can be laminated to paper; it has a decorative as well as a protective function, providing a high gloss, as well as resistance to water and grease.

Polyvinylidene chloride

Polyvinylidene chloride is derived from oil via ethylene. When polymerized alone it is infusible, but with 5–25% of copolymer a film can be produced. The higher copolymers can be used as coatings only when they confer exceptional resistance to water vapour. Softer copolymers give thicker self-supporting films which have the same overall order of protection as the thinner coatings but, because of their limpness, they are not easy to run on automatic wrapping machinery. Vinylidene chloride copolymer films are dense, and a film 25 μm thick has a yield of about 0·5 unit. The strength is 4 units and stretch 0·1. Hand-wrapping grades have a water vapour resistance of 3 units and oxygen resistance of 600 units. Copolymer films are widely used for hand wrapping (in styles 2, 3 and 5) where their surface tackiness and cling make it easy to keep the pack taut before sealing. Poultry, cheese and processed meat are wrapped in this film. Mechanical wrapping needs a grade of film where the slip has been improved (at the expense of halving the water-vapour and oxygen resistances), though even then the wrapping is restricted to the simpler styles (figures 3.14, 3.18, 3.20 and 3.21). The appearance of the pack can be improved further by shrinking with suitable grades.

There are other commercially available wrappings which do not fall into the above categories.

Water-soluble wrappings

Water-soluble wrappings have a limited use for packaging, laundering or dyeing materials. Detergents, water-softening powders or dyes can be handled conveniently in strip packs (figure 3.21) made from a film of polyvinyl alcohol, polyethylene oxide, or methyl hydroxypropyl cellulose. The packages break open fairly quickly in water to give up their contents and eventually dissolve completely. The first two dissolve more quickly in hot water than in cold. Toxic horticultural chemicals can be packed in this way. Strongly alkaline materials such as bleach should not be packed in polyvinyl alcohol. Thin films naturally dissolve most quickly, but to ensure reasonable strength of pack the yield of film chosen should not exceed 0·5 (21 m^2/kg).

GLASS AND PLASTICS-BASED PACKAGING

The surface of these films becomes tacky in all but the dryest atmospheres, and rolls of strip-packs are usually sold in a protective carton. Even then, prolonged storage in a damp place (kitchen or bathroom) is inadvisable.

Condensation polymers

Condensation polymers (polyamides, polyesters) are produced by a different chemical process from that used in the making of the addition polymers (polyvinylchloride, polystyrene, polyethylene, polypropylene). The process confers strength on the polymers, although the conditions of production may be more difficult to realize than addition polymerization. They therefore tend to be more costly and are restricted to specialized uses.

Nylon

Nylon films are prominent among those used in packaging. There are five or six "nylons" which can be produced as films, but of these only Nylon 11 combines good resistance to water vapour with reasonable workability, and it can be used for hand-wrapping (figures 3.1, 3.3, 3.4 and 3.5), as bags or pouches (figures 3.7 and 3.10) or in simple machine-wrapping (figure 3.20) of bacon or cheese. It can also be used for boil-in-the-bag applications, such as pre-cooked stews or kippers, because it does not allow much odour to escape during re-heating.

Polyester films

Polyester films are more expensive than nylon and are used only where strength is required. They are derived from oil and, because of their cost, they are limited to very-thin oriented films (up to $60\,\text{m}^2/\text{kg}$). These cannot easily be heat-sealed, but they are used as strong substrates for coating with, or laminating to, less-expensive water-vapour barriers such as polyethylene or vinylidene chloride copolymers. These laminates can be used for liquid or vacuum-packs. Thin films of polyesters with a layer of aluminium deposited under vacuum have decorative value, but the aluminium is much too thin to provide extra water-vapour resistance and they are most used for producing textiles. Some grades are suitable for shrink-wrapping irregular hardware (figures 3.20 and 3.21). Roasting bags are a convenience in cooking.

The properties of some of the other wrappings are collected in Table 3.6. Other developing wrappings, still expensive but with specific applications, are based on fluorocarbons (low friction and high chemical

PACKAGING WITH FLEXIBLE BARRIERS

Table 3.6. Other film materials

	50 μm polyvinyl alcohol	38 μm nylon 11	12 μm polyester
Yield	0·5	0·6	1·3
Strength	3	5	5
Stretch	0·5	0·3	0·3
Resistance to water vapour	0·03	0·4	0·3
Resistance to oxygen	100	30	80
Heat-sealing temperature	135°C	180°C	260°C
Shrink temperature	—	—	—

and weathering resistance) and on cyclic imides (high-temperature resistance).

Laminates

No single wrapping possesses all the properties of an ideal wrap when used at an economical thickness. It is natural, therefore, to seek to combine the complementary excellences of two or more wrappings by combining them together. In simple cases, an inexpensive thermoplastic can be extruded on to a stronger base, such as paper, the barrier properties of one complementing the mechanical strength of the other. Such coated papers are sometimes used for packing liquids. In more complicated cases the strength may be contributed by a paper, or a film of cellulose, or a polyester; the barrier properties by a thin aluminium foil and heat-sealing by a thermoplastic film. Such complex materials are more expensive to use, partly because of their relatively low yield, and partly because of the cost of the adhesive which is necessary to hold the plies together, but they can provide a degree of protection which justifies their use. Very sensitive dry foods—AFD or baked, particularly if a long life is necessary—are well packed in such laminates.

The properties of three of the simpler laminates are shown in Table 3.7. Broadly the properties are additive but, when liquid adhesive is used for laminating, it may augment or reduce the overall properties.

Table 3.7. Some simpler laminates

	Paper/9 μm aluminium foil/ polyethylene	Nylon 6/ polyethylene	Metallized MXXT/A polyethylene
Yield	0·2	0·3	0·3
Strength	4	5	8
Stretch	0·05	0·3	0·05
Resistance to water vapour	12	2	6
Resistance to oxygen	300	100	1000
Heat-sealing temperature	120°C	120°C	130°C

3.71

Incidental processes

In the previous sections the styles of packaging and the packaging materials have been outlined. There are various incidental processes which can be applied to improve or enhance the pack.

Preforming

Thermoplastic films which are rigid enough to hold their shape can be preformed for some kinds of package (figure 3.23). The sheet is heated by radiant heaters until soft, and then drawn down by pressure or vacuum over simple shapes placed on a porous base.[7] Cellulose acetate, polyvinyl chloride, polystyrene, high-density polyethylene and polypropylene can be preformed in this way, and automatic machinery is available for forming, filling and lidding. The "lid", which is sealed over the top of the package, is usually a thinner sheet of the same material as the base. Thermo-formed trays are also widely used for the mechanical protection they accord to loose or fragile wares.

Vacuum packing

Some "breathing" wares (page 3.59) last longer if the packets are evacuated before sealing, so that only small traces of oxygen are left to react with the wares, and this is the case with certain processed meats like bacon. The wrapping must be chosen for high resistance to oxygen and for efficiency of sealing. In fact a laminate is generally used, one ply, a strong film which resists oxygen and the other a thick layer of polyethylene for sealing. The package (figures 3.18, 3.20 and 3.21) is almost completely sealed. It is then evacuated through a small tube or inside a closed chamber before being finally sealed. Vacuum-packed wares are not perfectly protected, and the shelf-life attainable with perishable food products is strictly limited.

Gas packing

The reaction process with oxygen can be slowed down further in some instances (fruit and coffee) if the air in the pack is displaced with an inert gas such as carbon dioxide. The package is first evacuated and then flushed with gas. If the wares are too fragile to withstand crushing by the evacuated pack (flowers), the atmosphere can simply be displaced by blowing gas in. The same sort of laminate is useful for this purpose as for vacuum packing.

Shrink wrapping

As mentioned in the section on hand-wrapping, tight wraps can be made by hand more easily than by machine. Some kinds of wrapping can be applied loosely by machine, and afterwards tightened by shrinkage. The greatest shrinkage is achieved with the thermoplastic films which have been stretched in manufacture, these being heated to a temperature approaching their softening points to cause shrinkage. A brief passage through a tunnel at a thermostatically-controlled temperature is the usual way of finishing these packs. The various shrinkage temperatures are indicated in the appropriate sections.

BIBLIOGRAPHY FOR CHAPTER 3

1. Schrama, A. H., and Mot, E., *Emballages*, 1974–A'pl–92 and May–62.
2. Cairns, J. A., et al., *Climatic Protection*, Butterworth, London.
3. Platenius, H. J., *Mod. Packaging*, 1946, Oct., 139.
4. Tomkins, R. G., *Jour. Inst. Pkg.*, 1965, Sep., 27.
5. MacDougall, D. B., and Rhodes, D. N., PIRA/I. of Pkg. Conference Proceedings, Harrogate, 1969.
6. Briston, J. H. and Katan, L. L., *Plastics in Contact with Food*, Food Trade Press, Ltd., London 1974.
7. Dean, D. A., PIRA/IAPRI Conference Proceedings, London, 1972, Paper 14.

CHAPTER 4

Sacks made from Plastics Film

D. J. Flatman

Early developments

As early as 1957, some horticultural products were being distributed in France in 25-kg packs made in plastics film. These sacks were manufactured from calendered polyvinyl chloride (PVC) and part of the graphic design was devoted to explaining how the empty sacks could be converted into waterproof aprons as an "after-use" benefit for the user of the product.

Then in 1960, PVC sacks were produced in Italy by the "blow extrusion" process. These sacks were made from 0·25 mm thick lay-flat tubing, and used for the packaging of fertilizer in 50-kg units.

The first commercial uses of polyethylene (PE) as a sack material were in 1958 for the packaging of 25-kg (50-lb) quantities of fertilizer and of polyethylene resin. These developments took place in the United States, and the sacks for both applications were made from what was then known as 1000 gauge material (i.e. film of 250 micrometres in thickness). Because of technical problems, however, including those of polymer grade selection, progress was rather slow during the next two or three years.

By the early part of 1961, two large companies in Canada and the United Kingdom were using 1000 gauge (250 µm) polyethylene film sacks for commercial dispatches of fertilizer. The Canadians were packing 38 kg (80 lb) of product per sack, but in the United Kingdom 50-kg (1-cwt) units were employed, matching the standard weight offered in multi-ply paper sacks.

From then on, progress in the use of heavy-duty polyethylene sacks was fairly rapid, particularly in the United Kingdom, Canada, the United States and South Africa, and mainly for fertilizers. On the continent of

Table 4.1. Comparison of PVC and PE films for sacks

	PVC	PE
Raw material	More expensive than PE in the plasticized form required for sack making. The possibility of plasticizer migration must be considered for some products. Many plasticizers that would normally be selected are not free from odour. Formulations which do not give toxicity problems are necessary for animal feeding stuffs.	Less expensive than PVC, particularly in natural (transparent) form. Chemically inert. Comparatively odour free. Coloured sacks can be obtained at extra cost for the necessary polyethylene "Masterbatch" which obviously slightly narrows the margin between these and PVC sacks.
Processing	Calendering or extrusion (Extrusion method produces lay-flat tubing which is obviously easier and cheaper to convert into sacks than the flat sheet film obtained from a calender; PVC lay-flat tubing, however, is more susceptible to "blocking" problems, which cause the inner surfaces of the sack to stick together, and therefore filling rates may be seriously reduced.)	Extrusion only. (Standard print-treatment of lay-flat film overcomes printing difficulties with regard to "keying" of inks.)
Fabrication	High-frequency (HF) welded joints. (Radio frequency (RF) is a synonymous term.)	Heat-sealed joints.
Physical properties	Tensile strength in both machine and transverse directions very good. Slightly higher resistance to puncturing and snagging than polyethylene. Normally adequate moisture-barrier properties in thicknesses suitable for sack applications. Fairly good odour barrier. Poor low-temperature impact resistance. Sacks likely to deform if filled with product at a temperature higher than 50°C, particularly if stresses are applied immediately after filling. Not affected by ultraviolet light.	Tensile strength in both machine and transverse directions good. Very good moisture barrier. Poor odour barrier. Good low-temperature impact resistance, even at temperatures as low as $-20°C$. Distortion insignificant at temperatures up to approximately 70°C. In parts of the world where ultraviolet rays are more intense than the UK, outdoor storage periods longer than 6 months may cause significant degradation of unpigmented polyethylene sacks. For longer periods black pigmented polyethylene should be employed.

Europe generally, during this early stage of plastics sack development, and especially in Italy, PVC was used to a greater extent than polyethylene, largely due to a price advantage.

In the United States and Canada, polyethylene was always the preferred material, and after some early indecision on the part of some UK fertilizer manufacturers, the choice finally settled on polyethylene film which was, and still is, less expensive than PVC. It also offers some advantages in the UK climatic conditions during severe winters. It is claimed that all problems of filling and sealing, of transit and of climatic protection, permitting the storage of packed fertilizer in the open, are solved by both materials, but the poor low-temperature impact resistance of PVC can prove hazardous if sacks have to be moved from outdoor stacks during very cold weather.

A comparison of PVC and polyethylene films for sacks is given in Table 4.1.

Advantages of plastics sack for fertilizer

The claim that plastics sacks could be stored in the open during all weathers was the original reason for packing fertilizer in them, even when both polyethylene and PVC sacks were more expensive than the multi-ply paper sacks they began to replace. In addition to the higher cost of the plastics sack itself during these early days of transition from paper, as a plastics sack weighed only two-thirds of the weight of a multi-ply paper sack (even taking into account the heavy gauge of film then used) and since fertilizer was sold in "1 cwt gross" packs, there was a further extra cost to the fertilizer manufacturer in terms of the additional small quantity of product required.

But with an all-the-year-round production and packaging requirement, because of the overall tonnage demanded, and a largely seasonal market for purchases of fertilizer, the reduction in storage costs that could be achieved was very considerable. This advantage also applied to both merchants and farmers, who could relieve the pressure of seasonal demands and also derive benefit from "early delivery rebates"—a customary off-season price structure system operating within the fertilizer industry.

Even if outdoor storage was not used, the farmer was able to stack his fertilizer packed in polyethylene film sacks in types of barns or sheds that would be quite inadequate for storing this product in paper sacks, as most compound fertilizers are extremely susceptible to deterioration by moisture uptake.

Finally, during the season when the farmer uses the fertilizer, polyethylene film sacks may be taken to various parts of the farm and left out in the open, so that the fertilizer is available where and when it is eventually needed, and the farmer's handling and transport costs are consequently reduced.

With all these advantages, it is perhaps not surprising that by the end of the 1964 season, the fertilizer industry had changed from paper to polyethylene for nearly all its output of packed fertilizer, particularly as by this time the 800 gauge (200 µm) polyethylene film sack was less expensive than the conventional five-ply paper sack it had replaced. Today the most commonly used polyethylene film sack for 50 kg of product is still 200 µm in thickness, although some 175-µm sacks have been introduced. The main difficulty with these thinner sacks is not that they do not provide adequate strength, but that, because the thinner film conforms more closely to the shape of the granules of fertilizer, it becomes more easily abraded during transit, and this can result in failure to give adequate moisture protection or even spillage.

Naturally, even with 200-µm polyethylene film, care must be exercised in handling sacks during filling and conveying operations. Broken metal chutes and projecting bolts or wooden splinters on pallets in need of repair can cause snagging and punctures. Also vehicles must be in reasonable condition, and the decks and walls free from splinters or protruding nails. This is because punctures leading to leakage of product are obviously caused more easily in the single wall of a polyethylene sack than in the several plies of a multi-wall paper sack. Polyethylene film sacks have the advantage that, if small accidental holes are present, the high tear strength of the film prevents such punctures from running into larger holes or tears, and hence the spillage of granular contents is greatly reduced.

Another advantage of polyethylene film sacks over multi-wall paper sacks is their greater impact strength. Although it is not possible to simulate exactly the impact stresses encountered in normal transport and stacking by laboratory drop tests on sacks, these clearly indicate the greater strength of the plastics sack.

Valved polyethylene film sacks having all the established features of valved paper sacks, are available for fertilizer manufacturers equipped with valve-sack-filling equipment (see page 3.83). Woven plastics sacks are employed for the export of fertilizer to all parts of the world (see page 3.89).

Other applications for heavy-duty polyethylene sacks

The good impact strength and tear propagation resistance of polyethylene film permits the economic choice of slightly thinner plastics sacks than those of 200 µm film thickness common for fertilizers, to be used adequately for the packaging of products which are not abrasive or free-flowing. The thinner film may be more easily punctured during transit, but these products will not spill out seriously, even if small holes have been caused by mishandling, while the walls of the sack will not readily burst open.

In addition to permitting outdoor storage, polyethylene film sacks also provide protection against moisture gain or loss in the product. Furthermore, polyethylene film is a more suitable material for the packaging of certain chemicals which deteriorate paper. It is often economically advantageous to obtain product protection using a polyethylene sack rather than a separate film liner inside a multi-ply paper sack, or one with an inner ply made from polyethylene-coated paper.

Although the largest single use of polyethylene film sacks is for fertilizers, other products packed include horticultural peat, plastics moulding powders and masterbatches, granular and powdered chemicals, certain animal feedstuffs, such as mineral supplements and calf-milk equivalents, and some food products such as dried peas, ingredients for the manufacture of meat pastes and sausages, and confectionery mixes. Also, pre-packed coal for domestic use, insulation fibres and wadding materials, rubber underlay and other rubber products, and other miscellaneous products such as pipe fittings, cleaning cloths and wood chips.

An interesting application in the animal feedstuffs category, not included in the above list, is the replacement of polyethylene-coated fibreboard containers by less expensive polyethylene sacks (normally 200 µm in thickness) for the packaging of cattle feeding blocks. These are blocks or circular cakes about 480 mm in diameter and 100 mm thick designed for the animal to lick rather than to chew, to obtain a proper balance of vitamins and minerals. Different products containing various ingredients are distinguished by the use of opaque white and other coloured film sacks which are themselves printed in alternative colour combinations.

Such feeding blocks are individually machine-packed into the polyethylene sacks which are then automatically conveyed through a check weigher to a continuous band-sealing machine. As the strength of the vitamins may be affected if the blocks are not consumed within a period of 12 months or so, each sack is automatically date-coded during the heat-sealing operation, to ensure proper stock rotation through the distribution cycle.

The fully-sealed polyethylene film sack aids the final hardening process of the feeding block during the 48 hours immediately following the packaging operation, but the blocks are already sufficiently hard to be automatically palletized in the film sacks as the conveyor feeds them from the sealing machine to the palletizing unit. Each pallet load of 1 tonne of feeding blocks in their sacks is then covered with a hood of polyethylene

SACKS MADE FROM PLASTICS FILM

Figure 4.1 Outdoor storage of animal feeding blocks packed in polyethylene film sacks
Bakelite Xylonite Ltd. (BXL)

shrink film and passed into the shrink tunnel. The shrink-wrapped pallets are then stored in the open for several months if required (figure 4.1). But even when the pallet load is broken into units by a distributor or farmer, and the shrink-wrap cover removed, the individual sacks will protect the feeding blocks for a sufficient period until required by the cattle.

This is another market which traditionally uses an "early delivery rebate" scheme, and there is an advantage to farmers able to buy seasonal products of this kind during financially favourable periods without the expense of storage under cover. The product manufacturer also has lower storage costs, and a better opportunity to plan production on a consistently even basis throughout the year.

Summary of properties—heavy-duty polyethylene sacks

Good tensile strength
High tear resistance
Good impact resistance
Good bursting strength } Refer to BS 4932 for test methods and minimum values for, various thicknesses of sack film, etc.
Chemically inert
Odour-free (Film may be specified as not being capable of imparting objectionable odour or taint to foodstuffs, etc. Refer to BS 3755.)
Waterproof
Heat-sealable
Printable

3.79

Manufacture of heavy-duty polyethylene sacks

Heavy-duty polyethylene film sacks are made from unsupported single-wall low-density polyethylene lay-flat tubing not less than 125 μm in thickness and 380 mm in width. The sacks themselves will have a minimum length of 600 mm, and will be intended to carry not less than 25 kg (refer to "Definitions" in BS 4932).

Most manufacturers use resins in the density range 0·912 to 0·923 g/ml with a melt flow index below 1·0 (BS 4932 specifies a maximum density of 0·927 g/ml and a maximum melt flow index (MFI) of 1·40, determining density in accordance with method 509B or C, and MFI in accordance with method 105C of BS 2782.)

Polyethylene resins are classified according to their density and melt flow index. The melt index correlates with melt viscosity, which is related both to the processability of a polyethylene resin and to its mechanical properties. A low melt index results from a high molecular weight and, at a given density, the higher the molecular weight the greater is the chemical resistance of the polyethylene. For films of highest impact strength, a resin of the highest possible molecular weight (i.e. the lowest possible melt flow index) would be selected, limited only by processability becoming more difficult as the molecular weight increases (i.e. MFI decreases).

The manufacturing process consists essentially of the extrusion of a tube of polyethylene, which is then blown to the required diameter and cooled. The cooled film can be reeled up for further conversion into printed sacks, but often the extruder, printing machine and sack-making

Figure 4.2 Automatic heat-sealing of polyethylene film sacks

BXL

equipment are all linked together, so that a fully "in-line" process is employed. To ensure that the surface of the polyethylene film is suitable for printing, the standard method of print-treatment of the lay-flat tubing is carried out by passing the film between an earthed roller and a high-voltage high-frequency electrode separated from the roller by a narrow gap. The electrical discharge is accompanied by the formation of ozone, which oxidizes the surface of the film and thus renders it polar and receptive to inks.

The printing is carried out by the flexographic process, and the colour combinations possible are up to four colours on one side only, up to two colours on both the back and front of the sack, and three colours on one side and one colour on the reverse.

If gussets are required in the sack, these are formed either during the extrusion-blowing process, before the tubing is collapsed into a lay-flat state, or subsequently after the printing operation and before the sacks are cut and sealed by the sack-making machine. The latter is necessary if the sack is printed in the gusset areas.

Storage of empty sacks

Polyethylene film sacks are of single-wall construction, compared with the multi-wall construction of paper sacks, and one advantage of the plastics sack over the paper sack arising from this difference is the reduced volume of storage space required. Unfilled plastics sacks occupy one-third to one-fifth of the volume occupied by the same number of multi-wall paper sacks, depending on the number of plies in the sack.

Filling and closing

Open-mouth polyethylene film sacks are filled on the same type of equipment as is used for paper sacks and can also be closed by standard sewing machines (with or without over-taping). The best method of retaining the full protective qualities of the plastics material is by heat-sealing the top of the sack. Other closures suitable for some products are stapling, with or without the use of a board fitment, or bunch tying the neck of the sack with wire ties.

In heat-sealing, either jaw-type sealing machines or faster continuous-band sealing equipment can be used (figure 4.2). With continuous-band heat-sealing equipment, the speed of sealing must be synchronized with that of the conveyor from the filling machine, and this widely used method can handle 50-kg sacks at rates of 60 tonnes of product per hour (20 sacks a minute) without difficulty.

A problem sometimes arises with the packing of dusty products, because

3.81

the inner surface of the film becomes covered with dust, and this interferes with the formation of an adequate heat-seal. The problem can be solved by mechanically cleaning the bonding surfaces with rotating brushes, or by applying a polyethylene film tape which is folded and heat-sealed into position on both sides of the top of the sack.

Another method of cleaning the mouth of a sack is by a device which provides access for a vacuum extraction head, and as the sack passes into the band heat-sealer, any dust present is sucked away from the area of film to be sealed.

It is also possible to apply a simple date code, imprinted into the seal area or at the top of the sack as required.

An alternative method of heat-sealing uses a continuous hot-air sealing machine which automatically guides the top of the filled sack through a heating chamber, and trims off the surplus polyethylene film above the seal.

When polyethylene sacks are closed by heat-sealing, it is essential not to trap air within the sack, otherwise safe stacking may be impossible. The escape of residual air after heat-sealing is normally achieved through "micro-perforations" in the wall of the sack. The number and position of these needle-size perforations is important, and dependent on the product to be packed, as well as the size and shape of the sack. Larger-size perforations are punched out of the film walls of sacks for the packaging of peat. In the past, some products were packed without any perforations but, after sealing, a partial vacuum was drawn by means of a multi-point probe, following which operation a PVC self-adhesive patch label was applied. Another method to keep entrapped air to a minimum is the use of a maze-seal at the base of the sack.

Valved plastics film sacks

The earliest valved plastics sacks were made from PVC film by folding in one corner of the top of the sack and, in order to obtain sufficient penetration of the valve aperture into the sack, a projecting "turret" of film was left on the valve side edge of the top of the lay-flat sleeve. Then both top and bottom seals were applied, trapping the upper edge of the folded-in valve within the top seal of the sack. This type of valve construction was not successful in polyethylene film, because of the less pliable nature of this material compared with the plasticized PVC film formulated for sacks.

Attempts were made to design a polyethylene-film valved sack with an opening for the filling lance or spout on one side of the face of the sack which was covered with a patch of film sealed on three edges only. The main problem with this patch valve sack was that packing rates were reduced because of the difficulty in placing the sacks into position for

SACKS MADE FROM PLASTICS FILM

filling with product at the required speed.

Several other types of plastics valve sack were introduced, including square-ended designs, but as these were all based on semi-automatic production methods, none were commercially successful.

Some time elapsed before the automatic production of a block-bottom valved sack in polyethylene film was developed, providing a plastics equivalent of the valved paper sack. Indeed it was not until late 1969 that the first polyethylene film sacks of this type were manufactured in the United Kingdom, although they had been available from the Continent earlier. Special adhesives are used to secure the ends, which are reinforced with overcapping panels, so that when filled these valved sacks have a neat square-ended profile.

They are used by manufacturers of plastics resin and fertilizers who are equipped with valve sack-filling equipment (figure 4.3).

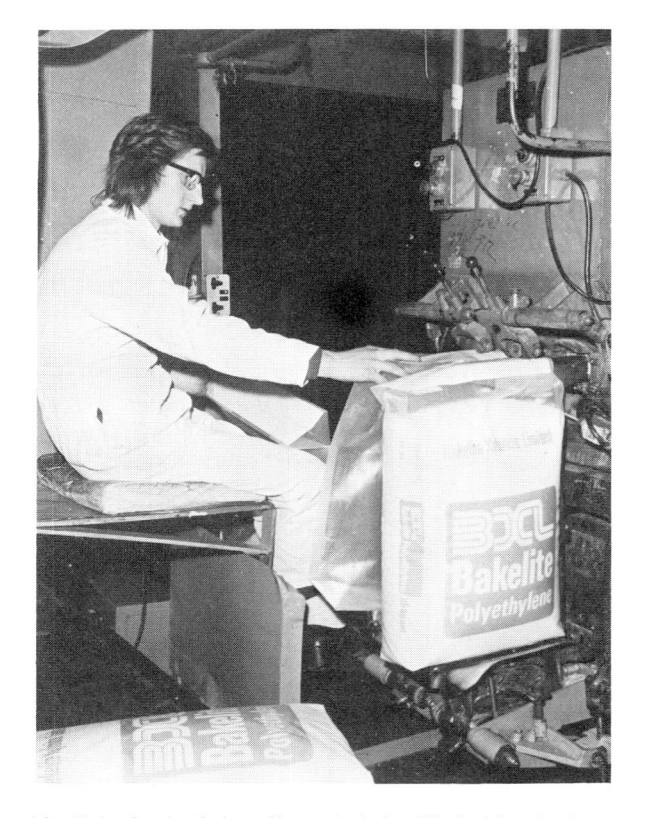

Figure 4.3   Valved polyethylene film sacks being filled with polyethylene resin
BXL

Laminated film sacks

During the past ten years or so, many attempts have been made to overcome the generally limp nature of low-density polyethylene-film sacks, and various blends of high-density polyethylene with other materials have been tried in an endeavour to obtain the stiffness and more paper-like handling characteristics of high-density polyethylene, whilst retaining the better functional properties of the low-density film. Whilst the tensile strength and the bursting strength of high-density film are very good, the impact strength and the tear resistance of low-density polyethylene film are more favourable for sack applications.

None of these attempts has been successful in producing a single-layer sack film that meets all the requirements, but a possibly better solution to the problem is the manufacture of a two-layer sack film by a lamination process from high-density and low-density polyethylene.

Co-extruded laminated sacks

Co-extruded laminates are produced by linking two (or more) extruders to a single die. In the United States film sacks have been manufactured by this process, using high-density polyethylene to achieve stiffness, and a blend of low-density polyethylene with ethylene vinyl acetate (EVA) to obtain the other properties required. The inclusion of ethylene vinyl acetate in the polythene blend widens the temperature range for heat-sealing.

These and other combinations of compatible materials could be co-extruded to provide the properties required for sack applications, but the higher costs involved in sack manufacture by this process are probably difficult to justify at the present time, in comparison with the existing types of sack that are adequate for the wide range of current packaging requirements. The process could become more important in future.

Cross-ply laminated sacks

Sacks made from a cross-ply laminate of two layers of high-density polyethylene film were first developed in 1968 in the United States and Europe.

The basic concept of cross-laminated films was conceived by a Danish engineer. Ole Benat Rasmussen. The Van Leer organization acquired the rights to this concept and, through a research and development programme, converted it into a workable industrial process. The registered trade name of the material is "Valéron" and two Van Leer manufacturing plants are now in operation, one in Essen, Belgium, and the other in

SACKS MADE FROM PLASTICS FILM

Houston, Texas. A contract has been signed allowing a Japanese company to produce "Valéron" cross-laminated high-density polyethylene film under licence in that country.

Manufacture

The material itself consists of two separate layers of high-density polyethylene films which have been produced in such a way that the lines of orientation run at an angle of 45° to the longitudinal direction of the film, and then lamination takes place so that the direction of the orientation of the two webs is opposed, forming a "cross" (figure 4.4).

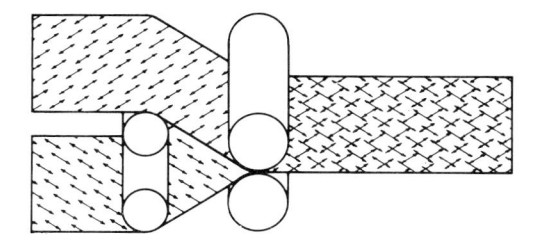

Figure 4.4   Cross-ply laminated high-density polyethylene "Valéron"[R] film
Van Leer (U.K.) Ltd.

The film produced by this process is an opaque white which provides a good background for printing. The same kind of print treatment as was described for low-density polyethylene film is carried out on both sides, and similar inks are employed, which provide very good resistance to rubbing or smearing on sacks printed by the flexographic process in up to four colours. An anti-slip lacquer coating can be applied if required. Special UVI grades of laminated film are available containing a UV absorber component.

It is possible to produce pigmented cross-laminated film sacks if the quantity required is sufficient to justify the special programme required to manufacture the film.

Standard web widths and thicknesses have been introduced to reduce the costs of film production and conversion. A slightly different range is available in the United States to that in Europe, for, although this type of sack has experienced slow development and growth on both sides of the Atlantic, greater interest and activity has occurred in the United States since 1973 where three thicknesses of cross-laminated film are available for the manufacture of sacks and bags (100 μm: 75 μm: and 63 μm). In the EEC the first two are manufactured, but only the 100-μm

3.85

film is used for both valved and open-mouth sacks. However, some European sacks are also supplied in a thicker (120 μm) material. The size range of "Valéron" sacks in the EEC is:

Width from 500 mm to 650 mm (in 50-mm increments)
Length from 460 mm to 1200 mm (up to 1500 mm for sewn open-mouth sacks)
Block-bottom width from 90 mm to 180 mm (with corresponding size valves)

In the United States sack converting companies have placed considerable emphasis on the supply of smaller sizes of sacks and bags (some with carrier handles) for consumer market packs where more use is made of four-colour printing than is usual for industrial packs.

In both the United States and Europe, the earlier sack-production methods were based on the use of modified paper-sack tubing machines, where a single web of laminated high-density polyethylene sheet was formed into a tube, and then an adhesive back seam applied. This method has now changed to provide an even stronger weld by extruding a copolymer melt for the back seam.

The folding and glueing required to produce the block-bottom and valve sections of the sacks were originally carried out by modified paper-sack bottomer machines. This confirms the paper-like characteristics of this type of laminated film which can be folded, creased, cut, sheared, punched and blanked with conventional tools. Again, however, the method has changed in that block-bottom and valved sacks are now manufactured on the same type of automatic sack-making machines used for low-density polyethylene film valve sacks.

Open-mouth sacks with block bottoms are produced on the same machine, but open-mouth sacks without block bottoms are also available with sewn ends, and these sacks may be of the flat-pillow type or side-gusseted. In the United States pinch-bottom sacks are manufactured, i.e. one face of the sack is longer than the other, and this longer portion is turned over and adhered to the opposing face. Valve sacks with side valves and pinch-bottom closure, top and bottom, can also be made.

Closure

Valved sacks are manufactured with a low-density polyethylene film valve which allows high-speed filling on conventional equipment, the valve closing automatically in the normal way for block-bottom valved sacks.

Because of the cross-lamination of the high-density polyethylene film, the sacks possess a very high tear strength, and this allows the use of conventional stitching machines with a crepe tape binder and cotton thread for closing open-mouth sacks. The needle should have a conical-shaped head, not one shaped like a pyramid with flat edges.

SACKS MADE FROM PLASTICS FILM

Heat-sealing of cross-ply laminated sacks is not recommended, because the orientation of the film layers of the laminate results in loss of strength in the seal area.

A more recent development has been the production of micro-perforated "Valéron" film which, when made into valved sacks, can be used with pneumatic filling equipment for fine powders. For very fine powders, the micro-perforated plastics sacks are provided with a single ply of light-weight kraft paper liner to act as a filter.

Improved filling speeds can be obtained with this type of micro-perforated plastics sack compared with paper sacks, and the strength properties of the cross-laminated film are not significantly weakened by the micro-perforation process.

Properties

High tear resistance/tensile strength.—As already stated, the manufacturing process involved provides cross-laminated high-density polyethylene film with very high tear and tear propagation resistance. It also results in an optimal balance of non-directional tensile strength and elongation in the material.

Good impact strength.—Sacks made from this laminate possess exceptionally good impact strength. Indeed, the normal test methods for low-density polyethylene film are inadequate, and a special testing method has been developed in which a free-falling metal ball is dropped on to a

Figure 4.5   Drop test on 5-ply Kraft Sack Wall, from a height of 1·2 m
Van Leer (U.K.) Ltd.

3.87

clamped circular specimen of material. The ball must be dropped from heights considerably in excess of that which punches a hole in five plies of paper sack material, or makes an indentation in 200-μm low-density polyethylene film before a 100-μm laminate specimen is damaged. The height from which the ball has to be dropped to damage the specimen is recorded, and this height multiplied by the weight of the ball gives an impact failure energy (see figures 4.5, 4.6 and 4.7).

Figure 4.6   Drop test on low-density polyethylene film (200 μm) from a height of 2·5 m
Van Leer (U.K.) Ltd.

Figure 4.7   Drop test on Valéron film (100 μm) from a height of 4·0 m
Van Leer (U.K.) Ltd.

Chemically inert/odour free.—As with all polyolefin materials, cross-ply laminated high-density polyethylene film sacks can be used in contact with foodstuffs for both human and animal consumption without risk of contamination.

Puncture/snagging resistance.—"Valéron" sacks are extremely resistant to puncture by sharp objects and possess very good snag resistance.

Waterproof/water vapour resistance.—The high-density polyethylene film construction ensures that the sacks are waterproof, and sacks made from the film have a low water-vapour transmission rate if good closures are made.

Applications

The manufacturing process for cross-ply laminated sacks results in their being more expensive than normal low-density polyethylene-film heavy-duty sacks, and the cost factor restricted development during the earlier years of availability.

The increasing prices of conventional sacks, particularly multi-ply paper sacks with polyethylene film liners, and the successful conclusion of long-term development programmes has now resulted in a growing use of cross-ply laminated high-density polyethylene-film sacks in both the USA and the EEC.

Applications include the packaging of chemical products for export; plastics polymers, especially where sharp granules are involved; food additives and special animal feedstuffs; herbicides and pesticides; and building products such as special plasters.

Woven plastics film tape sacks

Manufacture

Polypropylene or high-density polyethylene film is extruded from a slot die in the usual manner (although some processes employ an annular die) and the sheet of film is then multi-slit into narrow widths which are passed through a heating and stretching process which orients the film tapes in the longitudinal direction. The tapes are then wound up on cores or bobbins for the weaving operation, which can be carried out on flat looms or by a circular loom process.

Fabric made on flat looms has a "selvedge" at both sides of the woven material where the "weft" tapes are turned back on themselves. These are the tapes which are carried by the shuttle backwards and forwards across the width of the fabric, and pass alternately over and under the "wrap" tapes which run along the length of the fabric.

With circular-loom weaving, the "warp" tapes run along the length of the fabric, but the "weft" tapes are carried by rotating shuttles around the circumference of the woven tube of material, and there is no "selvedge".

The film tape is now normally specified by its *tex*, i.e. the weight in grams of 1000 metres of tape. In the past, before SI units were adopted, the term *denier* was more commonly used, i.e. the weight in grams of 9000 metres of tape. The number of warp and weft tapes per 100 mm of fabric are specified, and sometimes the weight of the fabric in g/m^2.

Although high-density polyethylene (HDPE) has a better low-temperature resistance, polypropylene (PP) has a higher creep resistance. Polypropylene is also 6–8% lighter than high-density polyethylene, and therefore produces a greater yield per unit weight.

In the United Kingdom, most woven plastics film sacks are manufactured from polypropylene tape made into fabric by the circular-loom weaving process, which, by obviating the necessity for a sewn side-seam on the sack, results in a stronger package. This type of sack was first manufactured in the United Kingdom in 1966.

The tube of fabric is cut into the required length by a hot-knife process which seals the raw edge of the woven plastics material. Indeed, the top edge of an open-mouth sack can be left unhemmed unless the customer specifies otherwise. If hemming is required at the top of the sack, this can be provided either as a single hem or a double hem. The bottom is folded over and stitched through all four thicknesses of material, normally with thread made from polypropylene filament yarn, but other sewing threads can be used. The advantage of polypropylene thread is that it matches the properties of the woven sack itself, in not being susceptible to rotting or attack by chemicals, etc.

Before any hemming or stitching operation is carried out, the sleeve of woven sack material can be printed, using special inks.

Liners

The use of a polyethylene film liner (made from low-density polyethylene film) in a woven sack provides a completely waterproof package and, because of the almost transparent nature of unpigmented woven polypropylene fabric, any printing required can be carried out on the liner. This method not only gives protection to the print, but also enables the woven sack to be left plain, and therefore re-usable for another purpose or product.

Polyethylene film liners can either be inserted loosely, held in position by a solvent-based adhesive at the mouth of the sack, or stitched into the manufacturer's base closure, using the "skirt" of the polyethylene liner for this purpose (i.e. the area below the base seal line which will be longer than on a normal film bag or liner). Obviously, this type of liner has to be ordered to specification.

Liner gauges vary according to the product packed and the degree of protection required. The normal range of thickness is 25 µm to 125 µm. Perforated film liners are used for some application.

Closure

After filling with product, the woven polypropylene sack can be closed by machine stitching or by bunch tying with wire ties.

When stitching, it is advisable to arrange that the stitchline passes through four thicknesses of fabric, although a stitchline through two thicknesses is satisfactory for some applications if the top of the sack has been hemmed, or if it is applied at least 50 mm from the top of the sack.

If a polyethylene film liner is used, it can be folded over and the sack stitched sufficiently close to prevent the film unfurling. When the liners are glued to the mouth area of the inside of the sack, this can be right at the top, so that the stitchline of the closure will go through the adhered section; or the adhesive can be applied about 200 mm down from the top of the sack, so that the liner can be heat-sealed to protect any hygroscopic products, and then folded down away from the final stitchline closure of the woven outer sack.

The woven plastics film-tape sack itself cannot be heat-sealed to provide a satisfactory closure, as the orientation of the tapes would be affected, causing a serious reduction of strength in the seal area.

Valved woven plastics sacks are available on the Continent of Europe. These are of the side-entry type, fitted with inner polyethylene film liners.

Properties

Woven polypropylene sacks are much lighter in weight than jute sacks and, weight for weight, polypropylene tape is significantly stronger than conventional sacking materials (see Chap. 6, Part Four).

Polypropylene tape is unaffected by water and does not absorb moisture. Woven polypropylene sacks do not rot in damp conditions. With the use of suitable UV-stabilized polymers, improved protection can be given against ultra-violet light degradation. These sacks therefore offer many advantages over traditional sacks for use in tropical areas.

GLASS AND PLASTICS-BASED PACKAGING

Woven polypropylene sacks are unaffected by most chemicals. Sacks made from polypropylene tapes are clean and do not impart any odour or taint to their contents. There is also an absence of contamination by loose hair-like fibres, a problem often associated with conventional sacking materials.

Applications

The largest packaging application for woven polypropylene sacks in the United Kingdom is for the shipping of British-made fertilizers and other chemical products to various export markets (figure 4.8).

Other main uses are for the packaging of potatoes, vegetable seeds, metallic abrasives, castings and other light-engineering products.

Sugar, coffee, cocoa, and other world crops are often imported into Britain in woven polypropylene sacks. The packaging of grain is another

Figure 4.8   Woven polypropylene sacks used for exporting fertilizer
Fairbairn Lawson Packaging Ltd.

obvious use. Also flour, particularly in Nigeria, where about 12,000,000 woven sacks are employed every year for this product. In Peru, woven polypropylene sacks are widely used in the fishmeal trade for their export markets.

Typical examples of woven polypropylene sacks used for the packaging of chemicals are as follows:

Sack 510 mm wide × 900 mm in length, 40 warps × 40 wefts, 110 tex, with 25-μm polyethylene film liner for 50 kg of sodium chloride.

Sack 510 mm wide × 820 mm in length in the same specification, with 125-μm polyethylene film liner for 50 kg of sodium hydroxide.

A typical lighter-weight woven polypropylene fabric would be 40 warps × 35 wefts, 100 tex. A heavy-weight fabric could be either 40 warps × 40 wefts, 120 tex; or 45 warps × 45 wefts, 110 tex. The more intersections per unit area provided by the 45 × 45 specification produces a tighter weave and therefore a slightly better sift-proof fabric.

CHAPTER 5

Plastics Drums and Crates

Bowater Packaging Ltd.

In this chapter plastics drums are considered to be any plastics moulded industrial container in the range 5 litres to 250 litres (1 gallon to 50 gallons). This definition includes containers known generally as drums, barrels, kegs, jerrycans, pails. Plastics drums are generally moulded in polyolefins, i.e. in thermoplastics materials produced by the polymerization or copolymerization of olefins. Examples of polyolefins are high-density and low-density polyethylene, polypropylene and copolymers.

Plastics drums are produced in a variety of styles and shapes. The styles are either *bung type* (also known as closed or tight head) or *lid type* (also known as open top). The shapes include straight-sided, barrel-shaped or tapered. Some plastics drums have circular cross-sections, others have square cross-sections, and there is a third variety of the so-called "square-round" cross-section.

The bodies of plastics drums are manufactured by one of three processes, namely rotational moulding, injection moulding or extrusion blow moulding. The plastics lids, bungs and handles of plastics drums are usually produced by injection moulding.

Moulding techniques

Rotational moulding, which is also known as sinter moulding or the "Engel" process (figure 5.1).

Basically this process or technique consists of three stages:

(a) Loading a metal mould with the polyolefin powder.
(b) Heating the metal mould containing the powder with gas flames, or in an oven with

PLASTICS DRUMS AND CRATES

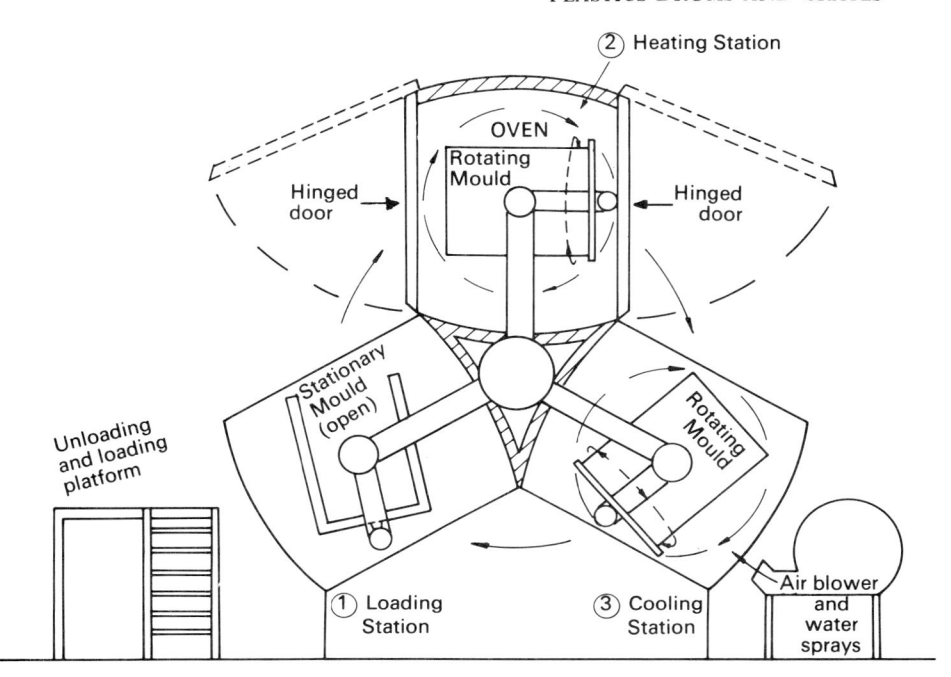

Figure 5.1 Rotational moulding

hot air, or with molten heat transfer salts, while rotating it simultaneously about two mutually perpendicular axes. In this stage the powdered polyolefin is melted and fuses as a coating of reasonably uniform thickness on to the inside of the metal mould.
(c) Cooling the mould at the end of the heating cycle, when the charge of powdered polyolefin has been completely fused, and finally extracting the moulding.

Injection moulding (figure 5.2)

This process consists of three stages:

(a) Plasticization or conversion of the plastics granules or powder to a molten state in an extruder. The extruder may be of the screw or ram type.
(b) Injection of the melt or plasticized material into a space defined by the male and female parts of the mould.
(c) Cooling of the plastics at the end of the injection cycle and extracting the moulding.

Extrusion blow moulding which is generally known as "blow moulding" (figure 5.3)

The process consists of five stages:

(a) Plasticization or conversion of the plastics granules or powder to a molten state in a screw extruder.

3.95

GLASS AND PLASTICS-BASED PACKAGING

1st step. The mould closes and moves up against the nozzle.

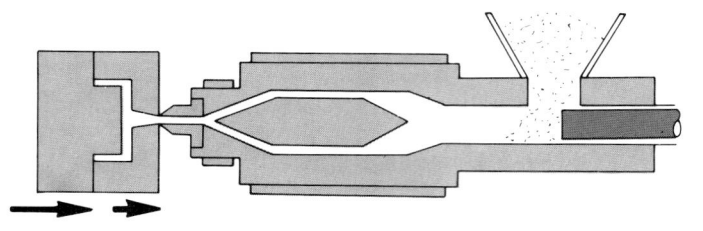

2nd step. The plunger moves forward and pushes raw material into the cylinder. At the same time it injects plasticized material into the mould.

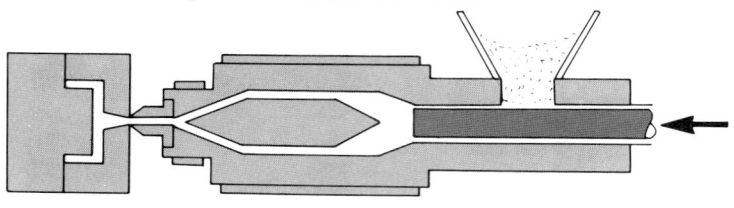

3rd step. The plunger stays in this position for some time. It still maintains pressure through the nozzle while the material in the mould cools and sets.

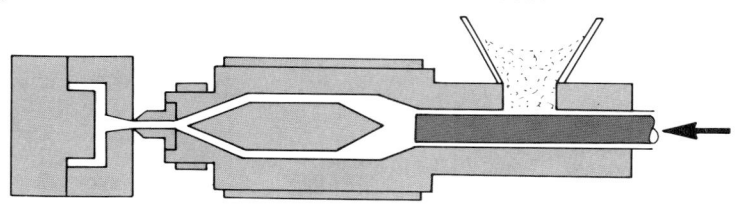

4th step. The plunger withdraws but the mould remains closed. A new lot of raw material falls from the feed hopper.

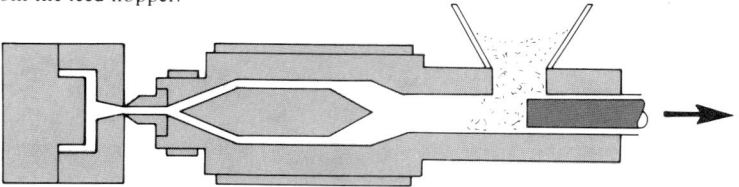

5th step. The mould moves away from the nozzle; the mould opens and the moulding can be taken out.

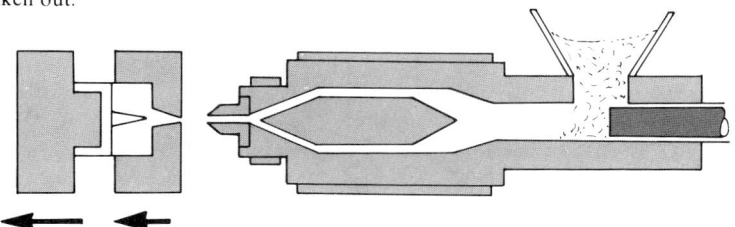

Figure 5.2   Injection moulding

PLASTICS DRUMS AND CRATES

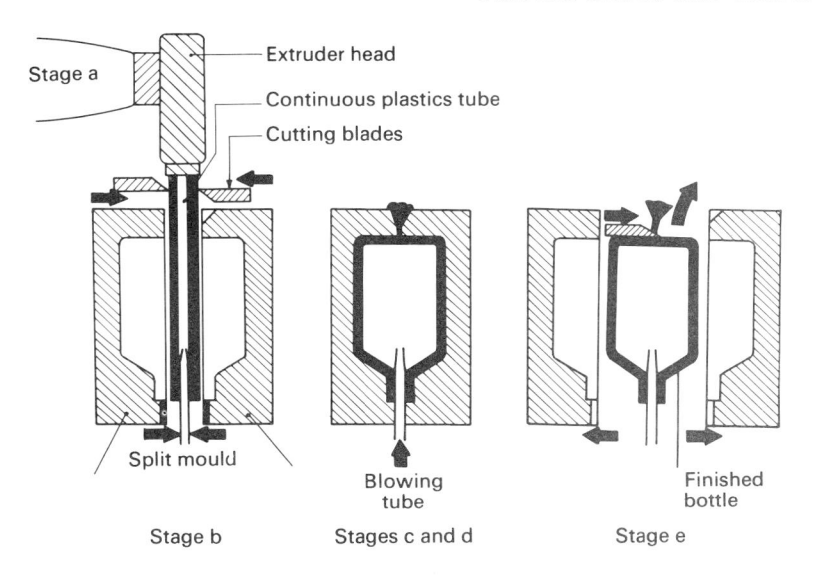

Figure 5.3 Extrusion blow moulding

(b) Formation of a parison or tube of melt which is extruded vertically downwards.
(c) Closing of the two mould halves around the parison and the separation of the parison from the extruder.
(d) Blowing of air or other gas into the parison whilst it is still molten, so that it is expanded to the shape of the mould cavity.
(e) Cooling of the moulding, mould opening, and finally the removal of the moulding.

Types of drums produced

The types of drums produced by the various moulding techniques are as follows:

By rotational moulding

Bung or cap type (closed or tight head) drums (figures 5.4 and 5.5) and lid (or open-top type) drums with straight parallel sides, tapered sides or barrel-shaped (figure 5.6). The cross-section can be circular, square or square-round, and the sides can be ribbed for extra strength.

By injection moulding

Normally injection-moulded containers or drums have tapered sides to facilitate removal of the moulded articles from the mould, but recent

3.97

GLASS AND PLASTICS-BASED PACKAGING

Figure 5.4 Types of small closed-end plastics drums

developments in mould design allow straight parallel-sided containers to be produced. Again, because of the limitations of the process, all single-piece injection-moulded containers or drums are open top, but recent developments enable two injection-moulded components to be welded together to form a tight head, i.e. a bung or cap type drum. The two components form the two halves or parts of the container, and one part contains the bung housing or the neck for the cap.

Injection-moulded pails normally have a circular cross-section, but other cross-sections are possible.

By extrusion blow moulding

The complete range of plastics drums can be produced by blow moulding, namely, tight-head and open-top types with any style or shape of sidewall (figure 5.7).

Advantages of the different moulding techniques

Rotational moulding

The production of a large number of small containers can be as economical as by injection moulding or blow moulding if the container required can be produced by the rotational moulding process. In addition, if relatively small numbers of large drums are required, then rotational moulding is more economical than injection moulding or extrusion blow

Figure 5.5 Examples of large closed (tight-head) drums

moulding. In general, large mouldings and drums with very thick walls can be produced easily by rotational moulding.

The rotational-moulding process has its limitations, due to the ease with which the polyolefin powder can be melted, and the way in which the melted powder will form a continuous layer on the inside of the mould. The ease with which the powdered polyolefin will melt and fuse is governed by the "melt flow index" (MFI) of the polyolefin. In general, a polyolefin with a high melt flow index, i.e. with a low viscosity, will fuse more readily into a continuous sheet to form the container. Unfortunately, polyolefins with high melt flow indices have poorer resistance to impact, particularly at low temperatures, lower resistance to environmental stress cracking, and lower resistance to chemical attack than polyolefins with low melt flow indices. The lower the melt flow index of the polymer, the more difficult it is to mould the polymer rotationally.

Other factors which govern the ease of rotational mouldings are the particle size and shape of the powder, and the density of the polymer. In general terms, rotational moulding is more difficult with high-density polymers than with low-density polymers, but high-density polymers provide advantages over low-density materials in terms of rigidity, and therefore stacking strength of drums, and chemical resistance.

Rotational moulding is also suitable for the production of thin wall liners for steel and fibre drums.

Injection mouldings

The advantages of injection moulding are that production speeds are high, granules or powder can be used, and good dimensional tolerances of the moulded article can be achieved. In addition, the complete range of

GLASS AND PLASTICS-BASED PACKAGING

Figure 5.6 Types of open-top plastics drums

polyolefins, i.e. materials with low and high melt flow indices and low and high-density polyethylenes, as well as polypropylene, can be processed. In practice, however, it is extremely difficult to produce large-capacity drums from stiff materials (high-density high-molecular-weight polyethylenes). Therefore the injection-moulding process tends to be used for the production of containers up to 25-litre capacity, although 50-litre containers have been produced by injection moulding. Blow moulding is usually more applicable for the production of drums above 25-litre capacity, particularly when low melt flow index materials are used.

Thus, for the smaller-capacity plastics drums, the injection-moulding process provides the opportunity to choose the most appropriate polymer to achieve the desired properties, i.e. grades of polymer can be selected to give good rigidity and high resistance to chemical attack and environmental stress cracking.

The limitation of the injection-moulding process is the high cost of equipment and moulds, which makes the process economical really only for long runs.

Although drums are normally produced in polyolefins, there has been

PLASTICS DRUMS AND CRATES

Figure 5.7 Plastics drums are blow-moulded from high-density polyethylene in two styles—tight-head and full-spintop—and in a range of sizes (50–120 litres). They provide a genuine alternative to costly lacquer-lined steel drums for the transport of many products, including chemicals to UN group 2.

a recent development of a barrel produced in acrylonitrile-butadiene styrene. The base and top of the barrel are injection-moulded separately and then welded together around the equator of the drum.

Injection moulding, because of the nature of the moulds wherein the outer and inner surfaces are completely defined, allows precise control of the wall thickness of the drums to be obtained.

Extrusion blow moulding

The main advantage of the blow-moulding method of drum manufacture is that one-piece tight-head (bung or cap type) drums can be produced from high-density high-molecular-weight polyethylene which provides an extremely rigid container of low tare weight with exceptionally good resistance to chemicals and environmental stress cracking. In addition, lid-type open-top drums can be produced by blow moulding with the same strength and chemical resistance properties as the tight-head drum.

GLASS AND PLASTICS-BASED PACKAGING

Drums in the complete range of capacities, styles and shapes can be produced by blow moulding.

The high cost of blow-moulding machines and moulds makes the blow-moulding process economical only for extended runs of standard-range containers. However, the relatively high cost of the moulds for blow moulding is partially offset by the fact that the moulds normally consist of three parts. The capacity can therefore be altered by replacing the centre section of the mould with another section of the same diameter but with a different height. Similarly, by changing the bottom section of the mould, either a tight-head (bung or cap type) drum or a lid-type (open-top) drum can be produced.

With the blow-moulding process it is difficult to produce bung housings or screw necks with fine threads, except by insert moulding or an extra turning operation. In practice, however, coarse threads which are readily produced in the blow-moulding operation have gained acceptance because of the strength of the threads which provide a strong closure.

As with rotational moulding, the blow-moulding process is suitable for the production of thin-wall liners for steel and fibre drums.

Identification of moulding method

The method by which a drum was produced may be readily determined by visual examination.

Rotational mouldings have a well-defined outer surface, a poorly defined inner surface, and show no sprue or parison weld.

Injection mouldings have good definition of both surfaces and a sprue mark.

Extrusion blow mouldings are similar to rotational mouldings in respect of surface definition, i.e. good outer surface and poor inner one. They also show no sprue, but the parison weld is visible, as are the mould split lines, normally down the sides of the drum.

Table 5.1. Differences between types of moulding

Type of moulding	Definition of surfaces Outer	Inner	Sprue mark	Parison weld	Mould split lines
Rotational	good	poor	none	none	—
Injection	good	good	present	none	—
Extrusion blow	good	poor	none	present	present (normally down the sides)

General properties of polyolefins used for drum making

*Polyethylene**

Polyethylene is available in two basic forms, namely, low density and high density.

Low-density polyethylene is also referred to as:

>   high-pressure polyethylene
>   branched polyethylene
>   flexible polyethylene

High-density polyethylene is also known as:

>   low-pressure polyethylene
>   linear polyethylene
>   rigid polyethylene

With an increase in the density of polyethylene, there is an increase in the following properties:

(a) tensile strength
(b) stiffness
(c) hardness
(d) resistance to solvent attack
(e) resistance to gas and vapour transmission

The molecular weight, i.e. the size of the molecule or the number of ethylene groups in the molecule, also affects the properties of polyethylene. An increase in molecular weight improves toughness, resistance to environmental stress cracking, and fatigue strength.

Polypropylene

Polypropylene was developed as an alternative material to high-density polyethylene for applications demanding high rigidity and good chemical resistance. Polypropylene is also noticeably different from polyethylene in that it does not exhibit environmental stress cracking.

Polypropylene can be used to produce either homopolymers or copolymers, propylene being most commonly copolymerized with ethylene.

Polypropylene homopolymers and copolymers have narrow density ranges. Changes in density have little or no effect on physical and chemical properties.

* Polythene is a generic term for polyethylene which has been used mainly in the United Kingdom; it is recommended that the word *polyethylene* is used rather than polythene to avoid any misunderstanding.

GLASS AND PLASTICS-BASED PACKAGING

The effect of molecular weight on the properties of polypropylene is similar to that on polyethylene, namely an increase in molecular weight improves toughness and fatigue strength.

The homopolymers of polypropylene are inherently brittle at low temperatures, and the copolymers were specifically developed to improve low-temperature impact performance.

Typical markets for plastics drums

Plastics drums are typically used for the transport and distribution of the following products:

(a) adhesives
(b) building products such as sealants and mastics
(c) chemicals such as weedkillers, hydrogen peroxide, disinfectants, synthetic resins, sulphuric acid, sodium hypochlorite.
(d) cleaning products such as detergents, bleaches and polishes
(e) dyestuffs and printing inks
(f) essential oils and flavours
(g) foodstuffs such as syrups, vinegar, fruit pulps and mincemeat
(h) hairdressing and cosmetic preparations
(i) mineral oils and greases

The choice of drum for any particular product to be packaged will depend upon the physical and chemical properties of that product.

Plastics crates

Plastics crates can be divided into three categories, and these three categories depend upon the type of product to be packed rather than the method of manufacture of the plastics crate. The type of product to be packed, and therefore the usage which the crate will have to withstand, determines both the style of the crate and the material from which it is manufactured. The main advantage of plastics crates over wooden and metal crates is that plastics crates are easily washable and therefore hygienic. There are no maintenance costs with plastics crates.

The three categories of product for plastics crates are:

(a) milk
(b) beer and soft drinks
(c) agricultural products and foodstuffs

Historically, crates for milk were originally of metal, while crates for beer, soft drinks, agricultural products and foodstuffs were originally made of wood.

Various methods of production have been developed for plastics crates, including rotational moulding, blow moulding, injection moulding and thermo-forming. However, although some half-depth cases have been produced by extrusion blow moulding, virtually all crates are now produced by injection moulding.

The thermo-forming (or vacuum forming) process consists of four stages:

(a) production of a sheet of plastics material
(b) heating of the plastics sheet so that it is soft and mouldable
(c) forming of the softened sheet in and/or around moulds by vacuum and/or air pressure, sometimes together with movement of the moulds
(d) cooling and removal of the moulding.

The material used in the production of plastics crates is determined by the specific requirements of each market. We will consider each of the three markets in turn.

Plastics crates for milk

Plastics crates for the milk market must be able to withstand caustic and detergent washes at 80°C, and must also have good impact strength, particularly at low temperatures (this is necessary as milk is frequently delivered at sub-zero temperatures). Hence high-density polyethylene is used for the production of milk crates, as this is the only material which meets the requirements.

Plastics crates for beer and soft drinks

Beer crates are stacked higher and for longer periods than milk crates, and hence a more rigid crate is required. In addition the sub-zero impact strength of beer crates is not as critical as that of milk crates. Polypropylene copolymers have been found to have the necessary balance of properties, and are used for this type of crate.

Plastics crates for agricultural products and foodstuffs

The crate market for agricultural produce and foodstuffs is extremely diverse and the usage very variable. For example, there are delicate lightweight high-cost products, such as flowers and hot-house salad produce, and weighty products such as meat, sausages and fish. Plastics crates for products such as flowers are handled extremely carefully, but plastics crates for fish receive considerable abuse and are also subjected to sub-zero temperatures.

The materials used and the design of crate employed for this sector are therefore equally diverse. For example, for light-framework crates for vegetables, polypropylene or toughened polystyrene is used, and the crates are single-trip. For bread, meat, sausages, etc., crates are produced in high-density polyethylene or polypropylene, and are of robust construction so that they can withstand many trips.

Plastics crates are now firmly established for a wide range of products and usages, since these injection-moulded crates can easily be coloured and embossed to provide instant identification of ownership and product.

CHAPTER 6

Closures and Dispensing Devices for Glass and Plastics Containers

E. P. Adcock

It has been said that "a bottle or jar is only as good as its closure"—but this is only part of the story, since a good seal is the outcome of a cooperative effort between the finish of the bottle or jar and its closure. The main duties that properly-marrying closures and bottle finishes are required to perform are as follows:

1. The closure must keep the bottle or jar sealed until the contents are required for use. This usually means that the contents cannot escape, and external environments cannot enter the container. The degree of tightness required by the seal is, however, dependent upon the nature of the product packed—many products, such as non-hygroscopic powders and tablets, do not need a completely hermetic seal.
2. It must be possible to open the bottle or jar without difficulty and to re-seal it properly with ease when only part of the contents is used at a time. Alternatively, the closure can be provided with a dispensing device, such as an orifice or spout, which is operated without removing the closure; or the closure may be fitted with a pierceable wad as is used for blood transfusion and injectable fluids.
3. The closure must neither affect nor be affected by the contents of the bottle or jar; it should be inert to any climatic conditions to which it may be exposed, and may need to withstand conditioning or processing treatment, such as pasteurization or sterilization, to which the pack is subjected.
4. The closure may need to provide a pilfer-proof device to show whether it has been removed prior to use, thus giving the consumer assurance that no-one has tampered with the contents. This is important when there might be a temptation to "water down" the contents, as with potable spirits; or to inspect the contents out of curiosity, as with certain food packs on supermarket shelves when, by so doing, the sterility of the product is destroyed.
5. The closure may need to provide a tamper-proof membrane adhered to the mouth of the container which must be torn to gain access to the contents. Such membranes are usually also designed to keep out moisture and/or oxygen, so as to preserve the freshness of the product. Membrane seals are mostly used on jars of instant coffee and dried milk powders.
6. Special-purpose closures may need to retain a vacuum or internal pressure within the container.

7 Not only must the closure perform its mechanical and protective functions correctly, but it must often blend with the graphics of the container, enhancing the appearance, and adding to the sales appeal of the pack.

The products packed will be liquids, creams, pastes, powders, granules or other solids. They will vary from completely innocuous to highly corrosive, from very penetrating liquids (e.g. oils, detergents, solvents and the like) to solids which have little or no penetrating properties. Even so, some solids (e.g. fine chemicals) are hygroscopic or even deliquescent, and these require particularly good water-vapour-tight seals.

Other products that are sensitive to the ingress of moisture include many pharmaceutical tablets, both plain and sugar-coated, as well as effervescent salt mixtures in both tablet and powder form. Many medicines (such as cough mixtures) contain small percentages of chloroform which must not escape. Nail varnish and nail varnish removers also contain volatile solvents.

Products such as bleaches, containing peroxides or hypochlorites, develop internal pressures on standing, and it is important to be able to relieve those pressures without risk of leakage when taken home in shoppers' baskets. This requires the use of a liner that vents when the pressure in the container reaches a predetermined level.

Some liquid products, of which whisky is an example, have high coefficients of expansion, and containers may develop high internal pressures on exposure to tropical temperatures. It is essential that the closures on such containers are able to contain the developed internal pressures without leaking.

Carbonated beverages must be sealed with pressure-tight closures, and the vacuum in vacuum packs must be preserved until the jars are opened for use.

The general requirements for a good seal

A seal is usually made by causing a resilient material to press against the sealing edge or rim on the top of the finish of the container. The pressure must be evenly distributed and maintained to ensure a uniform seal around the whole of the edge of the cushioning material in contact with the rim.

The resilient material is frequently a wad cut or stamped out of a composition-cork or pulpboard sheet; usually protected by a facing material against interaction with the contents. The combination of a wad and a facing material is called a *liner*. Hence, caps have to be fitted with liners in order to make good seals on bottles.

Liners consist of materials that combine the resilient properties of the wad with the protective properties of the facing material. Such liners can be discs of solid rubber, PVC, polyethylene, EVA, etc., or they can consist of purpose-designed fitments moulded from these or other similar resilient thermoplastics. Liners also can be flowed-in to form wads, sealing rings, or gaskets. A flowed-in liner results from injecting a PVC plastisol, or a natural or synthetic rubber colloidal dispersion, into a metal cap and causing the compound to set by stoving.

In addition to the requirements for the cap and liner, the third no-less-important aspect is the container finish, which must mate completely with its closure and liner. Fundamentally, the sealing edges or rims on the finishes of both glass and plastics containers must be seamless, smooth, even, and free from roughness or defects. Perfection in this respect is rarely attained in practice, but very good commercial finishes are now consistently obtainable from production runs, and the resilience of the liners then compensates for the small variations that still occur.

The mechanics of a good seal

To make a good seal, the liner is pressed against the sealing rim on the finish of the container with sufficient pressure, which must be maintained during the shelf-life of the pack. The higher this pressure (within reason) the more effective will be the seal, but it would obviously be self-defeating to increase the pressure to the point at which the cap can either break or deform, the bottle finish become chipped (glass) or deformed (plastics), or the liner break down by splitting or collapsing.

The tightness with which a cap is screwed on to a container is known as the *tightening torque*. Part of this torque is used in overcoming the frictional forces encountered in screwing on the cap, and the remainder is converted into a direct top-sealing pressure on the liner.

With rolled-on, crimped-on, and pressed-on caps, the effectiveness of the seal depends upon the direct pressure exerted on the top of the cap during the capping operation.

The sealing pressure ensures that the liner makes good contact with the sealing rim on the finish of the bottle or jar. The greater the area of the sealing surface on the rim of the bottle finish, the greater will be the area over which the load exerted by the closure is spread, and the less effective will be the seal for a given cap-tightening torque. It follows, therefore, that to obtain a good seal without having to use unduly high cap-tightening torques, the width of the sealing edge must be kept as narrow as possible, consistent with not damaging the liner or its facing. There are thus practical difficulties in maximizing the effective sealing pressure for low cap-tightening torques and, in practice, a narrow rounded sealing edge is particularly suitable for the finishes of glass bottles and jars, but there are also many completely satisfactory applications using flat sealing edges. Plastics bottles and jars are usually made with flat sealing edges, the best seals being obtained when these are kept reasonably narrow.

A further point to remember, when designing a good closure system, is that the impression produced in the liner by the sealing rim of the bottle finish must not be too near to the outside perimeter of the liner. If this occurs, it can cause the edge of the liner to collapse, particularly if the wad is of composition-cork or similar easily compressible material.

Both composition-cork and pulpboard wads make comparable seals up to diameters of about 40 mm. Above this, composition-cork wads are better able to take up any slight waviness that can occur in the sealing rims of bottles but, because of flexibility, composition-cork wads can be difficult to retain in large caps (say above 48 mm), unless the liners are glued-in. There used to be an advantage in using what was known as a "Rocdaw" liner (a stiffer liner of a faced pulpboard and composition-cork laminate). Nowadays rim waviness is a much lesser problem with glass containers, and it can be completely controlled on plastics containers. The "Rocdaw" liner has been virtually replaced by pulpboard, particularly in large plastics closures, whilst the larger metal caps tend more and more to be lined with flowed-in PVC plastisol sealing-rings.

The importance of thread engagement and thread pitch

Thread engagement is the number of turns given to a cap from the point of first engagement between the start of the thread in the cap with the start of the thread on the bottle finish to the point when the sealing edge on top of the finish of the bottle makes contact with the liner in the cap. In order that the liner should be pressed down uniformly around the whole of its circumference on the sealing edge of the bottle, at least one full turn of thread engagement is required. The greater the thread engagement, the better the cap stays on, and the more effective is a given cap-tightening torque in keeping it there.

Thread pitch is measured in terms of the number of turns of thread per unit of traverse. The pitch determines the slope or steepness of the thread. The lower the number of thread turns per centimetre, the steeper will be the slope of the threads, and the more rapidly will the caps screw on and off the bottles. Also the steeper the slope of the thread, the deeper the caps will have to be to achieve a given thread engagement.

It is thus clear that the performance of a screw cap depends both on the thread engagement and the thread pitch. In practice, the whole subject has been covered by BS 1918: Glass Container Finishes. This standard, first published in 1953, specifies the dimensions and tolerances for two series of external screw neck finishes:

(a) Shallow continuous-thread finishes—the R3/2 series
(b) Tall continuous-thread finishes—the R4 series

In the R3 series, the thread engagement of the cap on the bottle finish tends to fall a little short of a full turn, but two full turns of thread engagement are obtained with the caps in the R4 series of thread finishes. Recently, what is called the short R4 series has been used. This permits a thread engagement between those obtained with the R3 and R4 series of thread finishes.

There is no British Standard Specification for closures and, in practice, the major closure manufacturers endeavour to provide, where possible, a clearance of 0·1 mm between the maximum bottle finish and the minimum closure dimension.

There is no difficulty in producing plastics bottle finishes with a finer thread pitch than can be achieved satisfactorily with glass. Thus, while it is possible to design a shallower skirted cap having the requisite minimum of one full turn of thread engagement, plastics bottle finishes tend to follow the pattern of glass bottle finishes, if only for the reason that the same caps can then be used on both glass and plastics bottles and jars interchangeably. With plastics, of course, many speciality designs are available, where the cap and bottle thread forms are specific to each other.

Applying the correct tightening torque to screw caps

Having made sure that both the cap and the bottle finish requirements for a good seal have been satisfied, it remains to ensure that the caps are applied properly to the bottles. To do this we need a yardstick to measure the effectiveness of cap application.

This can be measured by means of a *torque tester*. To determine the cap application torque, the bottle is clamped on the spring-loaded table of the torque tester, the cap is screwed on to the bottle, and the torque applied is read directly from the scale. In practice, the torque tester cannot be placed under the capping head of a capping machine, and the torque applied by a capping machine must be determined indirectly by measuring the *immediate loosening torques* of the caps applied by the machine. The torque at which the caps should unscrew is first obtained by applying several dozen caps to the bottles, by hand, to the required tightening torque using the torque tester. The cap loosening torque for each is then measured. Having determined the range of values of the cap loosening torque for the required cap tightening torque, it is then only necessary to adjust the capping machine until the loosening torques of the caps applied by the machine fall within the same range of values as those of the caps applied by hand.

3.111

The cap tightening torque varies with the diameter of the cap, and each torque tester is supplied with a chart giving the minimum and maximum recommended cap tightening torques for the range of cap diameters in normal use. There are nevertheless occasions when the maximum recommended cap tightening torque may be exceeded, provided the caps are sufficiently robust. With thin plastics or aluminium caps, it is generally not advisable to exceed the recommended torque.

The importance of vacuity or ullage

Vacuity or ullage can be described as the amount by which a bottle falls short of being brimful with liquid. It is the volume of the air space in the head of the container above the level of the contents. It is also known as the *head space* (or *expansion space*) and is expressed (BS 3447:1962) as a percentage of the liquid contents of the bottle at the time of sealing.

If the liquid contents of a bottle expand as the temperature rises, then the higher the coefficient of expansion of the liquid, the greater the provision that must be made for vacuity.

For example, whisky and gin have high coefficients of expansion, and a 2% vacuity will result in an internal pressure of the order of $500 \, kN/m^2$ after a rise in temperature of about 22°C above the filling temperature. With a vacuity of 5%, the internal pressure would only reach about $140 \, kN/m^2$ after a rise in temperature of 35°C above the filling temperature. In practice, vacuities between 3% and 5% are used for spirits (such as whisky and gin) depending upon the type of closure used.

It is clear that the temperature of the product at the time of filling is important. If the product is at 5°C when filled, and no allowance is made to the height of fill, effective vacuity will be lost by the time the contents of the bottles reach room temperature. This means that a vacuity normally allowing a rise of, say, 30°C above a product filling temperature of 16°C would, if the same vacuity were used at a filling temperature of 22°C, only provide for a rise of 16°C above the ambient temperature of 16°C before leaking would occur.

A vacuity of 10% has to be provided for products like peroxides and hypochlorites which are liable to give off gases. In practice, these products are frequently sealed with closures that vent to relieve excess gas pressure developed.

The above considerations apply to all rigid bottles. Flexible and distensible plastics bottles could accept lower vacuities since the effective pressure developed against the insides of the caps is reduced as such bottles distend under internal pressure. But, if they distend too much, they can become unstable and fall over or burst.

Types of closures

When classified according to their primary function, all closures fall into one of four main groups:

(a) *Those that provide a normal seal*

These are primarily intended to keep the containers sealed and, although not specifically designed to resist pressure, will cope with normal rises in internal pressure resulting from increases in external temperature, subject to the use of proper vacuities (figure 6.1).

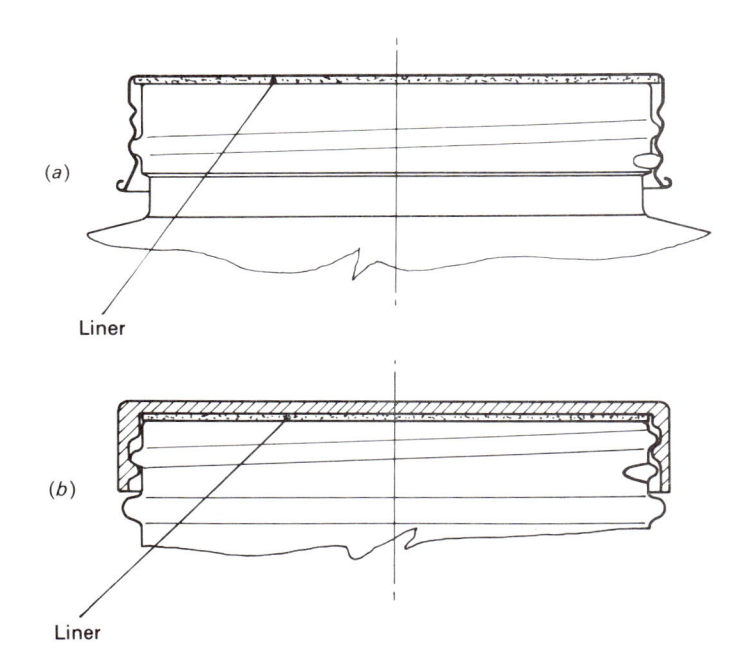

Figure 6.1 Conventional pre-threaded screw caps
(a) Cross-section of metal screw cap applied to the finish of a glass or plastics container
(b) Cross-section of a plastics screw cap applied to the finish of a glass or plastics container

(b) *Those that ensure a vacuum-tight seal*

Closures that are designed to make an air-tight seal when (at least) a partial vacuum has been developed in the head space of the container as a result of the sealing or processing conditions to which the packs are submitted—the maintenance of vacuum being necessary for the proper preservation of the contents.

3.113

GLASS AND PLASTICS-BASED PACKAGING

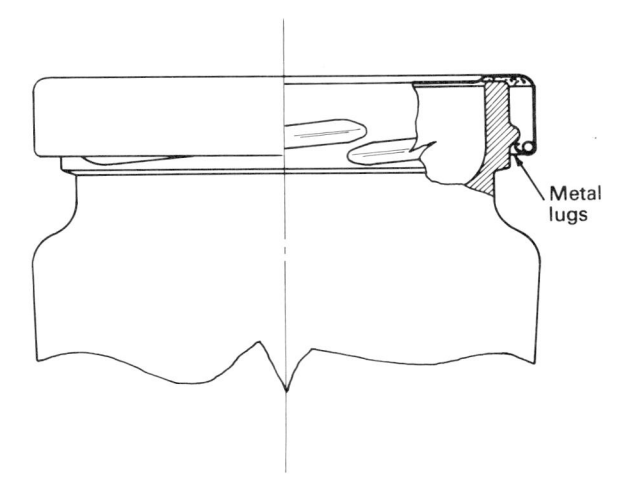

Figure 6.2 Lug cap applied to the finish of a glass jar

(c) *Those that are designed to contain high internal pressures of the order of* $700 \, kN/m^2$

These closures (figure 6.4) are commonly used for sealing pasteurized beers and carbonated beverages.

(d) *Those that have a venting feature*

Closures that allow venting (figure 6.7) when the internal gas pressure reaches a predetermined level, excluding vacuum seals (where venting takes place during the processing of packs that are required to develop at least a partial vacuum as they cool to room temperature).

Some overlap can occur between these groups. Certain pressure-tight and vacuum-tight seals can, for example, serve for both purposes or be used as a normal seal. Similarly, some normal seals can withstand quite high internal pressures.

Speciality features of closures

There are many closures that have novelty features such as: dispensing (figure 6.11), measuring, pouring (figure 6.12) and spraying devices; pilfer-proof, tamper-proof and non-refillable features; or are captive, wadless, or child-resistant.

Selection of closures

The main aspects to be taken into consideration when selecting a closure within the tabulated first three groups (*a*), (*b*) and (*c*) are as follows:

Normal seals

The basic requirements for a good seal are:

(i) The liner must be compatible with the contents.
(ii) The liner must be compressed evenly over the whole surface of the sealing edge of the container. With screw caps, a minimum of one full turn of thread engagement is required for maximum security of the seal but, in practice, a minimum thread engagement of three-quarters of a turn is usually satisfactory for general-purpose applications.
(iii) The sealing edge on the rim of the container should be seamless, smooth and reasonably narrow (about 0·6 mm to 1·2 mm wide) so as to ensure maximum pressure against the liner or gasket for the recommended cap tightening torque. For plastics containers, it is important to avoid sealing rims that are so narrow and sharp that they cut into the liner.
(iv) Sufficient vacuity must be allowed to provide for the expansion characteristics of the product. Even more must be provided for products such as ether, which develop high internal pressures for only moderate increases in ambient temperature.

Vacuum-tight seals

All closures used in vacuum-tight seals are made of metal, and, in practice, they are all fitted either with flowed-in liners, or gaskets, or with rubber rings. The reasons for this are as follows:

In any closure (metal or plastics) fitted with a conventional liner (such as a faced pulpboard or composition-cork disc), the vacuum in the container pulls the liner down on to the finish of the bottle. With a metal cap, fitted with such a conventional liner, the pull of the vacuum is likely to pull the liner out of the closure. With a plastics closure, the liner is usually glued-in, and the closure will then be so difficult to unscrew that the liner will shear before the vacuum is broken. Moreover, vacuum packs generally involve some form of processing or steam injection, and moisture entering the closure would seriously affect any liner based on a pulpboard or composition-cork wad.

The sealing medium must, therefore, be a flowed-in compound containing a slip additive to make the closure easy to unscrew or twist off. Flowed-in compounds do not adhere to plastics closures, which in any event will not withstand the temperatures at which flowed-in compounds have to be stoved. Hence, only metal caps are used for vacuum-tight closures.

A vacuum-tight closure is thus either a screw-off or a twist-off metal closure lined with a flowed-in compound containing a slip additive, or a prise-off metal closure lined with a flowed-in compound without a slip additive, or fitted with a rubber sealing ring. A prise-off closure is lifted

3.115

GLASS AND PLASTICS-BASED PACKAGING

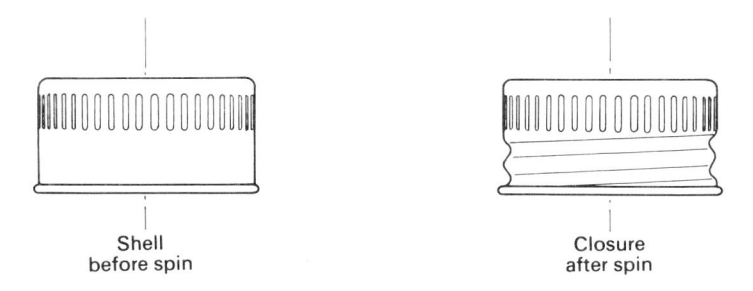

Shell
before spin

Closure
after spin

Figure 6.3a Roll-on closure

Breaking
bridges

Pilfer-proof skirt
spun under
bottle bead

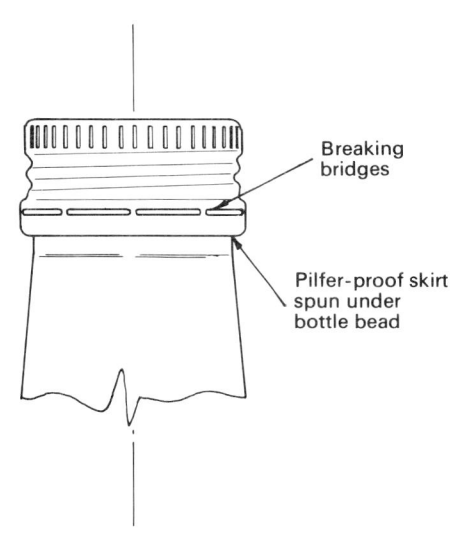

Figure 6.3b Roll-on or spin-on pilfer-proof cap applied to the finish of a glass container

directly off the finish of the bottle to break the vacuum and no twisting is required.

Pressure-tight seals

Pressure-tight seals are generally required to contain pressures from about $350 \, kN/m^2$ to over $1000 \, kN/m^2$. This means that consideration must be given to the strength of the closure in relation to the total force that will be exerted against it. For a closure about 25 mm diameter on the inside, the internal surface area is about $2\frac{1}{2}$ cm^2 and the force exerted against it, when the bottle is pressurized at 350 kN/m^2 will be 164 N or 500 N if the bottle were pressurized at 1000 kN/m^2. But, if the internal closure diameter is 50 mm, its internal surface area is about 10 cm^2 and the force then exerted against the inside of the closure is 670 N at a pressure of 350 kN/m^2 and 2000 N at a pressure of 1000 kN/m^2.

Thus, the larger the diameter of the cap, the thicker must be the plate from which the cap is made, or it will not withstand the internal forces that will be directed against it. But the thicker the plate, the less economical the cap becomes. Hence, in practice, the maximum diameter of commercial closures to withstand high internal pressures is likely to be about 40 mm.

The best pressure-retaining seal is the tinplate crown (crimp-on; lever off, figure 6.4) but the spun-on aluminium closure has good performance, particularly when fitted with a flowed-in liner that is made to hug the top and outside edges of the sealing rim (figure 6.5). Another closure, particularly in the larger diameters, that makes an excellent pressure-tight seal is the Eurocap which makes possible the replacement of cans by glass jars without altering the processing conditions. It is the only relatively large-diameter closure that will withstand autoclaving without counter pressure.

An internal screw provides the best pressure retention for plastics closures, but it is now expensive, and cumbersome. To contain pressures of the order of 400 kN/m^2 or 500 kN/m^2, an external screw plastics closure must have ample thread engagement, and the finish of the container has a downward and outward tapering sealing edge which makes a more

Figure 6.4 Crowns. Shallow fluted-skirt crimp-on tinplate closures with plastisol liners. These are the best pressure-tight seals but require an opener. They are the cheapest closures for beers and carbonated beverages. They withstand the brewery pasteurizing processes.

effective seal with the liner than a narrow conventional flat or rounded rim.

Secondary functions of closures

The secondary functions of closures, whether for use on glass or plastics containers, all have particular objectives such as inviolability, non-refillability, measuring and pouring facilities, child-resistance, sales appeal or two or more of these functions in combination. Additionally, advantage can be taken of the flexibility of some plastics to combine a squeezing action with a cap feature for dispensing small quantities of liquids at a time, or to provide a spray from a spray nozzle incorporated in the closure system.

The main secondary functions of modern closures for both glass and plastics containers are as follows:

Inviolability

A closure must demonstrate that unlawful access to the contents or their unwitting exposure to the atmosphere has occurred. Strictly, no closure is inviolable, since it must be possible for the consumer to gain access to the contents of the container, but tell-tale features can be introduced into the closure system that show up immediately the seal has been broken. Two such types of closures are:

Pilfer-proof closures

Here a pilfer-proof band is joined to the skirt of the closure by means of frangible bridges that fracture when the closure is unscrewed, leaving a tell-tale ring round the bottom of the neck of the bottle. These caps are normally represented by the roll-on pilfer-proof (ROPP) aluminium closure (figure 6.3b) that is spun on to the bottle finish, the tell-tale pilfer-proof ring being tucked under a bead on the neck of the bottle during the capping operation.

Plastics pilfer-proof closures (figure 6.9) are designed to snap over a security bead on the necks of glass or plastics bottles so that the closures cannot be opened until the band connecting the skirt to the pilferproof ring is torn away.

Tamper-proof closures

This system is designed to deter tampering with the contents, but the evidence may not become readily visible until the closure is removed from

CLOSURES AND DISPENSING DEVICES

the container. An example is the closure containing a membrane glued to the rim of the container (figure 6.6). It is not possible to check that the membrane remains intact until the closure is unscrewed from the container.

Non-refillability

Non-refillable closure systems are designed to prevent or at least strongly to deter bottles from part or complete refilling with spurious liquors by unscrupulous handlers. These closure systems usually consist of devices which are firmly locked on to the finishes of the bottles, and which contain valves and weights that only allow flow in one direction, namely out of the bottle.

Dispensing, measuring and pouring devices

These aids control the rate and manner of flow, or the method and/or amount of dispensing, so that the product is used to the best effect and convenience (see figure 6.12).

Child resistance

Child-resistant closures are designed to make it difficult for young children to gain access to the contents of bottles and jars while enabling adults to do so without difficulty (figure 6.10).

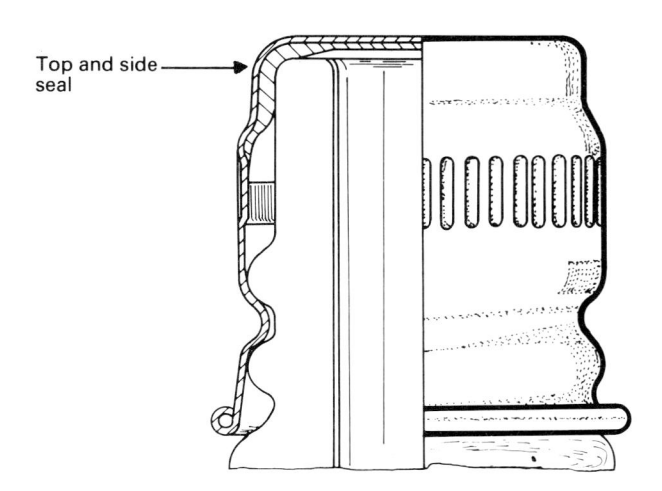

Figure 6.5 The sealing principle of "Flavorlok" and "Eurospin" closures

Sales appeal

There is virtually no limit to the versatility of craftsmanship in design, colour and decoration of plastics closures. There is much less scope in the design of metal caps—being limited by the constrictions of their method of manufacture—but they can be decorated in many colours and in an almost endless range of attractive printed designs.

The manufacture of closures

Closures fall into two main classes:

(a) metal caps
(b) moulded caps—(1) made from thermosetting materials,
                    (2) made from thermoplastic materials.

Metal caps

Metal caps are stamped out of sheets of either tinplate, aluminium or aluminium alloy, generally in a thickness of about 0·25 mm, depending upon the use for which the caps are required. The sheets are first coated by a roller coating machine with lacquers or enamels compatible with the product packed and resistant to any processing or climatic conditions the caps are called upon to withstand; they are frequently printed on their tops and/or their skirts with decorated designs. All lacquers, enamels and printed coatings are stoved immediately after application by conveying the coated sheets through an oven. This not only develops the hardness in the coating essential to withstand scuffing during subsequent tooling and handling operations, but also the required resistance to the product with which the coating may come into contact, as well as the processing treatment it may have to withstand.

For screw caps (figure 6.1a) the decorated sheets are fed to presses that stamp out the cap shells which are then automatically fed into the thread-forming, beading and knurling machines. The liners are automatically inserted into the finished caps but, in some instances (for example, when there is a possibility of damaging the edges of composition-cork wads) the insertion can take place after the cap shells have been knurled, but before the threads are formed. Other types of metal caps, such as lug caps (figure 6.2) and crowns (figure 6.4), will either be complete after pressing or will require one or two further tooling operations according to their design. Crowns and practically all lug caps are fitted with liners after they are otherwise complete, and the type of liner used in these caps is now almost universally a flowed-in plastisol gasket or sealing ring, although there is still some demand for flowed-in water-based compounds. There are also two-part metal caps in which each part is fabricated separately.

CLOSURES AND DISPENSING DEVICES

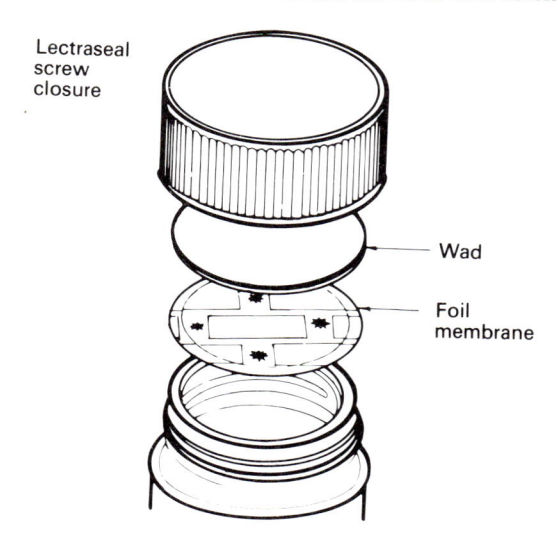

Figure 6.6 *Tamper-proof closures.* The most popular tamper-proof closure is the pre-threaded closure (metal or plastics) with a membrane seal. The membrane seal has the advantage that it provides extremely good moisture- and gas-tight seals; these ensure that the product is kept fresh and is seen to have been kept fresh during the shelf-life of the pack. A modern and effective membrane seal, known as "Lectra-seal" uses the heating effect of high-frequency eddy currents to heat the perimeters of aluminium foil membranes, coated with a thermoplastic resin mixture, and thereby causes the membranes to be peelably adhered to the rims of containers. The pre-capped containers are passed at high speed under a coil connected to a high-frequency generator. It is suitable for use on glass and on all plastics containers. Plastics closures are used for the "Lectraseal" process, since metal caps heat up when exposed to a high-frequency field. The diagram shows an exploded view of a typical "Lectraseal" closure.

In the case of spin-on or roll-on closures (figure 6.3) the manufacture of the cap shells is complete after the drawing operation, apart from any bead or pilfer-proof feature that may be required. The threads are developed in these closures when the cap shells are rolled or spun on to the threaded finishes of the bottles. The liners are fitted or flowed-in (depending upon the type of liner to be used) immediately after the caps are complete in their shell form.

Moulded caps made from thermosetting plastics

These caps are also known as *compression moulded* since they are formed in presses that compress by means of heat and pressure the moulding powder into the cap moulds.

3.121

Three types of moulding powders are generally used for thermoset screw closures, UF, WF, and PF.

UF is a paper-filled urea-formaldehyde resin
WF is a woodflour-filled urea-formaldehyde resin
PF is a woodflour-filled phenol-formaldehyde resin

There is a lesser-used fourth material, a melamine-fortified urea resin (urea-formaldehyde containing a proportion of a melamine resin) which is available in both paper-filled and woodflour-filled forms.

The manufacture of thermoset closures is based on the principle that urea and formaldehyde on the one hand, and phenol and formaldehyde on the other, will combine together under heat and pressure to form hard, insoluble and chemically resistant polymeric materials. In practice, the moulding-powder manufacturer starts the polymerization process in his plant and stops it at what is known as the "B" stage for phenol-formaldehyde powders. The resulting resin is compounded with the fillers (either paper or woodflour) and pigments and other minor ingredients in a steel roller mixing machine, such as is used in the rubber industry. The compounded mixture is then ground into a powder suitable for the moulding process.

The cap moulder either uses the moulding powder as received, or preforms it by initial compression into tablets (or pellets) according to the requirements of the moulding press. These are fed into the cavities of the moulding press, and the caps are removed by unscrewing as soon as the curing cycle is complete. The whole process is generally completely automatic.

In essence a mould consists of a steel cavity which determines the outside configuration of the cap, and a threaded pin which forms the threads on the inside of the cap. The cavity is loaded with a measured volume of powder or with a predetermined size of pellet. Both cavities and pins are heated to a temperature determined by the nature of the powder and the moulding conditions (generally of the order of 145°C for UF and WF powders and of 160°C for phenolic powders) so that, when the pin is pressed into the cavity under a pressure of the order of 1 to $1\frac{1}{2}$ tons/in^2 (15–20 MN/m^2), the moulding powder flows into the form and shape of the finished cap and the resin sets. Both pins and cavities are highly polished and also chromium plated, so as to resist corrosion and produce the best finish possible on the caps.

All thermoset caps have a light moulding flash where the excess material exudes between the cavity and the pin during the moulding process. This flash is removed either by rumbling the caps in a revolving barrel of wire mesh, or by passing them through a deflashing machine which automatically grinds the flash off the ends of the skirts. The caps are then

CLOSURES AND DISPENSING DEVICES

polished, either in a barrel with rags or by hand on rotating polishing mops. They can also be given decorative finishes by spraying, hot leaf stamping, metallizing, etc.

Finally the caps are automatically fitted with liners, which are almost always glued-in, before dispatch.

Moulded caps made from thermoplastics plastics

These caps are also known as *injection moulded* since they are formed in presses that inject the hot fluid plastics material into the moulds.

There is an increasing variety of thermoplastic materials from which caps can be made. The most commonly used are polystyrene, polystyrene

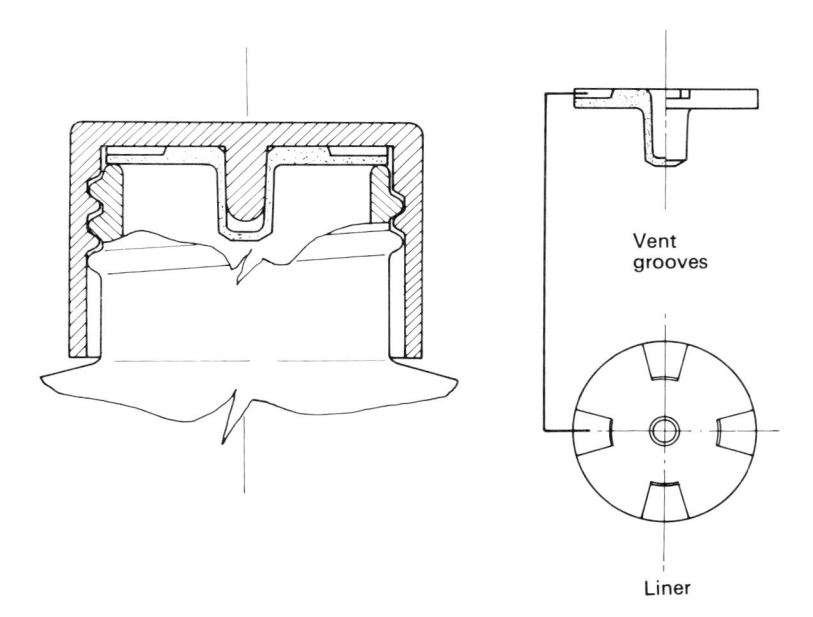

Figure 6.7a *Venting closures.* Developed originally for sealing bleaches (hypochlorites) and peroxides, venting seals are designed to relieve excess pressures that develop in bottles. A venting closure is basically a standard external-screw closure fitted with a liner that vents. The first venting liners consisted of rubber discs with central slits. Under the effect of internal pressure, the rubber disc arched into a dome in the top of the closure; the slit opened, and the excess gas escaped. Next followed discs in plastics and similar materials that were provided with grooves on their reverse faces, located radially so as to cross the sealing edges of the bottle finishes. As soon as excess pressure developed in the bottle, the grooves arched and the excess gas escaped. There are many variations on the groove method of venting.

The diagram shows an example of the grooved venting liner. The liner is moulded in suitable resilient material, such as low-density polyethylene, and is provided with grooves that arch when excess pressure develops in the bottle, thus allowing the gases that produced this pressure to escape.

3.123

GLASS AND PLASTICS-BASED PACKAGING

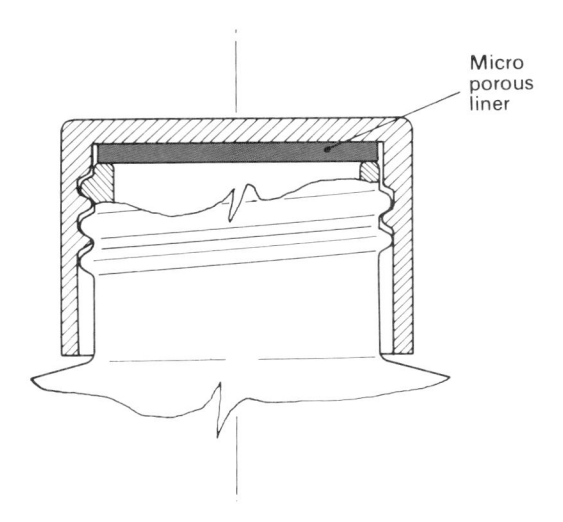

Figure 6.7b Many other venting devices are covered in the patent literature, but another that shows promise is a microporous plastics liner that allows gases to escape via the micropores when excess gas pressure develops in the bottle. It is, however, virtually impossible to devise a venting system that will vent only gases and not liquid when the bottle lies on its side, despite the claims that the dimensions of the micropores can be so accurately controlled that liquid cannot penetrate them, whereas gases do so at the venting pressure required.

The diagram shows a closure fitted with a microporous liner, such as microporous PVC, which allows gases to escape via the micropores when excess pressure develops in the bottle.

blends and copolymers, high and low-density polyethylene, and polypropylene.

Unlike thermoset powders supplied to cap manufacturers in partly polymerized form, all thermoplastics are completely polymerized when delivered to the cap moulders. All that the latter have to do is to heat the thermoplastic material to temperature and inject it under pressure into the moulds, which are then cooled to cause the thermoplastic material to set. The moulds are opened for the removal of the caps, and the process then starts all over again. As with thermosets, the whole moulding process is generally completely automatic.

Although the process is very simple, the tools for the injection moulding of thermoplastic materials are considerably more expensive than those required for the thermosets. This is because injection moulding tools consist of multi-impressions fed by runners from a central injection point, and the moulds also have to be made so that they can be water-cooled.

Once moulded, thermoplastics caps are complete and require no further finishing. Generally, too, thermoplastics caps have to be fitted with liners, but there are applications for linerless thermoplastics closures (figure 6.8).

Thermoplastics re-soften when heated and can, therefore, be recovered by grinding into granules for re-use. Thermosets, in contrast, are not reversible and do not re-soften when heated; reject thermoset mouldings cannot, therefore, be recovered and re-used.

Choosing the right type of closure

From the point of view of normal performance there is little to choose between the various materials used in the manufacture of conventional closures, and the choice is governed mainly by cost and appearance. However, aluminium will deform if caps are screwed on too tightly; tinplate is the most rigid of the metal caps, and the aluminium alloys are intermediate between tinplate and ordinary aluminium. But tinplate can rust at its edges, and both aluminium and aluminium alloys are free from this defect.

The strengths of moulded caps depend mostly upon the thickness of the sections in which they are made—caps in normal sections being comparable with tinplate caps. The material used does, nevertheless, have an influence: the phenolics (PF) being stronger than the urea-formaldehydes (UF and WF), and the medium and high-impact polystyrenes being stronger than normal polystyrenes. Caps in either high-density polyethylene or polypropylene are among the strongest, but screw caps in low-density polyethylene are a little too flexible and liable to jump the threads, particularly when applied to glass bottle finishes. Low-density polyethylene is consequently much more suitable for snap-on closures.

The outside appearance of metal caps is essentially functional, but moulded caps (whether thermosets or thermoplastics) lend themselves to an infinite variety of shapes. The external configurations and designs of plastics closures can be made to suit packs ranging from the most exotic for cosmetics to the purely functional pharmaceutical caps. Whilst the decoration of metal caps is, with a few exceptions, carried out on the metal sheets from which the caps are fabricated, moulded caps are decorated after manufacture. The principal methods consist of filling in designs with coloured paints, silk screen printing, hot foil stamping, decalcomania, the spraying of the outsides of the caps with gold bronze and other lacquers, or in metallizing them with a vacuum-deposited aluminium coating protected by a clear or coloured lacquer to produce gold and other highly reflective finishes. Even electroplating can be used to achieve particularly fine and lasting effects.

The fit of the liners in the caps is another feature that is determined by the material in which the caps are made. All metal screw caps have a recess, generally a knurl, in which the liner is retained, and the liners are consequently not glued into metal caps. Since moulded thermoset caps cannot be made with the same depth of recess as metal caps, the liners are almost invariably glued into compression-moulded caps. It should be added immediately that moulded thermoset caps often have shallow liner-retaining recesses, but these never retain the liners as well as the knurls of metal caps. In the past, moulded thermoset caps have been fitted with oversize liners retained in the caps as a result of an interference fit. This is not good practice, since it generally causes the liners to bow, and frequently to remain on the finishes of the bottles when the caps are removed. Thermoplastics caps can be injection-moulded, using collapsible core-pins that enable them to be produced with sufficiently deep recesses to retain the liners in position without having to glue them.

Liners

In its most general form, a liner consists of a cushioning material (i.e. the wad) with a facing material. The facing material isolates the contents of the bottle from contact with the wad, and is chosen principally for its resistance to the product packed. Its resilience or lack of resilience (i.e. hardness) is also of importance, particularly for products that seep easily for one reason or another.

The wad material will generally be either composition-cork or pulpboard, and the facing will be either a coated paper, a paper faced with a plastics film, a plain metal foil, a lacquered metal foil, or a metal foil faced with a plastics film or coated with a layer of a wax mixture.

Composition-cork consists of granules of cork bonded together with either a gelatine-type glue or a synthetic resin into blocks that are then sliced into the sheets from which the wads are cut. Glycerine is used as a plasticizer or humectant for the gelatine-glue-bonded cork, and diethylene glycol for the resin-bonded cork. The function of the plasticizer is to make the cork composition more pliable and resilient. The glue-bonded cork will support mould growth under damp conditions, and consequently resin-bonded composition-cork is always used whenever there is any risk that such growth can occur.

Pulpboard is produced from an 85/15 mixture of mechanical and chemical woodpulp. It is generally free from mould growth troubles, providing a suitable facing material is used to prevent contact with products such as emulsions and creams, which are susceptible to moulds.

Composition-cork is more resilient than pulpboard and is consequently better able to take up unevennesses in the sealing edges of the bottle finishes. Nowadays, however, the quality of both glass and plastics bottle finishes has improved to such a degree that pulpboard is replacing composition-cork more and more as a wad material. Also, composition-

CLOSURES AND DISPENSING DEVICES

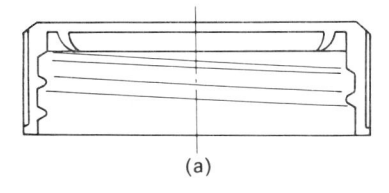

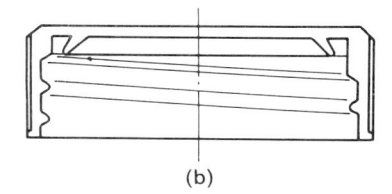

Figure 6.8 *Linerless closures.* A linerless closure is a closure that is rigid enough to function as a closure, yet resilient enough to make a seal on the finish of the container, without requiring any form of gasket such as a liner to do so. Polypropylene is a material that has these properties, and the diagrams show two typical designs for linerless closures. A continental design for a wadless closure illustrated in figure *c* is the VIMA closure that is provided with twin internal skirts that make a plug-type seal in the mouth of the bottle.

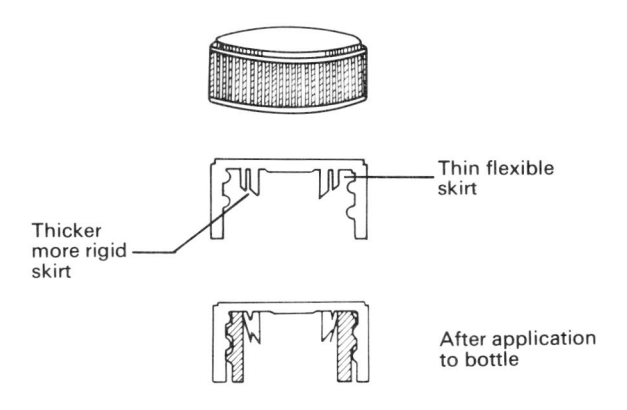

Figure 6.8c The VIMA wadless closure

cork wads develop a greater permanent set (i.e. they lose their resilience) than do pulpboard wads, and this generally results in lower cap-loosening torques after the bottles have been in stock for some time.

There are a number of materials used as complete liners that require no facings, since they possess in themselves both the resilience of the wad and the resistance of the facing to the products packed. Such materials are rubber, PVC, polyethylene, and EVA. Additional resilience can also be achieved by using plastics in expanded form. In this group also are included moulded liners in materials such as polyethylene and polypropylene, the shapes of these liners generally being such as to present an arch or the base of a groove against the sealing edge of the bottle, thereby increasing the effective resilience of the material used. Yet again, a moulded liner in one of the above-mentioned thermoplastic materials, particularly polyethylene, can be in the form of a disc made captive to a peg in a

3.127

GLASS AND PLASTICS-BASED PACKAGING

moulded cap and, if the cap also has a cored-out dome, the liner is made to seal against the edge at the entrance to the bore of the bottle, and a particularly good seal results. On the other hand, the cored-out shape can be moulded in a thermoplastic material and can then be fitted into a cap (British patent No. 981 716).

Both *natural and synthetic rubbers* are available in almost any degree of resilience or rigidity that may be required for particular applications. Natural rubber, however, has the disadvantage of a high coefficient of

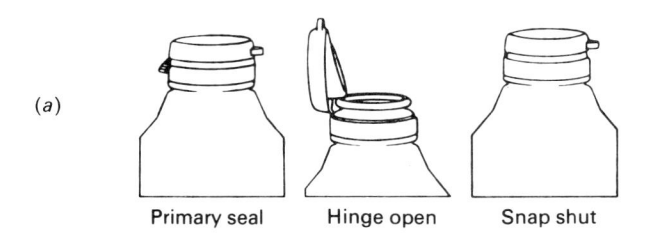

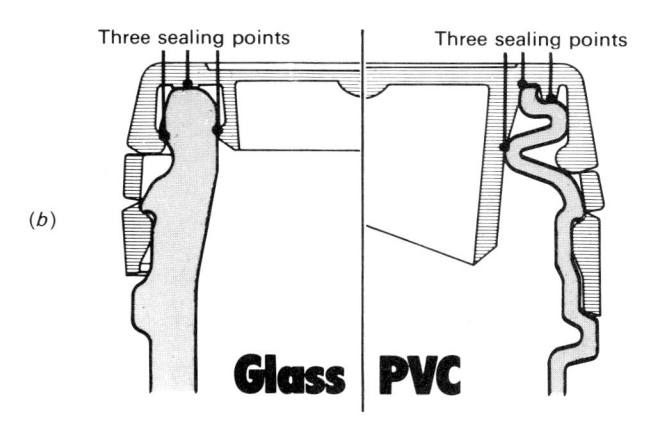

Figure 6.9 *Plastics pilfer-proof closures.* A plastics pilfer-proof closure is designed to snap over a security head on the neck of the container, so that it cannot be opened until the band connecting the skirt to the pilfer-proof ring is torn away.

One of the most common examples of this type of closure is the polythene Jaycap, currently used on both glass and rigid plastics containers. The Jaycap has a triple sealing feature—a plug seal inside the bore; a top seal on the rim; and an outside seal under the snap-over head. Between each sealing-point there is a capillary cavity that prevents seepage should the product penetrate the initial plug seal.

Jaycaps are applied to the containers by simple top pressure. Figure *a* shows the Jaycap as applied to a bottle, as well as in its in-use positions.

Figure *b* shows the sealing features of a Jaycap on both glass and rigid plastics (PVC) bottles.

3.128

friction that causes caps to spring back after tightening. This can be overcome by wetting the rims of the bottles or the wads before applying the closure. Two of the main advantages of rubber are that it will withstand high-temperature processing and that it is self-sealing after piercing with a hypodermic needle. It is, therefore, extensively used in pathological and blood transfusion closures. However, its increasingly high cost encourages users to seek alternative materials.

Plastisols and water-based (flowed-in) compounds are also liners that require no facings, since they possess both resilience and inertness.

Plastisols are dispersions of PVC resins in plasticizers. They are flowed into metal caps either in the form of annular sealing rings or overall lining discs. They are fluxed at a high temperature relatively rapidly to set the plastisols into resilient sealing-rings or discs.

Water-based compounds are colloidal dispersions of either natural or synthetic rubber in water. They are flowed into metal caps in exactly the same way as plastisols, but need prolonged stoving to drive off the water.

Flowed-in compounds are almost invariably used in vacuum closures as well, for all food processing applications, and are replacing the synthetic rubber (usually nitrile rubber) rings that were popular with a number of vacuum closures.

Compatibility between closures and the contents of containers

The product must in no way become tainted or affected physically or chemically by any of the closure materials with which it may come into contact. Nor must the product itself have any effect on the closures.

This compatibility must also include the sealing compound in the closure, namely, the wad, liner, or sealing ring. Pulpboard on its own has a degree of porosity; it is sensitive to moisture, and absorbs aqueous products readily. Gelatine-bonded composition-cork is sensitive to moisture and disintegrates in the presence of aqueous products; resin-bonded composition-cork is better in this respect, but it is still liable to taint aqueous products and will also disintegrate in them, although more slowly.

Consequently, pulpboard and composition-cork wads, even when waxed, are only used in closures for sealing dry moisture-insensitive products. In other instances these wad materials must be lined with an inert product-compatible facing.

When the resilient sealing compound forms a ring pressed against the sealing edge of the container, the product comes into contact with the inside of the closure that is encompassed by the ring. This applies particularly to metal vacuum closures for wide-mouth jars. The inside

3.129

GLASS AND PLASTICS-BASED PACKAGING

lacquer or enamel system with which the cap is protected must therefore also be inert to the product packed, not only at room temperature, but also under the processing treatment to which the pack is submitted.

Plastics closures are sometimes fitted with rubber sealing rings that expose part of the inside of the closure to the product, and this again requires that both the closure and product must be completely unaffected by each other. Plastics closures are not usually suitable for processing treatments, although some of the high-density polyethylenes, polypropylenes, and phenolic thermosets have at times been used for applications where some processing is involved.

Facing materials

Whenever a wad material does not provide the protection or inertness required, then it must be faced with a material compatible with the product packed. The facing material must also bed down well on the sealing edge of the container to ensure a good seal.

Almost all facing materials are based on paper (bleached sulphite or kraft) either coated with white pigmented synthetic resins such as vinyl copolymer or polyvinylidene chloride (PVDC) resins; or laminated to plastics films such as PVDC, polyester (Melinex), and polyethylene. Metal foils (aluminium and tin) make excellent facing materials where maximum

1	2	3	4	5
Press-and-turn	Squeeze-and-turn	Combination lock	Restraining ring	Press-and-lift

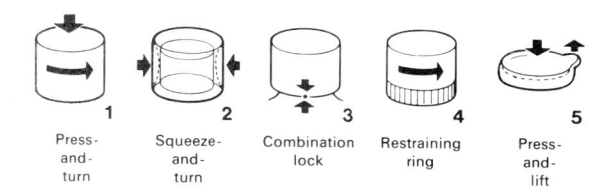

Figure 6.10a *Child-resistant closures.* A child-resistant closure is a closure that requires two co-ordinated actions, not generally possible by children, for removal. The five basic types into which child-resistant closures fall are as follows:

1. *Press-and-turn.* The cap must be pressed against the bottle and rotated at the same time.
2. *Squeeze-and turn.* A double cap with a free rotating outer which must be squeezed to engage the inner cap when the closure is to be unscrewed.
3. *Combination lock.* Two parts of the cap must be lined up according to the marks shown on the cap.
4. *Restraining ring.* A two-piece cap in which both parts (top and restraining ring), when screwed together, rotate freely on the bottle. Since the restraining ring is permanently attached to the bottle, the latter can only be opened by holding the restraining ring in one hand and unscrewing the top from it with the other hand.
5. *Press-and-lift.* Downward pressure relaxes the grip of the cap's edge on the bottle.

Some child-resistant closures require special bottle finishes, and some of these are suitable only for plastics bottles since the bottle finishes required are too complex to be suitable for glass bottles.

CLOSURES AND DISPENSING DEVICES

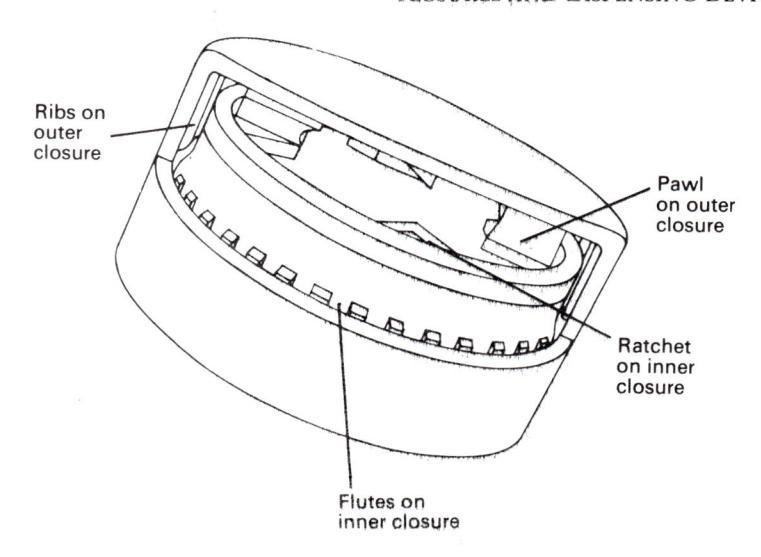

Figure 6.10b Of the child-resistant closures that are suitable for use on standard glass-bottle finishes, the most promising is probably the Clic-Loc closure. This is a "press down and turn" closure—two co-ordinated movements that almost always baffle a child under five—and the Clic-Loc closure has the further feature that it gives an audible alarm when anyone attempts to unscrew it by a normal turning action. Not only, therefore, is the mother secure in the knowledge that her child is most unlikely to be able to unscrew a Clic-Loc closure, but she will also hear an audible alarm if her child starts playing with it.

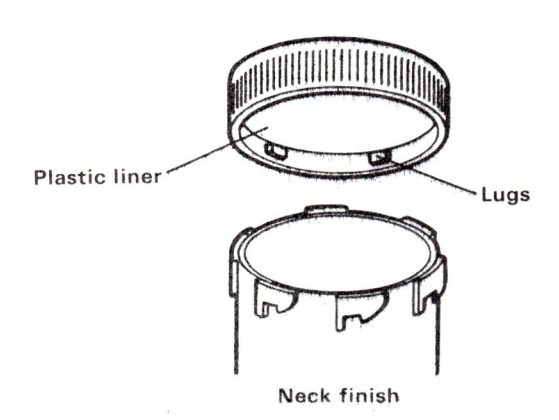

Figure 6.10c Shows the Screw-Loc "press and turn" closure which is more suitable for plastics than glass containers because of the special finish required to mate with it

3.131

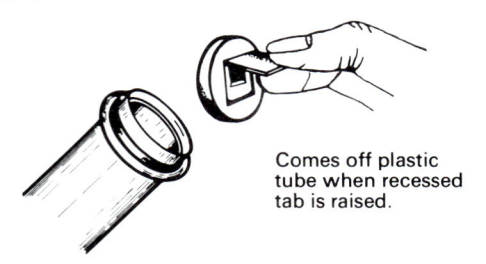

Figure 6.10*d* Shows the "Pop-Lok" closure, which is a cap that fits tightly on the container and cannot be removed by pulling the sides. Pressing the correct spot on the top of the cap releases a tab which is easily gripped to pull off the cap.

resistance to gases, water vapour or solvents is required. Lead foil is never used in the food, beverage, cosmetic, and pharmaceutical industries because of its toxicity. Metal foils are almost always backed with paper to facilitate their lamination to the wad material during the lining operation. Where there is risk of corrosion, the foil can be coated on one side with a lacquer or a plastics film.

The range of facing materials used to be more extensive than it is today, but an endeavour has been made to cut the range down in the interests of rationalization. Closure manufacturers have always sought to find a universal facing material or liner for sealing all products, but this ideal is not yet in sight.

Experience is usually the best guide in recommending a sealing system for any particular product, but extensive sealing tests could be required on a new product, or a significant alteration in the composition of an existing product before sound advice could be given on the best sealing material(s) to use.

Closure design

A closure is designed for its appearance and performance. Its appearance will depend upon its external shape, the material in which it is made, and the decoration required to achieve a particular effect or purpose. Its performance will depend upon ensuring that the specification of its bottle finish engaging features match the specification of the glass or plastics bottle finish; that the cap is fitted with the correct sealing system for the product packed, and that it will withstand any processing treatment involved.

Methods of application of closures

There are four principal methods (see Table 6.1, p. 3.134).

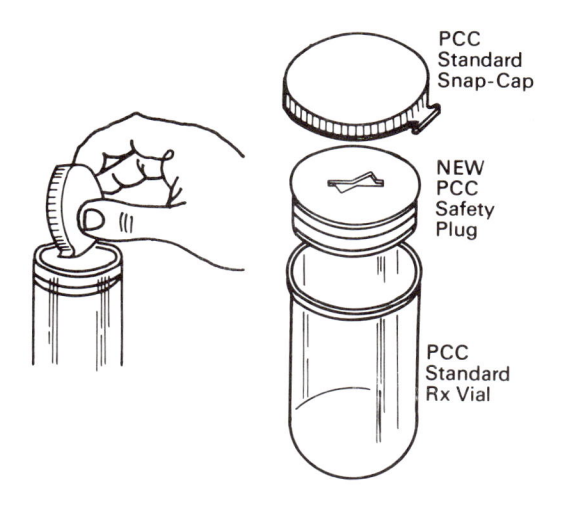

Figure 6.10e Shows the PCC "Safe-Key" cap, consisting of an outer snap cap and an inner flush-fitting plug seal. The tab of the snap cap is provided with lateral projections that fit into a moulded-in keyhole opening in the top of the plug. The user first removes the snap cap and then inserts the projecting tab ends into the keyhole to pull out the plug.

There are many more ingenious devices for producing child-resistant closures, and a whole chapter could be devoted to this subject. Child-resistant closures are usually more suitable for rigid plastics containers than for glass containers (with a few notable exceptions such as the Clic-Loc closure) since it is easier to mould in plastics the special finish configurations that are so often required for the child-resistant feature than to reproduce these configurations on the necks of glass bottles.

Screw-on caps mainly have an external screw-thread cap which can be applied by hand; but this is a slow operation, and consistency of tightness falls off as the operatives become tired.

All standard screw closures can be applied automatically by a rotating head having a chuck that fits over and grips the loosely pre-applied cap and tightens it to the required torque which is regulated by means of an adjustable friction clutch fitted to the head. Hand-fed single-head semi-automatic machines can apply caps at speeds of up to about 60 per minute whilst multi-head fully-automatic rotary machines are capable of operating at speeds of 250 per minute or more. Caps can also be tightened by conveying the loosely capped bottles in a straight line between rollers or bands applied to their sides. Really fast speeds in excess of 400 caps per minute are possible with such a straight-line capper.

Screw-on lug caps are pressed into position on jars by a horizontal moving belt while the jars are moving along a conveyor; the necessary twisting action is obtained by dividing the belt in two lengthwise, and

Table 6.1. The four principal methods of application of closures

	Method of application	Types of closures
1	Screw on Screw in	Internal and external screw thread caps and lug caps
2	Push in Push on	Driven corks and flanged-top cork stoppers; plugs; snap-on caps and other forms of press-on friction-fit caps
3	Crimp on	All closures which are secured to the rims of bottles by squeezing the skirt of the cap into a groove—under a bead on the side of the neck of the bottle; or on to screwing-off threads round the outside of the finish of the bottle.
4	Roll on Spin on	All caps in which the screw threads are formed in the skirts by means of spinning rollers that shape the skirts to the contours of the threads of the bottle finishes. Also all threadless closures in which the skirts are rolled under retaining rings on the necks of the bottles.

running one half faster than the other. A normal capping speed is 250 per minute, but speeds up to 750 are possible.

Screw-in caps are now almost entirely superseded by either pre-threaded external screw or spin-on closures. The old internal screw stopper is the main example here, and it was either entirely applied by hand, or loosely fitted into the bottle and tightened by means of a rotating head. Its rubber washer made an excellent seal, and re-sealing was just as effective when replaced by hand. The main disadvantage was an unattractive appearance, and the lack of protection afforded around the rim over which the liquid contents of the bottle are poured.

Push-in corks were originally and still are largely represented by the driven cork, now mostly used for sealing vintage wines. A good many corks are still inserted by hand or mallet, but mechanical methods are also available. The modern example of a push-in cork is the flanged-top cork stopper extensively used for spirits and fortified wines, and which can be applied by mechanical means.

Push-in plugs and shives are pushed into the mouths of bottles prior to the application of the closures, and mechanical means are available to do so when the output warrants it.

Press-on plastics caps are represented by a wide variety of snap-on plastics closures. There are various means for mechanically applying

CLOSURES AND DISPENSING DEVICES

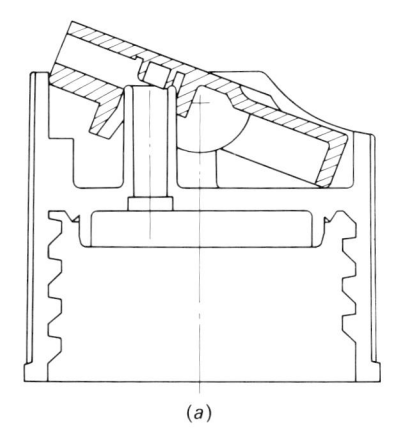

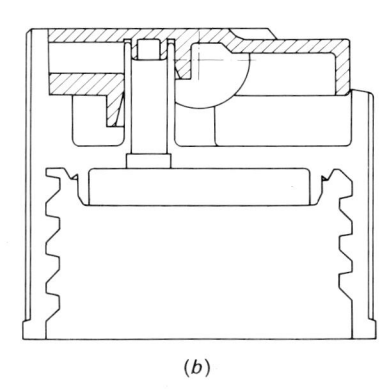

(a) (b)

Figure 6.11a *Dispenser closure systems.* A dispenser system is a means for controlling the flow of product out of a container without removing the closure.

Figures *a* and *b* illustrate the Toggle Dispenser closures in the open and closed positions. The closure makes a plug seal in the neck of a rigid plastics container, and the contents of the bottle are dispensed after depressing the disc so as to open the spout. Light pressure on the walls of the bottle help to ensure an even flow of liquid through the spout.

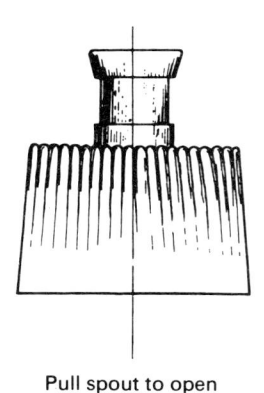

Pull spout to open

Push back to close

Figure 6.11c The Tip-Top dispenser closure. Again this closure makes a plug seal in the neck of a rigid plastics container. The contents of the bottle are dispensed by pulling the spout out, and the bottle is sealed by pushing the spout home again. The spout is provided with an anti-drip nozzle.

3.135

GLASS AND PLASTICS-BASED PACKAGING

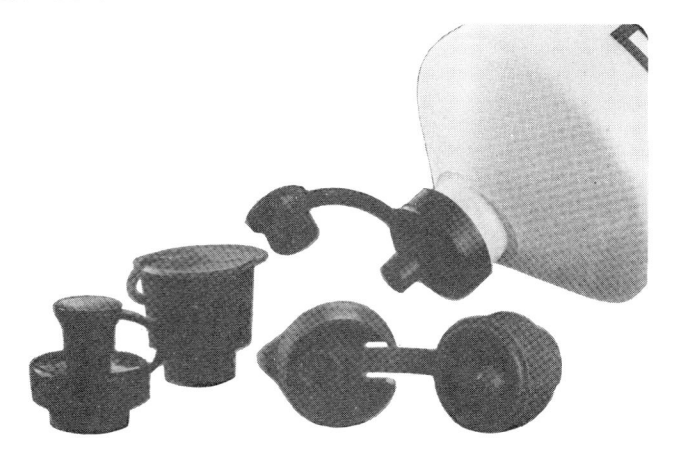

Figure 6.11d Flip Top. Low-cost one-piece fixed snap-on polythene closure with dispensing spout and captive re-seal push-on cap. It is widely used for sealing detergents and similar products that are dispensed from plastics bottles by a squeezing action.

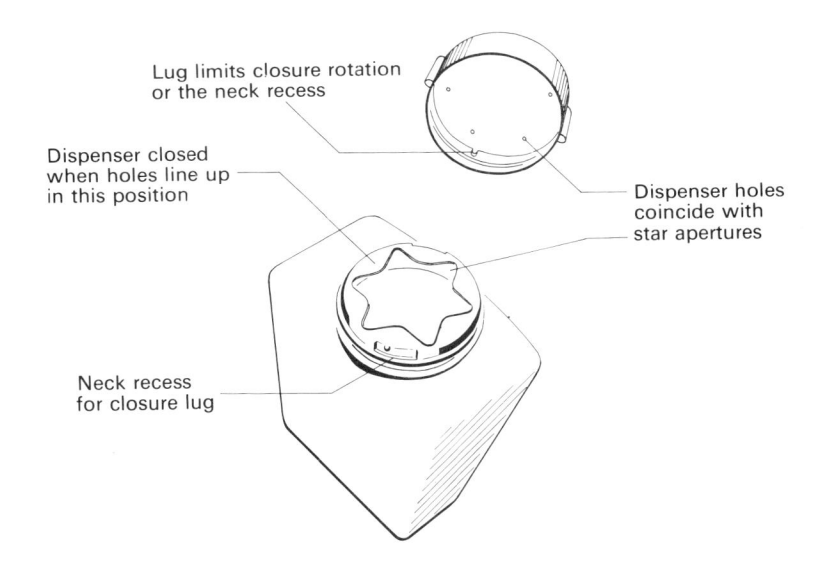

Figure 6.11e Shows a one-piece talc dispenser snap-on closure. This closure takes full advantage of the moulding possibilities presented by thermoplastics. The finish of the bottle is provided with star apertures that coincide with dispenser holes in the cap. When the cap is snapped on to the bottle finish, a lug limits the rotation of the closure to the open and closed positions of the dispenser holes.

3.136

CLOSURES AND DISPENSING DEVICES

them—the simplest being to pass loosely-capped containers under a moving pressure belt which presses the caps home.

Press-on metal caps are represented by the prise-off cap applied by a pressure belt using downward pressure only. The cap is retained on the jar by an in-pack vacuum developed by the use of steam flushing in the capping machine.

Crimp-on metal caps are represented by the crowns used for pressure-retaining seals and by the larger-diameter vacuum-tight closures commonly seen on jams, preserves, and similar products. The crimping-on operation is effected by a shaped chuck that, when applied to the cap under top pressure, squeezes the skirt on to the cap retaining feature (groove, bead, or threads) on the neck of the container. Sealing equipment starts with a single-head semi-automatic machine and extends to multi-head fully-automatic rotary machines capable of applying closures at speeds up to about 750 per minute for crowns.

Roll-on or spin-on metal caps are represented by all those aluminium caps supplied to the bottlers in shell form, and in which the threads are developed in the skirts during the capping operation. During application, the liner in the closure is compressed by top pressure exerted by the capping machine head, while spinning rollers rotating round the skirt of the cap develop the threads in the cap using the thread on the bottle finish as a former. In the pilfer-proof versions, the pilfer-proof tell-tale ring is, at the same time, tucked under its corresponding bead on the bottle finish. Capping machines range from single-head to multi-head fully-automatic rotary machines capable of applying caps at speeds of up to 400 per minute.

Sealing methods

There are a number of methods for sealing the mouths of containers in addition to the rim top seal. These other methods tend to be associated with the types of closures for which they are used. There are five main methods.

(a) *The rim top seal*

The rim top seal is described in detail under "The mechanics of a good seal" (page 3.109). It is the standard seal in which the sealing component in the closure, namely the liner, is pressed downward against the top of the sealing rim on the container finish.

3.137

GLASS AND PLASTICS-BASED PACKAGING

Pourer fitment remains in bottle when cap is unscrewed

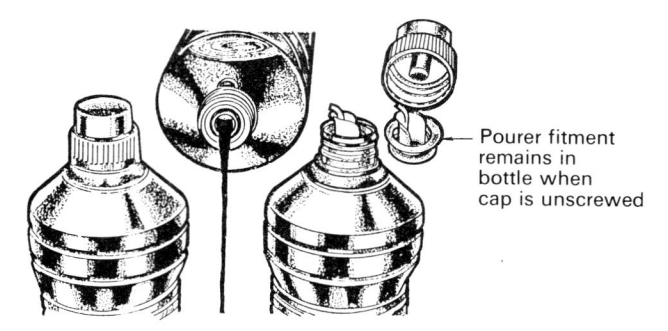

Figure 6.12a *Pourer systems.* There are a number of devices that improve the pourability of bottles and that often also act as anti-drip or non-drip devices.

Figure *a* illustrates the Rohill oil-pourer closure. The pourer plug fitment is supplied already fitted in the closure, which is applied to the bottle by a conventional capping machine. When the cap is unscrewed, the pourer plug fitment remains firmly fixed in the mouth of the bottle, and the oil contents are dispensed via the pourer spout, which also acts as a non-drip device. The closure is fitted with a central tube that seals in the orifice of the spout when the closure is screwed back on to the bottle finish. This closure is also available with a pilfer-proof tear-off band.

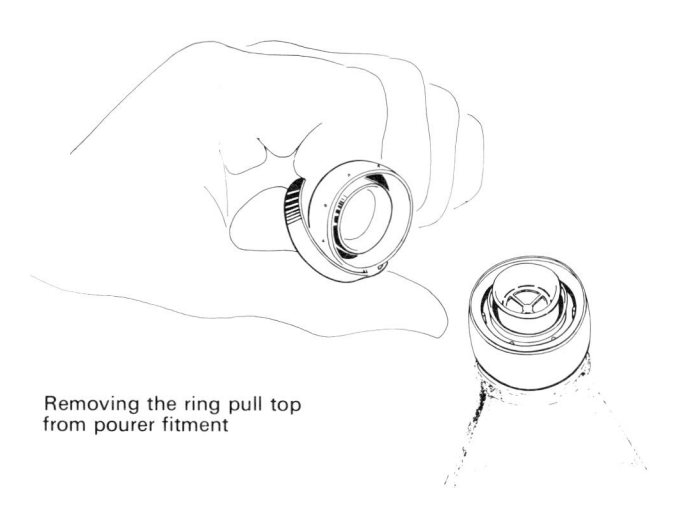

Removing the ring pull top from pourer fitment

Figure 6.12b The Ring Pourer cap. This cap is in two parts, generally in contrasting colours. The pourer base is a plug fit in the mouth of the bottle, and its skirt snaps over a bead on the outside of the neck of the bottle, thus retaining the base portion firmly on the neck of the bottle. The top portion is a tamper-proof ring closure.

CLOSURES AND DISPENSING DEVICES

(b) *The side top seal*

In the side top seal, a rubber or synthetic rubber band, fitted on the inside of the skirt of a metal lug-type closure, mates with a slightly outward tapering seal edge on the outside wall of the top of the rim of the container finish. This type of seal is essentially a vacuum seal, since it relies almost as much on the external atmospheric pressure as on the downwards sealing pressure exerted by the lugs. It was at one time popular, particularly in the United States, for sealing baby food jars.

(c) *The combined rim and side top seal*

The most important examples of this type of seal are The United Glass "Eurospin" (figure 6.5) and Metal Closures "Flavorlok" roll-on aluminium closures. They consist of cap shells lined with an overall flowed-in plastisol gasket that rides up slightly on the inside of the skirt and which increases to a thicker annulus in the angle between the crown and the skirt of the cap. In the course of spinning the cap on to the bottle finish, a specially designed pressure chuck reforms the top of the outside knurl section of the cap shell, so as to cause the plastisol liner to seal on both the top and side of the rim of the bottle finish before the threads are formed in the skirt of the cap by the standard spinning process. By this means, the pressure that a relatively weak aluminium closure can contain in a bottle is very much increased, and both the "Eurospin" and "Flavorlok" closures are extensively used for sealing carbonated beverages.

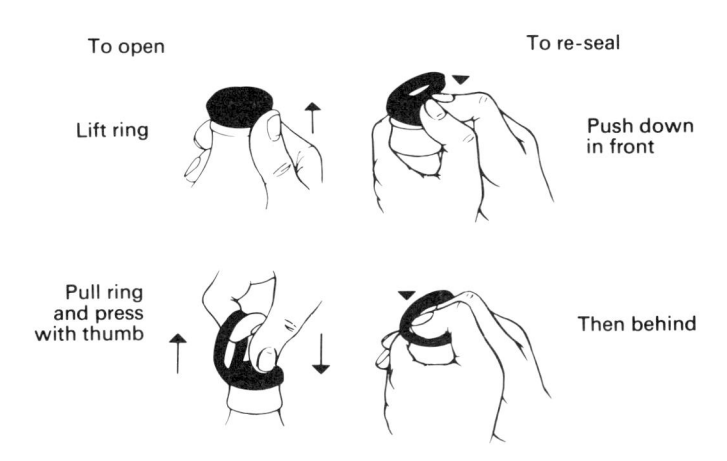

Figure 6.12c   Opening and re-sealing a ring closure

3.139

GLASS AND PLASTICS-BASED PACKAGING

Table 6.2. Normal seals

	Closure	
Method of performance	General classification	Materials in which made
Pre-threaded screw on screw off generally single-start threads	compression-moulded plastics	thermosets
	injection-moulded plastics	thermoplastics
	vacuum-formed plastics	thermoplastic sheet
	metal	tinplate
		Hi-Top plate
		aluminium
		aluminium alloy
Two-piece pre-threaded screw on screw off	metal	tinplate
		aluminium
		aluminium alloy
Lug type screw on twist off	metal	generally tinplate
Roll on or spin on screw off	metal	aluminium
		aluminium alloy
Press on prise off	metal	tinplate
Crimp on prise off	metal	aluminium
Lever type crimp on lever off (now obsolete)	metal	generally tinplate
Crimp on screw off	metal	tinplate
		aluminium
Push in pull out	driven corks	cork
	stoppers	corks with either plastics tops or wooden tops inside decorated aluminium covers
	stoppers with ribbed or finned hollow plugs	all thermoplastics, generally polyethylene
Push on pull off	Snap on	thermoplastics, generally polyethylene
	Push on	

3.140

CLOSURES AND DISPENSING DEVICES

Normally it is not good practice to design closures to seat on two (or more) sealing surfaces at the same time since, almost invariably, mating occurs first on one sealing surface and, since any further sealing movement of the cap is then blocked, it prevents the other mating surface from bedding down properly. With the "Eurospin" and "Flavorlok" closures, however, the top of the cap is reformed while the whole cap is under top pressure, thus ensuring maximum sealing effectiveness on both the rim top (top pressure) and the side top (reforming action) of the bottle finish.

(d) *The bore or plug type seal*

In this type of seal an interference-fit cork or stopper is inserted into the mouth of the container. It is the most ancient method of sealing a bottle—natural corks having been used for over 300 years for this purpose. They are still used today for sealing vintage wines and champagnes in glass bottles. The modern counterparts to natural corks include hollow plastics plugs, with or without fins, or rings designed to conform more positively with bore dimensional variations and thus ensure a more effective seal. Plastics shives are also used for making a plug-type seal in the larger-diameter mouths of (principally) glass containers.

(e) *The snap-on seal*

With the advent of resilient thermoplastics, such as low-density polyethylene and, to a lesser extent, polypropylene, increasingly popular closures have been designed to make a snap-on fit on the finishes of containers. A snap-on cap is forced over a cooperating bead on the mouth of the container, so that the cap hugs the bead tightly and uses the whole surface of the bead as a sealing area. The plastics cap is the resilient component; the tighter the fit it makes on the bead, the more effective is the seal.

Combination closures

In certain applications double closure systems are used for sealing (mostly) glass containers. Lead-tin, and pleated aluminium foil capsules, viscose bands, and heat-shrinkable plastics bands are applied over the closures on to the necks of bottles, not only to dress them in an attractive manner, but to protect the primary closures from contamination and provide evidence of pilfering, since the closures cannot be removed without destroying the capsules or bands.

The plastics shive, mentioned above, which is not a pilfer-proof device in itself, is usually used with an additional screw closure which can be

GLASS AND PLASTICS-BASED PACKAGING

used on its own (without the shive) to re-seal the container after removal and disposal of the shive.

The tamper-proof membrane adhered to the mouth of a container (see "Tamper-proof closures on page 3.118 and figure 6.6) is also a combination closure in as much as, once the membrane has been torn off the rim of the container, the cap itself, be it a screw cap or a snap-on cap, provides a re-seal during use.

Table 6.3. Vacuum seals

Closure		
Method of performance	General classification	Materials in which made
Screw on / twist off } lug caps	metal	Almost invariably tinplate, but aluminium and its alloys are possible
Press on prise off	metal	tinplate
Press on twist off	metal	tinplate
Two-piece screw on screw off	metal	tinplate
		aluminium
		aluminium alloy
Two-piece roll on screw off	metal	aluminium
		aluminium alloy
Crimp on prise off	metal	tinplate
		aluminium
Crimp on screw off		aluminium alloy

3.142

Table 6.4. Pressure seals

| | Closure | |
Method of performance	General classification	Materials in which made
Screw in (internal screw) screw out (now almost obsolete)	moulded plastics	cold-moulding (ebonite)
		thermosets
		thermoplastics
Screw on screw off	moulded plastics ("Estseal") (now obsolete)	thermosets
		thermoplastics
	metal ("LoTork")	tinplate
Crimp on lever off	metal	tinplate
		aluminium
Crimp on screw off		aluminium alloy
Roll on or spin on screw off	metal ("Eurospin" and "Flavorlok")	aluminium
		aluminium alloy
Crimp on pull off or rip off	metal	tinplate
		aluminium
		aluminium alloy
Crimp on or roll on with tab-locked band undo band	metal (Phoenix) (gradually going out of use)	tinplate

3.143

PART FOUR

WOOD AND TEXTILE-BASED PACKAGING

CHAPTER 1

Wood and Wood Products in Packaging

F. A. Paine

Statistics on forestry[1]

The total annual world production of wood is more than 2000 million cubic metres in round or log form and is steadily increasing. Table 1.1 shows data for 1959, 1963 and 1968 and gives a breakdown of the 1968 production by regions. Forecasts for the future are that both production and consumption will increase.

Table 1.1. World production of wood and wood products

Year	Round wood (million m^3) Total	Industrial	Sawn wood (million m^3) Softwood	Hardboard	Total	Ply-wood (million m^3)	Fibreboards 000' tonnes
1959	1900	1010	266	69	336	15	4100
1963	1960	1040	273	77	350	20	5400
1968	2130	1200	301	88	389	28	7100
Region (1968)							
Europe	310	241	58	17	75	3·6	2700
USSR	380	290	94	17	110	1·8	550
N. America	450	418	96	17	114	15·0	2700
C. America	51	10	2	1	3	0·1	40
S. America	229	35	4	6	11	0·4	140
Africa	247	33	1	2	3	0·2	80
Asia	438	152	43	25	68	6·8	620
Pacific	26	19	2	3	5	0·1	200

4.3

WOOD AND TEXTILE-BASED PACKAGING

The wood produced is grown on some 4126 million hectares* or 32% of the total world land area. (13,333 million hectares.)

Except for Asia and the Pacific areas, most regions have at least one-quarter of their land surface classified as forest. South America, North America and the USSR, each have over one-third. The world as a whole possesses some 238,000 million cubic metres of standing timber, made up of 114,000 million cubic metres of conifers or softwoods and 124,000 million cubic metres of broad leaved trees or hard woods. Tables 1.2 and 1.3 show the areas of forest land in the various parts of the world and the timber harvest from them.

Table 1.2. Annual timber harvest

Area	Standing volume of timber (million m³)	Annual harvest (million m³)	Annual harvest as % of standing volume	Harvest per head (m³)
N. America	44,000	434	1·0	1·9
Asia	17,000	419	2·6	0·2
USSR	79,000	395	0·5	1·6
Europe	12,000	353	3·0	0·7
Africa	3,800	216	5·0	0·7
S. America	78,000	199	0·3	1·3
C. America	800	47	5·2	0·7
Pacific	3,800	27	7·0	1·6
Total	238,000	2090	0·9 (average)	0·7

Table 1.3. Areas of forest

	Land area (million hectares)	Forest land (million hectares)	% under forest	Forest area per head of population (hectares)
USSR	2144	910	42	3·8
S. America	1760	890	51	4·9
N. America	1875	750	40	3·4
Africa	2970	710	24	2·1
Asia	2700	550	20	0·3
Europe	471	144	31	0·3
Pacific area (incl. Australasia)	842	96	11	5·2
C. America	272	76	28	0·9
Total	13,033	4126	32	1·1

* 1 hectare is roughly 2½ acres.

WOOD AND WOOD PRODUCTS IN PACKAGING

If we adjust the annual harvests and relate them to the standing volumes, although wide regional differences appear, the average annual cut does not exceed 1% of the standing volume.

In spite of the amount of coal, oil, gas and electric power available, nearly half the wood felled in the world today is still used for fuel. This occurs mainly in the less industrialized countries and amounts to almost a thousand million cubic metres annually.

Wood utilization[2]

The main primary products from the forest are round-wood products, sawn-wood products, plywood and laminated constructions, particle boards, fibreboards, wood pulp and paper. In addition to these, there are a number of mechanically derived products, such as fishing rods, golf clubs, archery bows, cricket bats and the like, and a number of chemically derived products, such as acetic acid, charcoal, wood alcohol, vanillin. Apart from pulp wood used in the manufacture of pulp for papers and boards, and a certain usage of fibreboards and plywoods, the majority of packaging material is derived from sawn wood.

The principal timbers are those generally referred to as softwoods, and Table 1.4 summarizes the wood species used for case making grouped in similarity of case making characteristics. Some of the properties of certain woods are given Table 1.5.

Timber for wooden packaging should be well seasoned, either by air

Table 1.4. Wood species (grouped in similarity of case making characteristics)[5]

Group 1	Group 2	Group 3	Group 4
spruce	Douglas fir	beech	sweet chestnut
pine (except pitch pine)	pitch pine	birch	elm
	larch		
fir			oak
hemlock			
			Order of preference decreasing →
willow	gaboon	African mahogany	afzelia
alder	lime		guarea
magnolia	tupelo		
poplar			
obeche			

Group 4 should not be used, except in special circumstances, and never for metals or metal parts.

4.5

WOOD AND TEXTILE-BASED PACKAGING

Table 1.5. Properties of certain species of wood[5]

Species	Density* (g/cm³)	% volume shrinkage	Static elastic bending modulus** (kg/cm²)
Oak (*Quercus alba*)	0·71	15	125,000
Beech (*Fagus sylvatica*)	0·68	18	160,000
Birch (*Betula verra cosa*)	0·61	14	165,000
Pine (*Pinus sylvestris*)	0·49	12	120,000
Spruce (*Picea abies*)	0·43	12	110,000
Balsa (*Ochroma lagopus*)	0·13	7	26,000

* oven-dried wood
** based on small clear specimens (12% moisture)

drying or kiln drying, and contain somewhere around 15% to 20% moisture content. Drying is an essential preparation of all kinds of wood for further use, because it reduces the magnitude of dimensional changes due to shrinkage and swelling, protects the wood from microorganisms, reduces weight, and prepares the wood better for most finishing and preservation methods, and increases its strength. The object of air drying is to reduce the moisture content of the wood to the lowest value obtainable under the weather conditions in the shortest possible time without producing defects. Naturally, the level to which the moisture can be reduced will depend upon the temperature and the relative humidity. Wind will reduce the time required, but direct precipitation (rain or snow) wets the wood and hinders the progress of drying.

An air drying yard must have reasonable air movement, unobstructed by tall trees or buildings, and a ground surface which is free from debris and vegetation. Gangways must be provided for the working areas and to permit air movement, and the first row of any timber being dried is kept about 40 cm above the ground to allow space for air circulation. As layers are added, spaces are also left for this purpose. Some sort of roof, usually of low-grade timber or panelled material, will be placed on top of each pile, and the time required to air dry from green to 20% moisture will vary from about 20 to 300 days for $2\frac{1}{2}$-cm thick material, depending upon the species, the place and the time of year.

The other method of drying, kiln drying, is conducted in a closed chamber under artificially induced and controlled conditions of temperature, relative humidity and air circulation. This permits much more rapid reduction of moisture content to levels required, independent of the weather condition. A reduction of moisture from green to 6% will be accomplished in 2 to 50 days for $2\frac{1}{2}$-cm thick material. Heating is usually by steam circulating in pipe coils, and some control of the relative humidity by allowing steam to enter the chamber is necessary in order to regulate the exit of moisture, to avoid warping and so on.

In addition to being well seasoned, timber for packaging should also not contain defects which materially reduce the strength, i.e. it must be reasonably sound, free from bad cross-grain, contain no knots which are more than one-third of the width of the piece concerned, and none of these knots must run into the edges. Generally, freedom from wane (the original rounded tree surface) is also required.

Plywood

Plywood is a panel product manufactured by glueing together one or more veneers to both sides of a veneer or a solid wood core (figure 1.1). The grain of alternate veneers is crossed (in general at right angles); the species, the thickness and the grain direction of each layer are matched with those of its opposite number on the other side of the core. Usually the total number of layers is odd (3 or 5 or 7) and after assembly the panels are brought to presses for cold or hot pressing. The glues employed for pressing at room temperature (cold pressing) are either the natural or synthetic resins, while for hot pressing, synthetic resins only are used.

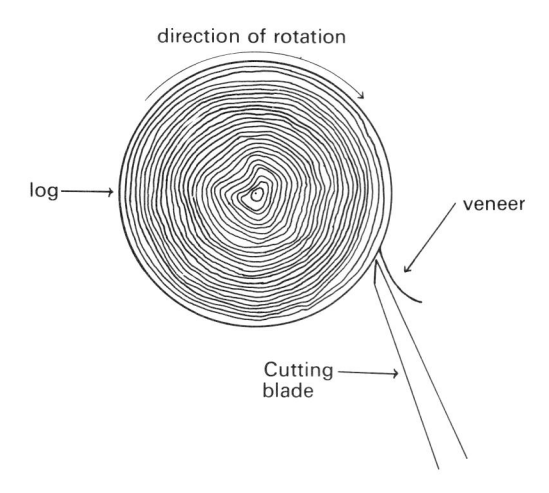

Figure 1.1 Rotary cutting of veneers

Plywood has several advantages over natural wood, namely, greater dimensional stability and greater uniformity of strength. It is, however, somewhat more expensive than ordinary timber. Fibreboard is a product made from fibres of wood and, as with particle boards, wood of low

4.7

quality is utilized and put through a pulp preparation, sheet formation, pressing and finishing, treatment not unlike that of manufacturing paper and board. The pulp is usually of a mechanical or semi-chemical nature, and the processes of sheet formation are wet felting and dry felting. Certain materials may be added to the pulp before felting to improve water resistance, strength, etc.

Glue is added in the proportion of one to four per cent on dry fibre weight in the dry process in order to bond the material together. The air felting process can confer some advantages with regard to water consumption and pollution, and for this reason is sometimes preferred over the wet process. According to the pressure applied and the resulting density, fibreboards are classified into insulation boards, medium-density boards, and hard boards. Wet-felted hardboards are the ones most commonly used in packaging, and these may sometimes be subjected to heat treatment to improve strength and water resistance, and to oil tempering which, in addition, gives better resistance to abrasion and weathering. Finally, it should be mentioned that wood is one of the primary packaging materials and is (or was) most commonly used first in packaging developments in any particular country. In most developing countries it is going to remain one of the most important packaging materials for a long time to come.[3] The most obvious reasons for this are:

Wood is often readily available as an indigenous raw material.

The construction and assembly of wooden crates and boxes is a labour-intensive operation and most developing countries have plenty of cheap labour.

Wooden packages have a comparatively favourable ratio of cost/strength, especially for shipment by sea freight.

Wooden crates and boxes are easily tailor-made in small-quantity shipments, whereas other shipping packs, e.g. corrugated boxes, require longer runs to be economical.

There are, however, a number of negative factors to consider in connection with the use of wooden packages:

The customers in industrialized target markets are less and less prepared to receive, handle, open and dispose of wooden packages.

Wooden packages have an unfavourable ratio of cost/weight and shipping space required.

The effective use of wooden packages requires good technical know-how concerning methods of constructing and assembling wooden boxes and crates, the right material specifications to use, etc.

A very important aspect is frequently overlooked by exporters in developing countries. Most business is done on an f.o.b. (free on board) basis and not enough attention is paid to the effect of packaging weight and volume on the final landed cost of the products sold, unless the buyer specifies the exact method of packing. The result is often excessive freight costs, with subsequent negative effects on the competitive position.

It is therefore felt that much more attention should be paid to the structural design of wooden packaging in general.

Being a replenishable product, good husbandry in silviculture will, in fact, result in a continuing supply to meet all our needs.

REFERENCES

1 *Enc. Brit.*, Vol. 19.
2 *Enc. Brit.*, Vol. 19.
3 "Wood as a packaging material in the developing countries," a study by B. Hochart for UNIDO (1972, 111 pp. from UNIDO, Felderhaus, Rathausplatz 2, A-1010 Vienna, Austria or from distributors of UN publications).
4 BS 1133.
5 *Enc. Brit.*, Vol. 19.

CHAPTER 2

Timber and Plywood Cases and Crates

M. Rawson

The very earliest types of shipping containers were bales and bilged casks, made to facilitate manhandling. The Industrial Revolution, and the resulting improvements in transport, the building of railways and better roads led to the development of wooden boxes and crates as the first modern shipping containers. Timber was then plentiful, and so inexpensive that wooden box makers, like most other users of wood, demanded high grades and ignored inferior material. Little consideration was given to wastage, either in forestry or in box making, and even less to questions of design to give maximum strength for minimum use of material.

Today the cheaper grades of timber are used for box making, and questions of design and cost are of major importance. Wooden shipping containers are used only when their strength characteristics are needed to carry their contents safely to their destination, or when the value of the contents warrants the extra protection afforded.

This chapter will deal only with cases and crates used for overseas shipments of industrial products, thus enabling the information in **BS 1133** to be supplemented by consideration of established export practice, and the principles of packing economy.

Basic types

Wooden shipping cases came into general use after the introduction of lifting appliances at the major ports in the United Kingdom and overseas. Four main types were developed, which form the basis of today's wider range, and which can justly be called the foundation-stones of export packaging practice.

TIMBER AND PLYWOOD CASES AND CRATES

Crate (skeleton crate) (figure 2.1*a*)

Open slatted construction, affording purely physical protection to one or more self-supporting items not liable to climatic deterioration in the course of transit.

Flush-side case (merchant case) (figure 2.1*b*)

Close-boarded box construction without battens, combining physical and climatic protection (with kraft-union liner) for "easy-type" distributed loads up to 100–150 kg (250–300 lb).

Battened case (belted case) (figure 2.1*c*)

Close-boarded assembly reinforced with end and girth battens (belts or cleats) for "average" and "difficult" loads beyond the safe capacity of the flush-side case in terms of volume and/or weight.

Skid-base case (figures 2.1*d* and 2.2*d*)

Closed case with crosslaid base-boards or planks and longitudinal skids or "runners" fixed underneath to support heavy loads, and (originally) to enable movement of the case on rollers or snow if necessary.

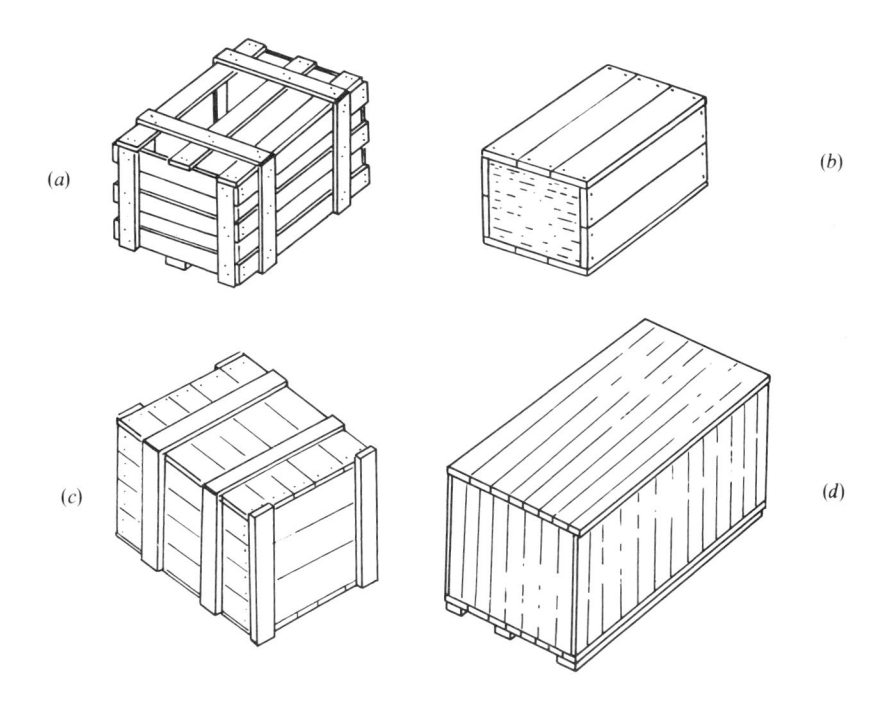

Figure 2.1
(*a*) Skeleton crate
(*b*) Merchant or flush-side case
(*c*) Battened case: end and girth bands
(*d*) Sheathed wood case with interior framing

WOOD AND TEXTILE-BASED PACKAGING

Present-day range

Just as lifting appliances brought about the change from casks to cases, so other factors have intervened to influence export packing-case design. Thus the shortage of imported softwood in the Second World War and the increasing cost in post-war years have compelled economy in its use, and paved the way for the wider use of plywood or other sheet materials forming the whole or part of the case. In addition to this, the increased use of Container Traffic has reduced the need for handling and therefore the strength needed in the pack, when these are being delivered on a door-to-door basis. It should, however, be remembered that like airfreight, these packs often face severe journey hazards across country before they reach their final destination, and goods should be protected accordingly.

Most export manufacturers seek to neutralize rising production and freight costs by comparable economies in packing, and this has helped to accelerate the use of lighter and cheaper packaging materials. It is important to note that shipping charges are based on the actual gross weight of case and contents, or on weight which used to be calculated at 40 cubic feet per ton, whichever is greater. The metric equivalent is 1000 kg (= 1 tonne) per cubic metre. Volume is reckoned on "outside dimensions", so that a battened wood case normally adds four thicknesses of timber to each dimension—length, width and depth.

Taking the four traditional types in turn, an abridged list of containers in current use shows how many of them have been developed from the parent type, and the particular purpose which they serve.

Crate (skeleton) (figure 2.1)

The term *crate* is used in the United Kingdom to indicate an open slatted construction as distinct from a closed *case*. Purely physical in function, it does not offer much scope for improvement, and little for economy except by widening the spaces between the crate members. Lining with waterproof paper does not convert a crate into a closed case, but large crates may be lined with thin plywood, hardboard or fibreboard as a cheaper alternative to a closed wood case when the load is suitable (example: large open tank) (figure 2.2b).

Types based on flush-side case

(a) *With battened ends only* (figure 2.2a(1))

Vertical end battens enable greater depth and loading capacity with relatively small increase in shipping dimensions and cost. If the end battens can be fixed inside instead of outside, space and volume is conserved (example: cans and drums).

(b) *With framed or panelled ends* (figure 2.2a(2))

4.12

TIMBER AND PLYWOOD CASES AND CRATES

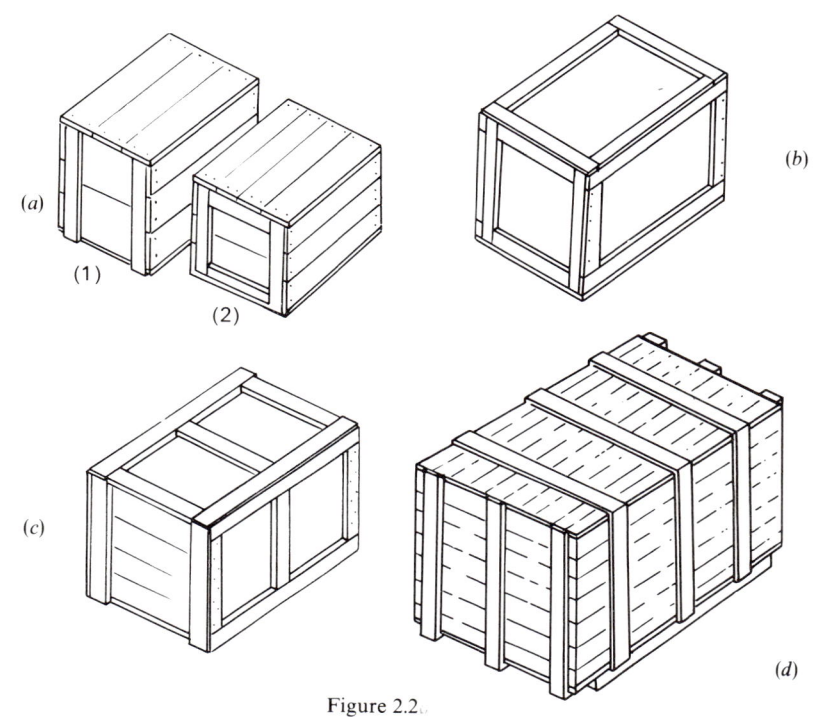

Figure 2.2
(a) (1) Battened ends (2) Framed ends
(b) Battened plywood case
(c) Plywood composite case with wood ends
(d) Skid-base case, outside battened type; with sling boards on underside of skids to assist lifting

Fully-framed ends allow some increase in both width and depth, with more-secure nailing around the ends into the two thicknesses of timber. These cases permit greater convenience in handling, except by slings, but are liable to be "grabbed".

Note: The application of tensional steel strapping girthwise to flush-sided cases permits some addition to the net weight load (alternatively an increase in the size of the case or a reduction in the thickness of the timber used).

(c) *Metal-edged plywood case*

Plywood panels assembled by means of riveted metal edges instead of wooden battens form a light but tough flush-side case. They provide maximum economy of freight space for evenly distributed loads, normally prepacked or cushioned. They are often used in conjunction with softwood-base, skid-base or pallet-base for easy handling, and are particularly suited for Container Traffic on a door-to-door basis.

(d) *Collapsible plywood case*

Economy of weight and shipping volume is combined with the ability to be stored flat and erected at the point of packing. Various constructional methods (mainly patented) include interlocking sections, metal fasteners and flexible hinge materials. They provide overall economy for "easy/average" standard packed for regular shipping, and can be used with softwood and pallet bases.

WOOD AND TEXTILE-BASED PACKAGING

(e) *Wirebound wood box*

This is a three-piece container supplied in flat case-sets, comprising wired body ("mat") and separate ends. Its wirebound construction strengthens relatively thin boards used in assembly. There are two main types, both convenient for storage prior to packing.

(f) *Frame-end fibreboard case*

This is a three-piece lightweight case consisting of two wood-framed solid fibreboard ends and wrap-around solid fibreboard body fixed to end frames with coated lath nails. It is suitable for export of cartoned products packed in unit quantities up to 40/45 kg weight.

Types based on battened case (figure 2.1c)

Because of their strong joints and rigid construction, together with great flexibility in size and load-bearing capacity, outside-battened cases are used for a wide range of export goods. The girth battens ensure that the body of the case is above ground-level, thus encouraging safer handling and stowage.

Extra girth battens and/or diagonal braces are introduced to stiffen the assembly where necessary and, when the nature of the contents permit, current practice frequently uses thicker battens and reduces the thickness of the sheathing timbers, thus reducing the overall timber content and the weight of the case. With the coming of "stress-graded" timber, this may well become more common, for, although the timber may cost more initially, it may be possible with good design and knowing accurately its structural strength to reduce the amount required. Large cases require the addition of lid bearers (crusher bars) to resist the compressive action of slings. With the exception of fork-lift battened cases the other types developed from the battened wood case are modifications designed to secure economy of cost, tare-weight and shipping volume.

(a) *Fork-lift battened case*

A traditional case with battens encircling the body, but the bottom battens are thickened to enable handling with fork-lift trucks. Side battens are lengthened to permit nailing into ends of base battens for extra strength.

(b) *Battened plywood case* (figure 2.2b)

The battened plywood case provides a lighter and cheaper alternative to the battened wood case for "easy type" loads preferably cartoned or cushioned. It has wood-framed plywood panel construction, using 4 mm to 6 mm plywood according to size and weight. Styles vary for loads up to about 180 kg. (The use of thicker plywood enables heavier loading, but cost advantage over wood is usually lost unless the cases can be produced from standard plywood sheet sizes with little or no waste.)

(c) *Plywood composite case* (figure 2.2c)

The plywood composite case combines the sturdiness of battened wood ends with the economy of framed plywood lid, bottom and sides. Double nailing into the wood end and vertical end battens gives more rigidity than panelled plywood ends, thus enabling safe and economic use for "average type" loads such as gas and electric appliances. The load range is from 135 to 270 kg with 4–6 mm plywood.

Types based on skid-base case

There is a sharp distinction between battened wood cases for loads up to 250 kg and 2 m^3 and skid-base cases for heavier loads. In practice, however, skid-base battened cases (figure 2.4) are often built for loads of 5 tonnes or more, and are seldom used for loads under 1 tonne, but the following types are constructed on engineering principles, and are therefore used when the cost and the need is justified.

(a) *Interior-framed (sheathed) wood case* (figure 2.2d)

A large wood case consisting of substantial frame members designed to withstand the load to which the case is subjected. Vertical wood sheathing is applied to give strength and complete covering on the outside of the frame.

(b) *Interior-framed plywood sheathed case*

This is constructed with internal framing members in the same way as the sheathed wood case, but plywood is substituted for the sheathing. The use of large panels, as opposed to narrow boards laid vertically, enables quicker assembly, gives better moisture resistance because of the absence of joints and, when 9-mm plywood is used in lieu of 19-mm boards, there is some reduction in volume.

Both these cases may be constructed with either a skid-base or a sill-type base, depending on the nature and disposition of the load. Skids are fixed underneath the base in conjunction with a flat deck or floor; sills are raised above floor level to support the load and permit the use of light rubbing-strips beneath instead of skids. The factors governing the need for each type may be defined thus:

Sill-type.—Whenever a supporting point or points of the object being packed are above the floor line, and proper bearing can be made so that the cubic displacement is conserved, a sill-base should be used providing that no weight is borne by the floorboards (figure 2.3).

Skid-type.—The skid-type base should be used for heavy loads when the supporting area bears on a large flat surface or is concentrated at a few points on the floor line (figure 2.4).

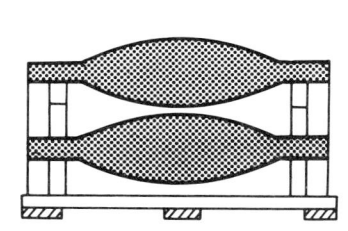

Figure 2.3 Typical load application for sill-type base, end view

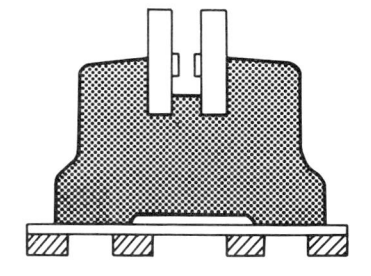

Figure 2.4 Typical load application for skid-type base, end view

Waterproof linings

Few export cases can claim to be waterproof in their own right. Shrinkage occurs across the boards of butt-jointed wood cases, and the joints of battened plywood cases may be loosened by various forms of handling during transit. It is therefore necessary to introduce a physical barrier strong enough to give "raincoat" protection to the contents, more especially if they are liable to corrosion or climatic deterioration.

Many special barriers are employed for packing goods, and for lining wooden cases and other types of shipping containers, particularly where packing for the Defence Services is involved. These are produced to the Defence Specifications and are referred to by the appropriate number. The most usual ones are to DEF 1242—a waxed wrapping paper and DEF 1236—a complex laminate of foil, cotton scrim, cellulose film and paper. The laminating material is waterproof (usually a hot melt) and the foil is coated with a heat sealing agent. A creped kraft paper coated with a special wax compound on both sides (DEF 1238) can be moulded around irregularly shaped articles, as can a mouldable waxed grease-resistant wrapping made by coating a laminate of cellulose film and cotton scrim with wax (DEF 1237). Kraft paper coated on one side with grease-resistant lacquer (DEF 1316) is also used. In addition, various types of coating are applied to kraft paper and used instead of union kraft when desired. The most common coatings are based on polyethylene, polypropylene and PVC. However, for commercial use the lining materials most commonly used are:

(a) *Waterproof paper (kraft union or pitchpaper)*

Two thicknesses of heavy kraft paper with a centre core of bitumen, stapled or adhered to all interior faces of packing-case (laid between framing and sheathing of interior-framed cases).

(b) *VCI—coated waterproof paper*

Adds chemical action of inhibitor to the physical protection from water afforded by the case liner.

(c) *Bag liners*

Bags with sealed closure, which conform to the inside dimensions of packing-case; made in various materials including creped and plain waterproof papers, special laminates and polyethylene.

(d) *Tinplate linings (tin shells)*

Enclosure shells with hermetically sealed lid and soldered joints, usually made with tinplate, terneplate or zinc and fitting snugly inside the packing-case. Mainly used for tropical cargo, moisture-sensitive goods, scientific instruments and chemicals liable to seepage. The use of desiccant material within the shell needs to be considered.

TIMBER AND PLYWOOD CASES AND CRATES

Closure of cases

Returnable wood and plywood cases are often fitted with closure devices to avoid damaging the lid in conditions of re-use. Methods include the fitting of captive-screws, palm bolts, strap bolts and various types of fastener, many of which are patented.

The lids of export cases are normally nailed (using nails two and a half times the thickness of the lid), and care should be taken that battens are "locked" where they butt. Lids should be screwed if the case is to be produced to H.M. Customs at time of shipment.

Principles of packing economy

(a) *Use of timber*

In the past it has often been necessary to produce case-boards by "deep cutting" from 2″ or 3″ imported deals, with a consequent slight loss in thickness arising from the saw-cut. Thus four boards resawn from a 3″ deal would be $\frac{11}{16}$ in thick, and insistence on the use of full $\frac{3}{4}$″ boards would cost correspondingly more. Normally the "scant" sizes serve equally well. Now that the majority of imported timber used in the packing-case trade is being supplied in metric sizes, it is advantageous to make the cases to metric sizes. For example, it is difficult to get four 100 mm (the metric equivalent of 4 in) boards, to cover 16″ in width, and the same problem often occurs on the lengths where the metric equivalent is often shorter than the old "Imperial" equivalent.

(b) *Arrangement of contents*

Wood or plywood cases of near cubical shape usually have a lesser superficial area (and use less material in their construction) than shallow elongated cases of the same capacity. By applying the three packing-factors in reverse order to the three dimensions of the units being encased, the cost and weight of the case can often be reduced, as the example (Table 2.1) worked out in feet and pence will show:

Table 2.1. Case-load of 144 cartons 6 × 4 × 3 in

Packing factors	Case dimensions (internal)	Cost at 10p per ft²	Weight of case
6/6/4	36 × 24 × 12 in	£3·20	47 lb
6/4/6	36 × 16 × 18 in	£3·00	45 lb
4/6/6	24 × 24 × 18 in	£2·80	42 lb

Comparison can usually be made by adding together the three internal

4.17

WOOD AND TEXTILE-BASED PACKAGING

case-dimensions, in this instance 72, 70 and 66 in respectively. It should be remembered that this principle does not apply to one-piece fibreboard cases, where the superficial area is affected by the relative size of the top and bottom flaps.

(c) *Economy of larger case-load*

An increase in the number of units packed per case will bring about a reduction in the cost, providing that the case can be enlarged without the need for thicker timber or plywood. The most graphic way of demonstrating this principle is by considering two cases each $0.50 \times 0.50 \times 0.50$ m which require two lids and two bases, and one case $0.50 \times 0.50 \times 1.0$ m which requires only one lid and one base of identical size (figure 2.5). The capacity has been doubled by deepening the walls.

The same principle holds good when extra units are packed in any direction (length, width or depth), and has equal truth when, for example, two appliances are packed side by side with a partition between them using three "ends" instead of four. The economy also extends into the realm of ocean freight charges, as the double-unit case incurs the addition of only one set of battens for volumetric reckoning instead of two (Table 2.2).

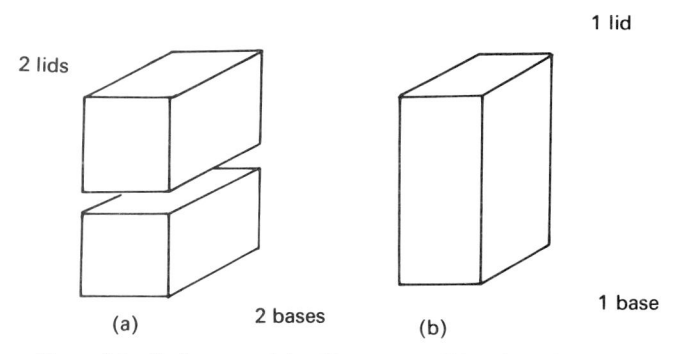

Figure 2.5 Saving material. (*b*) saves one lid and one base

TIMBER AND PLYWOOD CASES AND CRATES

Table 2.2. Physical data (main types)

Type and thickness of export container	Shipping dimensions Addition—mm (in) to			Average loading kg (lb)	Type of load
	Length	Width	Depth		
16 mm (⅝ in) skeleton crate	64 (2½)	64 (2½)	64 (2½)	75/160 kg (170/350 lb)	self-supporting
19 mm (¾ in) skeleton crate	76 (3)	76 (3)	76 (3)	90/180 kg (200/400 lb)	self-supporting
25 mm (1 in) skeleton crate	100 (4)	100 (4)	100 (4)	160/270 kg (350/600 lb)	self-supporting
19 mm (¾ in) merchant case	38 (1½)	38 (1½)	38 (1½)	50/90 kg (120/200 lb)	compacted
25 mm (1 in) merchant case	50 (2)	50 (2)	50 (2)	70/135 kg (150/300 lb)	compacted
19 mm (¾ in) frame-end case	75 (3)	38 (1½)	38 (1½)	70/135 kg (150/300 lb)	compacted
25 mm (1 in) frame-end case	100 (4)	50 (2)	50 (2)	90/160 kg (200/350 lb)	compacted
16 mm (⅝ in) battened case	64 (2½)	64 (2½)	64 (2½)	110/160 kg (250/350 lb)	difficult
19 mm (¾ in) battened case	75 (3)	75 (3)	75 (3)	110/205 kg (250/450 lb)	difficult
25 mm (1 in) battened case	100 (4)	100 (4)	100 (4)	205/365 kg (450/800 lb)	difficult

In practice 25-mm boards are often used for cases carrying 1 tonne.

Plywood cases

4 mm battened	50 (2)	50 (2)	50 (2)	70/110 kg (150/250 lb)	distributed
6 mm battened	56 (2¼)	56 (2¼)	56 (2¼)	90/205 kg (200/450 lb)	distributed
4 mm ply-composite	75 (3)	50 (2)	50 (2)	110/180 kg (250/400 lb)	distributed
6 mm ply-composite	80 (3¼)	56 (2¼)	56 (2¼)	180/270 kg (400/600 lb)	distributed

NOTE (for Table 2.2). The average loading figures are representative of normal (civilian) exporting conditions, under which proper transit, handling and port facilities are envisaged. Reference should be made to BS 1133 or DEF 1303 when transport, freight or storage conditions are known to be abnormal.

4.19

CHAPTER 3

Wooden Casks and Plywood Kegs

F. A. Paine

Wooden casks and plywood kegs include three main types of construction:

1. Wet cooperage—intended for the transport of liquids. The wooden beer cask is probably the best known example.
2. Dry cooperage—made from wooden staves and suitable only for dry or near dry products.
3. Plywood barrels and veneer casks—straight-sided cylindrical drums manufactured from sheet plywood; and from a double row of veneer staves respectively. Veneer casks are no longer manufactured in the United Kingdom.

The trade terms are often quite distinctive, and the following definitions will assist in understanding the subject.

Wet and dry cooperage

A *wet cask* is a round bilged coopered wooden container made of staves, heads and hoops, and constructed to hold liquids without leaking.

A *dry cask* is a round bilged coopered wooden container made of staves, heads and hoops, but constructed to hold powders and commodities other than liquids.

A *stave* is the name given to the arched vertical components forming the walls or sides of a cask, which are held together with *hoops*. The *hoops* are made of steel or wooden strips passing round the circumference of the cask.

The wooden disc forming the end of a cask is called the *head*. The *bilge* is defined as the largest circumference of the cask, located at the horizontal plane through its centre.

Plywood barrels

Plywood barrels are cylindrical containers, each made from a plywood sheet butt-jointed and held in position either by means of a plywood strip on the inside, or a metal strip on the outside. The head and bottom are discs of plywood.

The reinforcing plywood bands riveted to the outside of the barrel, normally at each end, are known as *body bands*. Intermediate bands may be used if necessary.

The plywood band riveted inside the body wall and set down from the end to form a seating for the head is called a *lining hoop*. They are generally positioned one at each end of the body.

After inserting the bottom (a disc of plywood which is seated on the lining hoop) another plywood band is riveted inside the body wall against the ouside of the bottom head. This plywood band is called the *closing hoop*. The top closing hoop is supplied loose, for fixing by the customer after the barrel has been filled.

A lifting ring of metal is fixed to the outside of the lid to enable it to be removed. The ring is normally secured by a clenched staple.

Production methods

Wet and dry cooperage

The first operation is to shape the staves using the axe, knife, hollow knife and jointer in that order. Having shaped the requisite number of staves, they are placed side by side in a raising hoop,* then further truss hoops are placed on the staves to hold them firmly, these then being placed over a cresset* of wood chips and timber until the staves are warm enough for bending. When the staves are bent and while the cask is still warm, the chimes are made by using the adze to slope the tops of the staves, the inside then being levelled with a chive* so that a perfect groove may be cut with a croze.*

---

* A *raising hoop* is usually a thick ash hoop of the required circumference for the finished cask, i.e. it acts both as a tool and a template.

A *cresset* is an open fireplace rather like a watchman's fire.

A *chive* is a tool with a curved blade which is swung around the inside of the chime and cuts the staves level.

A *croze* is a tool which looks similar to the chive but which holds three small blades, two of which cut the wood while the third clears the groove.

The next operation is to make the heads, which is done by flattening and jointing the heading timber on the heading jointer, after which the pieces are dowelled together forming a rough square. Next the grooves

WOOD AND TEXTILE-BASED PACKAGING

are compassed, and the rough heads are marked for sawing into a circle. When sawn to the round shape, a bevel is cut with the heading knife. The heads are then placed in position, the hoops are made and driven into place, and the cask is completed. This method is used with minor variations for the making of all bilge casks in Great Britain, whether for wines, spirits, beer, cider or dry goods.

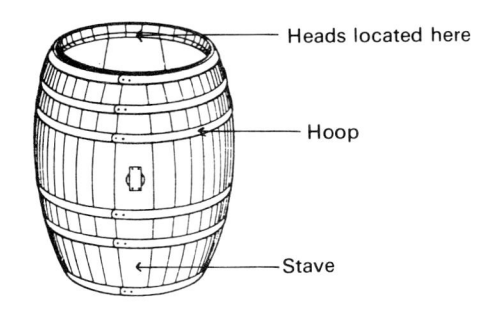

Figure 3.1 A wet cask.

Plywood barrels

The components are cut to size by power-driven saws. It is therefore possible to produce the containers in an almost unlimited variety of sizes, without special tool charges. Broadly speaking, the upper limit of the size range is determined by handling and weight considerations, and the lower limit by the minimum diameter to which the plywood can be bent.

The components are first dimensioned by power-driven saws, and bent to shape after steaming. They are then ready for assembly. The vertical body joint (or joints if more than one panel is used) is secured by a metal or plywood jointing strip, to form the *body cylinder*.

Assembly after bending is by riveting. Bifurcated rivets (which have two legs which penetrate the plywood and clench on the inside) are normally used. This makes mass production quick, easy and simple, and enables construction to be altered as necessary to meet specific consumer needs. The strips of plywood which are to act as lining hoops and body bands are then placed into position. The bottom is inserted, followed by the bottom closing hoop. The barrel is then closed by the lid, after the loose top closing hoop has been placed inside. Apart from the top closing hoop which is left loose for use by the customer after packing, all components are riveted.

4.22

Types and styles of container

Wet and dry cooperage

Wine casks.—Wine casks are all made in the country of origin from native or imported timber, and it would be fair to say that almost all the better-quality wines are shipped in oak casks, and the cheaper wines in chestnut casks.

Ninety per cent of port and sherry is shipped in oak casks and 90% of vermouth in chestnut. Bordeaux, Burgundy, Rhone wine, etc., vary, the better wines being shipped in casks of oak and the cheaper in chestnut casks. The casks in general use for sherry are the butt (110/112 gal capacity = 500 litres) and the hogshead (54/55 gal capacity = 250 litres). For port the corresponding sizes are the pipe (116/118 gal capacity = 53 litres) and the hogshead (58/59 gal capacity = 265 litres).

Vermouth, Bordeaux, Burgundy and other wines travel in hogsheads, that for Vermouth holding 54/56 gal (250 litres) and the others 46/48 gal (215 litres).

Spirit casks.—All spirit casks are made of oak and with the exception of brandy, for which French oak is used, gin, rum and whisky are all filled in American oak.

Whisky and gin casks have the same measurements as sherry casks. Rum is usually imported from the West Indies, principally from Jamaica in puncheons (110/115 gal capacity = 520 litres) and in hogsheads (54/58 gal capacity = 255 litres) and more recently in 40-gal barrels (180 litres).

Brandy is almost always imported in hogsheads of 62/63 gal capacity but occasionally in quarter casks of 31/32 gal capacity.

Beer casks are made in these capacities, butts (108 gal capacity = 500 litres), hogsheads (54 gal = 245 litres), barrels (36 gal = 165 litres), kilderkins (18 gal = 82 litres), firkins (9 gal = 41 litres) and pins ($4\frac{1}{2}$ gal = 20 litres).

Dry casks are made to any measurement required, some to hold a specific weight, some a particular volume. It is worth noting that very few dry casks are made nowadays, and that beer casks are being replaced by metal containers.

Plywood barrels

There are two types:

(a) *Those which employ an unsealed vertical joint.* These have either a metal body jointing strip on the outside of the body wall, or a plywood jointing strip on the inside. There is a tendency toward the former if bag linings are to be used, and the latter if direct packing is intended.

WOOD AND TEXTILE-BASED PACKAGING

(b) *Those which have a sealed vertical joint and have a sealing gasket between the body and the ends.* These are mainly employed for direct packing of powders and similar products which would "sift" through unsealed joints.

In each instance, the number and thickness of the components may be varied according to the size of the container, its content weight, and whether it is for home or export use.

Types of closure and methods of closing

Wet and dry cooperage

A wet cask is always closed by a wooden shive* driven home with an adze, and smoothed off flush with the bung stave.

For solid products the two top hoops are removed so that the head is loose enough to be lifted. After packing the head and hoops are replaced.

Plywood barrels

The type of closure generally employed is the closing hoop. This may be secured by: (a) nailing; (b) wing nuts (for returnable use); or (c) staples (driven by a hand-stapling machine).

Other methods of closing which are employed according to circumstances are:

(a) Griffiths fasteners—a special fitting, sometimes demanded by Government Departments.
(b) Toggle fasteners—these are used in conjunction with double lids, the under lid being to the internal diameter of the drum, and the top lid to the external diameter.
(c) Swivel bars—metal bars are fixed to the lid, and located to engage corresponding slots in the body wall. If required, they can be secured in the locked position by a screw.
(d) L-shaped lugs—secured by a bolt through holes in the body band.

In all of these instances, the lid must be firmly and evenly seated when the drum is closed.

(e) Body slots—slots cut in the body wall to enable tension strapping to fit flush over the heads.

* A *shive* is a wooden stopper for a bung hole; a cork stopper is called a *bung*.

4.24

CHAPTER 4

Hardboard and Softboard

M. J. Ford and H. P. Mostyn

Introduction

Hardboard and softboard—known generally as *fibre building boards*—have been relatively unexploited as packaging materials until quite recently. Detailed assessment of their potential in the packaging industry has, however, established their particular capabilities and is leading to their use on a wider scale.

These wood-based sheet materials are relatively cheap (occupying a position on the cost scale between timber and plywood on the one hand and corrugated board on the other), yet offer a useful range of properties. They are appropriate for use in situations calling for strength and protection, but where economy is an essential factor. This chapter summarizes the main types of board and their properties, indicates a range of handling and use techniques, and gives examples of packaging applications.

Board types and manufacture

Fibre building boards, defined in BS 1142,[1] are a family of sheet materials manufactured from wood fibres, the basic strength and cohesion of the boards being derived from the felting together of the fibres themselves and from their own inherent adhesive properties. (Bonding, impregnating or other agents may, however, be added during or after manufacture to modify particular characteristics.) Thus fibre building board is wood which has been reconstituted from its fibres, as opposed to the reassembly of timber by the bonding together of veneers or particles.

This process offers the significant advantages of over-all consistency, absence of surface defects, and a higher resistance to fungal attack and timber pest infestation (since most of the sugars and starches are extracted from the fibres during manufacture). Furthermore, laboratory tests in which samples of standard hardboard have been placed in contact with milk powder show that food is not tainted by odours from the hardboard. These test results are supported by practical experience of tea shippers who often use hardboard tea chests to package their product.

The manufacturing process is described in detail elsewhere.[2] Briefly, however, forest thinnings and sawmill waste are cut to chips. The chips are softened by steam under pressure, then reduced to fibres, either by grinding between disc knives or by ejection from a steam-heated pressure vessel. The fibres are added to water to form an aqueous pulp, and additives such as rosin size and aluminium sulphate may be added at this stage. The pulp is laid on to a slowly-moving Fourdrinier-type wire mesh and, as water is removed by drainage, suction and thicknessing rollers, the fibres interlock or "felt" before approximate sheet lengths of this "wet-lap" are cut by a rotary knife. The wet lap is fed on to racks through a series of drying ovens, when the remaining water evaporates to leave air cells within the board, thus producing the *low-density softboard*. Because it is chiefly used in the construction industry for its thermal insulating properties, this material is commonly called *insulating board*. It has a density normally about $200 \, kg/m^3$ and is commonly produced in 13 or 19 mm thicknesses. Thicker boards (up to 50 mm) can be produced, although this may involve the lamination of separate sheets. Common sheet sizes are 1200 mm × 2400 or 2745 mm. Softboards, by virtue of their low density, tend to be absorbent and have a low bending strength. It should be noted, however, that they have a high resistance to compression in the plane of the board.

Hardboards and medium boards are much denser, and hence considerably stronger. During manufacture, the cut lengths of "wet-lap" are transferred to wire mesh carriers, which are then loaded into a multi-daylight press (which may be as long as 7200 mm and as wide as 2135 mm). A heat and pressure cycle forces the remaining water out through the wire mesh (thus impressing the familiar "screen" pattern on one face of the board), while the polished surface plates produce the very smooth glossy face. Further heat treatment in drying ovens follows, and the final stage before trimming and cutting involves humidity treatment to re-establish an appropriate moisture content in the board. Variations in the nature and length of the fibres used, and in the variables of the press cycle, significantly affect the physical properties of the finished product.

Thus, hardboard from one manufacturer may be better suited to applications calling for high strength characteristics or industrial painting,

HARDBOARD AND SOFTBOARD

while that from another producer may be better suited to die-stamping or moulding techniques.

Standard hardboard is normally in the density range 800–960 kg/m^3, with a colour ranging from straw to dark brown. It is widely available in thicknesses from 2 mm to 6·4 mm and in sheet sizes 1200 mm × 2240 or 2745 mm. Cut sizes—even of quite small pieces—can be obtained in large quantities at costs which compare reasonably with those for full sheets. Standard hardboards have high bending strength and are particularly strong in their own plane, the denser boards having the highest strength levels and the most consistent machineability. The smooth face has a higher resistance to water absorption.

Tempered hardboard has high strength characteristics and greater resistance to water and abrasion. It is produced by impregnating the board, after pressing, with drying oils followed by further heat treatment. Thicknesses range from 3·2 mm to 8 mm, and sheet sizes are generally as for standard hardboard.

Medium board comes in two density ranges. The lower (Type LM) lies in the range 350–560 kg/m^3, while the higher (Type HM, or "panelboard") ranges from 600–800 kg/m^3. Although less dense than standard hardboard, it is normally thicker, being available in thicknesses from 6 mm to 12 mm.

Processing and use techniques

Cutting

Hardboard and softboard are relatively easy to work by normal woodworking processes. For low-density softboards, sharp knives are recommended for clean cut edges. Hardboard and medium boards can be cut with most common panel-sawing equipment, jig saws and routers. Softboards and some hardboards can be guillotined, whilst irregular shapes or curved outlines in hardboard can be obtained by die-stamping or punching.

Most of these techniques produce cuts at right angles to the surface of the board. Many useful effects can, however, be produced by cutting hardboard with bevelled edges. A simple version of this, to produce mitred corners, consists of using a saw blade canted at an angle of 45°. More sophisticated mitred edges can be produced by the use of V-grooving techniques. In this process a V-shaped notch is cut to a carefully-controlled depth into the board by using a specially-shaped cutter block. This block removes material part of the way through the board, leaving a very thin layer of fibres (or in some cases a laminated layer of another material such as PVC film) on the side remote from the cutter. This final layer maintains the integrity of the board and acts as a hinge when the board is folded along the line of the V-groove. Before folding, glue is applied to the groove to produce a permanent rigid joint (figure 4.1a). The creation of such a change of plane gives a significant improvement in stiffness without the need for a supporting framework.

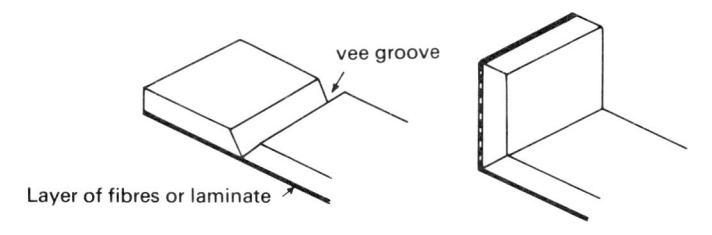

Figure 4.1a Controlled machining by V-shaped cutter block leaves constant thickness of fibres or flexible laminate, which acts as hinge whilst panel is glued and folded

Bending and moulding

Hardboard can be bent through quite small radii by a number of simple techniques. The simplest consists of bending dry board by pressure alone, but this cannot produce radii much below 300 mm (for 3·2 mm thick board), and the curved board requires fixing to a former. The addition of moisture effects an improvement. Soaking hardboard in water for several hours facilitates the formation of bends with radii of curvature down to about 125 mm (3·2 mm thickness).

Even better is the application of heat and pressure during the bending of a soaked board; this enables permanent right-angled rigid bends to be made. Using a heated platten press and suitable moulds, the bending technique can be extended to produce complex shapes such as inner door panels for the motor industry. The same moulding process can be used for the production of domestic and industrial trays.

Creasing and folding

During 1974/75 it was appreciated that some grades of standard hardboard could be creased to form a flexible hinge which can be folded. The process is similar to that used for many years to crease solid fibreboard. The fact that a rigid material such as hardboard can be creased, yet retain considerable strength[4], is of considerable importance for packaging applications.

Research and development is underway both by FIDOR/Pira and in several industrial companies. Packages incorporating the technique are being marketed but it is too early to assess the commercial potential fully.

Jointing

When panel products are used to make containers, the method of fastening the sheets together is a major consideration. The simplest methods for hardboard are the use of frameworks and mechanical fixings. Nailing or stapling of hardboard to softwood structures is common, as also are methods of riveting hardboard sheets to preformed angle strips of materials such as tin-plate. A further technique being increasingly used for produce trays and boxes is the use of angled corner stitching in which a wire staple is applied to two panel sections held at right angles to each other—the stitching being possible either with or without a timber corner support (figure 4.1b). All the above methods use through fastenings, and do not rely on the inherent strength of the bond between individual fibres in the hardboard. However, the perforation of the sheet by mechanical

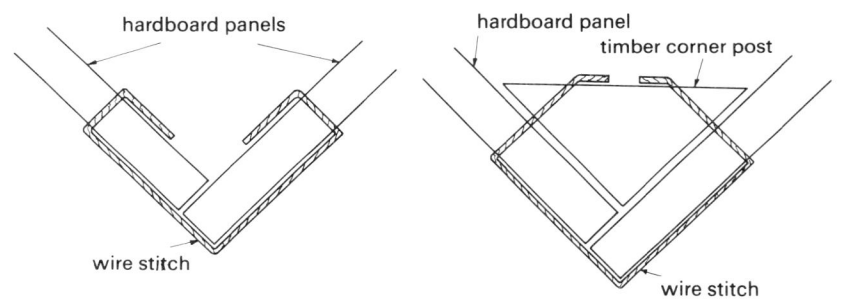

Figure 4.1b Hardboard panels fixed with corner stitch—with and without corner-post attachment.

fasteners may weaken the board. Furthermore it is advisable that the equilibrium moisture content of both the hardboard and framework be taken into consideration.

Surface fastenings make use of the strength of hardboard without producing such perforation weaknesses. In addition, if the fastening is itself flexible, a fold-flat property can be designed into the finished article. Surface fastenings employ adhesives in conjunction with metal angle strips, fibre board angle strips, or tapes of reinforced paper, textile or plastics. The metal strips are usually rigid, whilst the fibre board and the tapes are flexible.

Packaging applications of hardboard and insulating board

Since an appreciation of the characteristics of fibre building boards as packaging materials is a relatively recent development, it is clear that their application in this field will be subject to continuous and rapid change. A considerable body of information is, however, available and

WOOD AND TEXTILE-BASED PACKAGING

reference may be made to a range of established uses. Currently packaging applications fall into a number of well-defined areas.

Case and pack reinforcement

The strength and rigidity of hardboard can be easily employed to provide reinforcement to conventional styles of case. Unless the contents are themselves sufficiently rigid (canned goods, for example) the case may require considerable strength to withstand stacking in the warehouse and during transit. Hardboard or softboard are cost-effective materials for reinforcing corrugated cases, both to protect the products and improve stack stability. Reinforcement can be in the form of side panels or dividers. Tests[3] and service experience show that these can provide considerable improvements in compression failing load (CFL) and enable relatively lightweight cases to be used.

Table 4.1. Effect on compression failing load (CFL) of corrugated fibreboard case by addition of reinforcing pieces made of hardboard
(internal case dimensions 460 × 280 × 380 mm)

Case make-up	CFL at 23°C and 50% relative humidity kN	lb	Estimated cost per 50 kg CFL (pence)
CORRUGATED FIBREBOARD ONLY			
(1) "B" flute 125 gsm kraft facings	1.8	400	1.5
(2) "B" flute 400 gsm kraft facings	4.5	1000	1.2
(3) "BC" flute	5.8	1300	0.9
(4) "AA" flute 400 gsm kraft facings	17.8	4000	0.9
CORRUGATED FIBREBOARD REINFORCED WITH HARDBOARD			
(5) Case of Type 1 above with two end panels of 3.2 mm standard hardboard	9.3	2100	0.4
(6) Case of Type 1 above with end and side panels in 3.2 mm standard hardboard	15.6	3500	0.4

Table 4.2. Effect of temperature and humidity on compression failing load of corrugated fibreboard case reinforced with 4.8 mm thick hardboard pieces in "H" configuration

(Internal case dimension 520 × 250 × 275 mm)

Case style	CFL
(1) Corrugated fibreboard case with corrugated board dividers	9.7 kN (2200 lb) at 23°C, 50% relative humidity
(2) Corrugated fibreboard case with 4.8 mm thick hardboard dividers	20.3 kN (4600 lb) at 23°C, 50% relative humidity
(3) Corrugated fibreboard case with 4.8 mm thick hardboard dividers after conditioning for 7 days	20.3 kN (4500 lb) at 38°C, 90% relative humidity

HARDBOARD AND SOFTBOARD

The details given in Table 4.1 show the degree of improvement and the relative costs. Table 4.2 demonstrates that the high CFL can be maintained under wet conditions.

It is essential that the reinforcement is designed to suit the pack, the stacking pattern used, the type of pallet employed, and the journey conditions. The panels must be positively located so that they cannot be easily displaced. In some instances the contents of the pack will locate and hold the panel, and this method is particularly suitable in boxes containing a small number of bulky articles. An example is a case holding four 5-litre polyethylene bottles (figure 4.2a) where rigid hardboard panels

Figure 4.2a Hardboard and insulating board used as compression pieces in corrugated cardboard packing cases. Such compression pieces improve the stacking performance of the outer cases, and are being used for this purpose in the packaging of chemicals in plastic bottles.

located between the bottles provide the reinforcement. A single dividing piece of 3·2 mm hardboard can raise the Compression Failing Load of a 350 mm cube-shaped corrugated box from 500 to 1000 lb. Where such location by the contents is not possible, spot glueing, taping or stapling the hardboard to the case may be required. Alternatively, it may be located by sandwiching it between the case flaps. A further development envisages the production of a case made from both corrugated board and hardboard.

WOOD AND TEXTILE-BASED PACKAGING

This would have four faces of corrugated and two faces of hardboard (in a style similar to the "Ratio-Pak" system shown in figure 4.2b).

Figure 4.2b "Ratio-Pak" style of box with hardboard end panels to provide good stacking performance

Softboard (insulating board) may also be used for reinforcement, and in some situations may be preferable. It provides a cushioning effect. Moreover, should failure of the pack occur, the edges of the board deform rather than the panel fracturing. Consequently, even at "failure", a panel will still substantially support the load, reducing the risk of complete collapse of the stack. On the other hand, softboard has a lower resistance to moisture than hardboard: a factor which may be important in high-humidity situations.

Reinforcement of this type is not restricted to corrugated cases alone. For instance, a shrink-wrap pack containing four plastics bottles can be given added strength by the incorporation of a cruciform hardboard separator. In these circumstances a load-spreading device or collating tray is usually incorporated, and this may also be of hardboard.

As well as their use to provide vertical support, hardboard and softboard have a number of uses in horizontal reinforcement. A typical example is in cases containing heavy equipment, such as addressing machines or typewriters, where point loadings could damage the pack. A simple base plate of hardboard or softboard effectively distributes such point loads. With hardboard, locating and fastening holes enables the base plate to be positively fixed to the equipment packed, whilst softboard also provides cushioning.

HARDBOARD AND SOFTBOARD

Fruit and vegetable trays

The fruit and vegetable trade has for many years used wooden trays. Rising costs, however, have forced an examination of other materials, and hardboard offers a suitable alternative at reasonable prices. In these trades, the trays are required to permit considerable air circulation and have a degree of moisture resistance.

Hardboard trays and boxes provide performance comparable with timber in these respects, whilst being free from the associated blemishes and splits. Hardboard trays are made either by nailing hardboard to solid timber end pieces or, if hardboard is used for all side panels and the base, by using a corner stapling technique. In the latter instance, timber corner posts are usually included to provide the required stacking pattern (figures 4.2c and d).

Figure 4.2c Apple box made from hardboard and triangular timber corner posts

When changing from one packaging material to another, it is often advantageous to start with the pack requirements and examine how they can best be met with the new materials. So far hardboard has been used as a replacement to timber in traditional types of tray designed for timber. While this has been successful, development work on forming and jointing processes is facilitating the design of new styles of tray which are expected to have advantages over current styles in convenience and cost.

4.33

WOOD AND TEXTILE-BASED PACKAGING

Figure 4.2*d* Dutch trays with hardboard sides and bases wire-stitched to timber corner posts to give the required stacking pattern

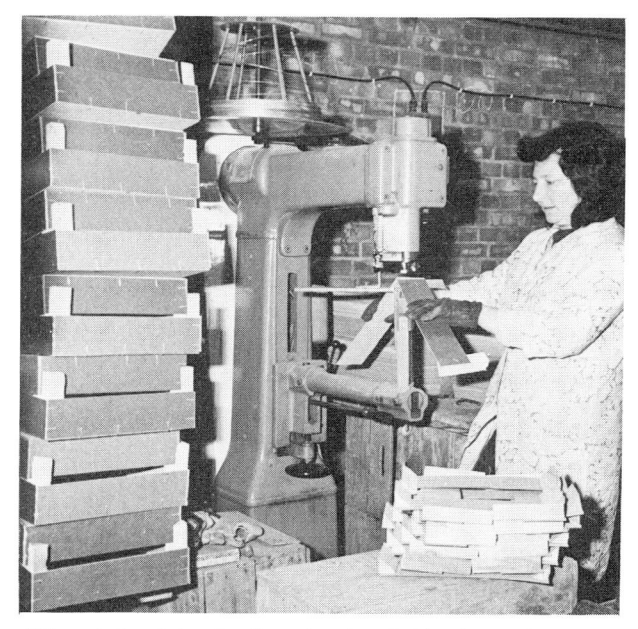

Figure 4.2*e* Wire stitcher fixing hardboard end and side pieces of produce trays to the timber corner post. Finished frames (stack at left) are then similarly stitched to hardboard base.

HARDBOARD AND SOFTBOARD

Hardboard cases

A number of styles of case have been developed based on hardboard or medium board panels jointed in various ways.

Large wood-framed hardboard cases are quite common. The hardboard is nailed to the frame, producing a flush-sided case, as shown in figure 4.3a. Medium-density hardboard panels, which are of thicker construction, can often be used with advantage for very large cases.

Figure 4.3a Flush-sided case

In smaller cases, the timber frame can be limited to the ends, or even eliminated altogether, joints being made with a riveted or stapled metal strip. Although care has to be taken that perforating the hardboard with the rivet holes does not provide a line of weakness, a case made from 3·2 mm standard hardboard and preformed tin-plate strips with overall dimensions 670 × 400 × 300 mm is suitable for 50 kg of machine tool parts (figure 4.3b).

As mentioned in a previous section, mechanical "through fastenings" may be replaced by surface fastenings. For this, solvent-based or pressure-sensitive adhesives are used in conjunction with jointing materials such as paper, textiles or plastics tapes, or with metal angle strips. In this situation, reliance is placed on the tensile strength perpendicular to the board surface (Z-strength) of the hardboard, and the rigidity depends entirely on the hardboard panels.

4.35

WOOD AND TEXTILE-BASED PACKAGING

Figure 4.3b Export pack for machine tool parts (50-kg load) consists of hardboard sides and back joined with riveted tinplate strips "tea-chest" fashion

Satisfactory transit performance depends on both the strength of the jointing material and on suitable methods of enabling shock to be distributed between adjacent panels. A style of pack which particularly meets the latter requirement has been evolved as the result of detailed study and practical testing. This "PIRA-style" fold-flat case (figure 4.3c) consists of six hardboard panels joined with flexible tapes. The edge of each panel is bevelled at 45°, enabling the production of mitred joints which both provide a self-locating feature during erection and also perform the function of distributing shocks received by one panel to the adjacent sides. This style has been found to be especially suitable for long thin packs (figure 4.3d) and for other applications where pack rigidity is particularly required.

Hardboards in 3·2 and 4·8 mm thicknesses are suitable for many cases of this type. As the dimensions increase, however, the use of medium board (panelboard) in 9–12 mm thickness becomes more appropriate.

HARDBOARD AND SOFTBOARD

Taped cases with medium board panels and the mitred corner techniques are extremely robust, the ultimate strength usually being related to the transit stresses that the tape can withstand. Furthermore, this type of box provides very good water resistance, the homogeneous sheet material and the completely sealed taped edges giving an unusually high degree of protection against water penetration to the goods being transported.

Figure 4.3c   This fold-flat case, developed by FIDOR and PIRA, consists of six pieces of hardboard with bevelled edges joined by adhesive tapes

Rigid boxes

Small rigid boxes made from hardboard panels nailed or stapled to timber end pieces are used in the fish (figure 4.4a) and other trades. The V-grooving technique is used to produce a complete box from a single piece of hardboard; plain hardboard may be used, or flexible PVC laminate, or tape may be added at the fold lines to provide more robust hinges during assembly (figure 4.4b).

4.37

WOOD AND TEXTILE-BASED PACKAGING

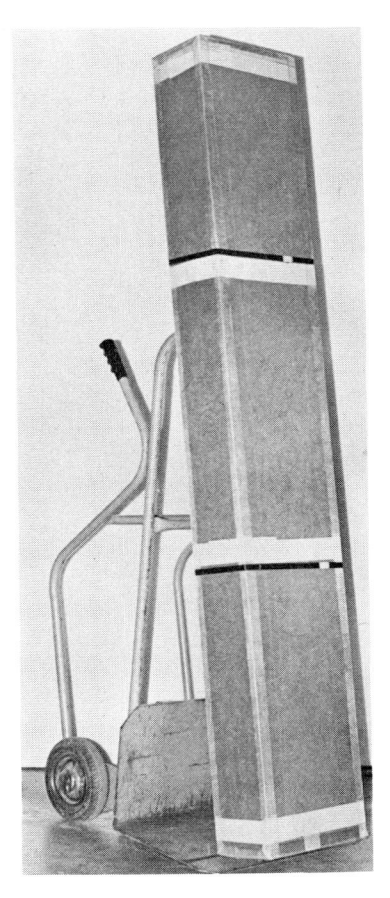

Figure 4.3d PIRA-style case with long panels, each cut with bevelled edges and joined by tapes to produce a long thin pack for rotary printing screens

Drums

Hardboard is also used for the construction of drums. A simple bend without the use of heat or moisture is adequate for the larger drums, where a small radius of curvature is not required. In the most basic situation, the fastening mechanisms tend to be very simple, with overlapped and metal-stitched side seams. The bottom is held in place by a simple metal chimb, also stitched into place. Such a drum is used for the transport of natural asphalt, where the hot liquid is poured into the drum and then allowed to set solid (figure 4.5a). It is not uncommon for the contents of this type of drum to weigh more than 200 kg, but the bitumen producers also require smaller-capacity drums. As bitumen is thixotropic, completely-sealed drums are sometimes required, and development work utilizing the techniques of adhesives and metal strips and tapes is in progress. A sealed drum is suitable for many hot-poured liquids and powders.

Figure 4.4a   Fish cake boxes—timber sides with hardboard top and bottom panels

Figure 4.4b   Multi-trip transit and storage pack. The sides and base of the box are hardboard, V-grooved to form corners, and channelled to accommodate hardboard side pieces.

WOOD AND TEXTILE-BASED PACKAGING

Figure 4.5a Range of hardboard drums with stitched seams and metal chimbs. When used for asphalt, lids are unnecessary.

Cable reels

Cable reels are regularly constructed from hardboard, both the flange and the core being produced in this material (figure 4.5b). To obtain the shape of the flange it is usual to die-stamp the hardboard, this process also facilitating the inclusion of the necessary perforations to accept the locating lugs of the core. The thickness of the flanges can be as much as 8 mm. Because of the small radius of curvature of the core pieces, soaking the board in water is necessary before bending is carried out.

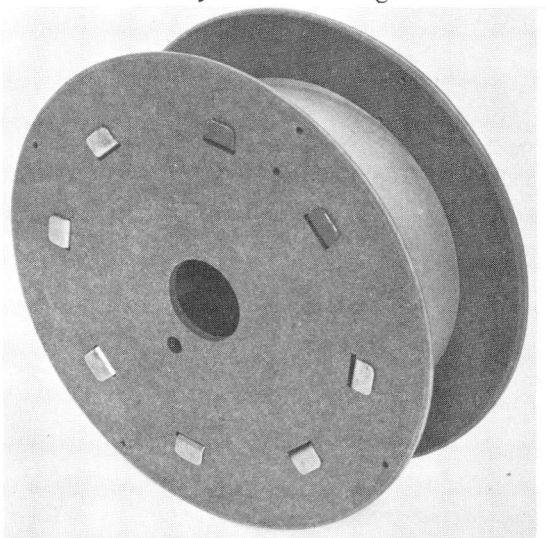

Figure 4.5b Cable drum with flanges made from hardboard

HARDBOARD AND SOFTBOARD

Pallets

Sheet materials such as hardboard and medium board have a number of applications in pallet construction. These include the maintenance of true "squareness" where this is required, e.g. in automatic palletizing and depalletizing machines, and certain types of pallet racking where out-of-square pallets will jam. Current commercial styles of pallet include:

(a) Conventional timber pallets incorporating a sheet of hardboard either above or beneath the top deck boards. The function of the hardboard may be to provide squareness, a continuous surface, or to repair a damaged pallet.
(b) Top and bottom decks of hardboard fastened to timber blocks (4-way entry) or bearers (2-way entry) giving a fully reversible pallet as in figure 4.6.
(c) Top deck of hardboard with plastic feet. 3·2 mm hardboard is used when continuous or secured loads are involved, and the pallet is often expendable. Thicker medium boards are used where the loading is more severe, or the pallets are returnable.

Figure 4.6 Four-way-entry fully-reversible pallet

Development work is in progress on a number of other styles. These include "sandwich" top decks (similar in construction to flush doors with a core between two sheets of hardboard). These are extremely stiff and, when tempered hardboard is used, can be highly weather-resistant. Moulding and forming techniques may also have application in pallet construction.

Palletized boxes

Fold-flat palletized boxes using the techniques discussed earlier can be produced. An example of such a box capable of carrying approx $\frac{1}{2}$ tonne is shown in figure 4.7a. The pallet itself can be made from timber or hardboard as is appropriate. The side panels and top of the box are hardboard with bevelled edges and flexible joints. However, for rigidity, the bottom edges of the side panels can have square edge boards and rigid through fastenings. Rigid containers on palletized bases are illustrated in figures 4.7b and c.

Figure 4.7a   Palletized box made from hardboard panels with bevelled edges and a heavy-duty tape undergoing testing in the laboratory

Miscellaneous uses

Packaging uses of fibre building boards are varied. Discs of insulating board are used as end protecting pieces in paper reels. Reeded hardboard is used as an overwrap for coils of stainless steel. Standard hardboard is used for protective side pieces in loads of sacks and bales. Strips of softboard have been used for many years in the packaging of plate glass. Laboratory studies of the cushioning performance of softboard indicate that its effect would be similar to that of bonded cork, and that it would

Figure 4.7b Rigid box made from 9-mm thick medium-density hardboard fastened mechanically to a timber base

Figure 4.7c Timber frame and hardboard panels form a 1-ton leakproof case into which material in liquid form can be poured. The container is used for liquids which solidify before distribution.

be most suitable for the cushioning of dense and heavy items, as in the postal pack for a house name plate made in slate.

REFERENCES

1. "Fibre building boards"—British Standard 1142: Part 1: 1971—Methods of test; Part 2: 1971—Medium board and hardboard; Part 3: 1972—Insulating board (softboard).
2. "Board manufacture"—Information Sheet IS/2, Fibre Building Board Development Organisation Ltd.
3. Mostyn, H. P., "Hardboard gives Cases Better Stack Strength at Low Cost", *Packaging News*, June 1973.
4. Mostyn, H. P., and Ford, M. J., "Creased Hardboard, preliminary report on case performance," Joint FIDOR/Pira Study, December 1975.

CHAPTER 5

Pallets and Unit Loads

Harri P. Mostyn

Introduction

Whenever applicable for distribution and storage it is advantageous:

(a) to handle goods mechanically rather than manually.
(b) to group a number of items into a single larger unit for mechanical handling.

Broadly speaking, these Unit Load Concepts have been applied for centuries—ever since man started unitizing items onto a base and using a windlass to lift them. Awareness of the value of aids of this type in modern distribution systems started with the development of pallets and fork-lift trucks in the 1939–1945 war.

A *pallet* has been defined as

a flat portable platform constructed to sustain a load and permit handling by mechanical equipment.

Many pallets fall within this definition, but there are types which are not flat and/or not properly described as a platform.

A *unit load* has been defined as

one or more packages secured to a pallet or skid in such a way that the entire unit may be handled and stored by mechanical equipment.

Again many unit loads come within this definition, but there are methods of unitization which do not employ a pallet or skid.

For our purposes therefore the wider definition which is linked with the principal means of handling will be used. This is:

WOOD AND TEXTILE-BASED PACKAGING

a palletized or unit load is one in which one or more items can be handled as a single unit by fork-lift trucks or similar equipment.

Modes of conveyance may be classified into

>road
>rail
>water
>air
>other (such as pipeline)

Large quantities of products in liquid, powder or granular form may be conveyed by pipeline, bulk tanker or intermediate bulk container (IBC). These are all outside the scope of this chapter, although an IBC is a form of unitized load. This chapter is concerned with liquids, powders and other products in conventional packs, such as cases, drums and sacks; also with other items, such as some engineering products, which may not be packed, but where unitization is a method of preparing them for delivery. The three principal methods of conveyance are Break-bulk; Unitized Loads; and Freight Container. A unitized load may in some instances be an alternative method to freight containers; in other instances it is complementary when unitized loads are transported in freight containers.

Methods of palletization and unitization

There are a large number of methods for unitizing loads. Most, but not all, are based on a pallet. The pallet type used in any given situation depends on a number of interrelated factors. The following are generally the most important and are basic to a consideration of pallet types.

1. The nature of the goods to be handled, and whether the pallet is to be used for one specific purpose or as a general-purpose pallet for a range of packs and products.
2. The distribution system(s) and particularly
   (a) The number of directions from which fork-lift equipment will need to enter the pallet. Two-way and four-way-entry pallets are common. Occasionally an eight-way (entry on the four sides and the four corners diagonally) entry pallet is used.
   (b) Whether the pallets will be handled by pallet trucks. These have small diameter rollers on the fork ends which need to be able to come through the pallet base and bear on the ground.
   (c) General handling and storage methods, including type of fork-lift or similar equipment, slings, spreaders, pallet racking, stack height, etc.
   (d) Whether the pallet "footprint" is significant in the downward loading applied. This may be important for aircraft floors and for corrugated fibreboard cases on the top of the pallet load below.
3. Cost of the pallet in relation to the expected service life.

PALLETS AND UNIT LOADS

Figure 5.1 illustrates by sketches and notes some of the principal types of pallet. Figure 5.2 illustrates some methods of unitization where a pallet is not used.

In addition to those shown in figure 5.2, there are other ways of unitizing loads without conventional pallets. The slip board system uses a thin flexible base board in place of a pallet; this can be gripped by an attachment on a special truck and pulled with the load onto a rigid platform on the truck. The SCULL system employs a "pallet" of wooden slats and wire made up with the load and which can be dismantled for return and re-use.

Figure 5.1 Principal types of pallet

(a) 2-way-entry pallet illustrated by a single-deck flat timber pallet.

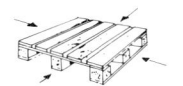

(b) 4-way-entry pallet.

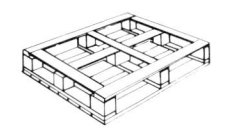

(c) Full-perimeter-base pallet sketched inverted. The loading area of the base is normally 40% or more of the total, giving an acceptable footprint for many applications. The cruciform base is compatible with pallet trucks.

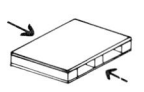

(d) Reversible pallet with similar top and bottom decks. The decks may be of sheet material such as plywood or hardboard (as drawn) or timber. They may be 2-way-entry as drawn, or 4-way-entry, typically using 9 wooden blocks instead of 3 bearers.

(e) Wing pallet. As illustrated, both decks may project beyond the outer bearers to facilitate the use of lifting slings. In other designs, for instance to run in live pallet racking systems, only the upper deck projects.

(f) Timber and steel stillage with uninterrupted entry for a stillage truck which cannot enter a pallet with centre bearers or bottom deck.

4.47

WOOD AND TEXTILE-BASED PACKAGING

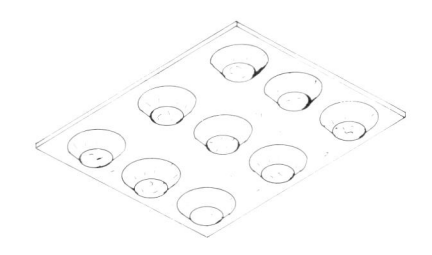

(g) Moulded pallet (drawn inverted) made in one piece of one material such as plastic or wood-fibre bonded with adhesive. Pallets of similar appearance but made of two materials may have a top deck of corrugated fibreboard or hardboard with plastic feet. Both types may be designed to nest when empty.

(h) Injection-moulded plastic pallet.

(i) Box pallet typically has 3 or 4 sides and may have a lid. Vertical sides may be fixed, removable or collapsible.

(j) Collapsible pallet.

Figure 5.1 Principal types of pallet

4.48

PALLETS AND UNIT LOADS

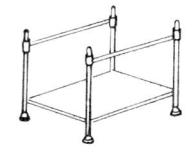

(k) Post pallet, usually made in steel with fixed or detachable posts, plus possibly rails or sides.

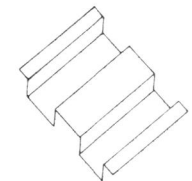

(l) Trough pallet. If the dimensions are compatible with the units (say cases or sacks) making up the load, one layer of these can lie in the troughs reducing the volume of the total loaded pallet.

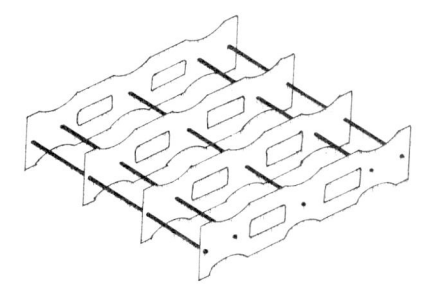

(m) Keg pallet. An illustration of one of the many types of special-purpose pallet. This one is designed for kegs loaded on their sides.

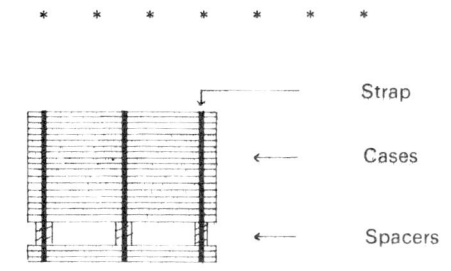

* * * * * * *

Strap

Cases

Spacers

(a) A load of heavy-duty corrugated cases folded flat and unitized for fork-lift handling with spacers (typically short lengths of spirally wound fibreboard tubes) and straps. No pallet is used, and the lowest cases are placed directly onto the warehouse and vehicle floors.

Figure 5.2 Unitization for fork-lift handling without a pallet

4.49

WOOD AND TEXTILE-BASED PACKAGING

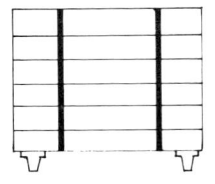

(b) Unitized load of sheet materials requiring feet and strapping only.

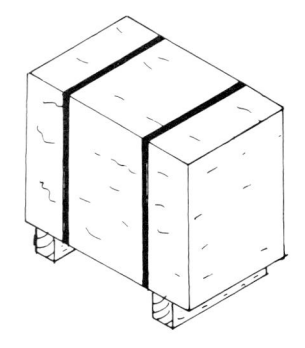

Figure 5.2

(c) Unitized load consisting of a single wooden case with skids.

Equipment for handling palletized and unitized loads

There is a great variety of mechanical equipment to handle palletized and unitized loads. Two common types are illustrated in figure 5.3.

Figure 5.3 Some mechanical equipment for unit load handling

(a) Counterbalance fork-lift truck conveying unitized load two high. Occasionally trucks are used with the forks considerably shorter than the pallets which can result in the latter being unduly stressed. (See figure 5.6e.)

4.50

PALLETS AND UNIT LOADS

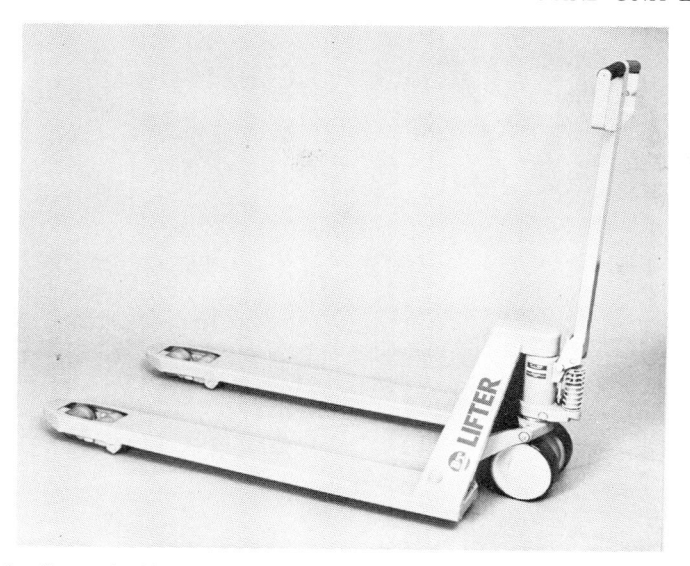

(b) Hand pallet truck. After the forks are pushed beneath the pallet, the two small wheels are moved downwards hydraulically to lift the pallet and allow it to be readily moved. If these wheels rest on the lower deck boards they can be forced off during the lifting operation.

Figure 5.3

Wooden pallet construction and specification

The terms used to describe the components of wooden pallets are shown in figure 5.4.

Most wooden pallets are made by specialist manufacturers, although some user companies make their own. Pallet manufacturers generally produce a range of types in various dimensions; timber of relatively low grade is usually used which often has considerable dimensional variation. These two factors make fully automatic manufacture impractical, even on the largest scale. Production rates in excess of 25 pallets per hour are, however, achievable by semi-automatic methods. Pallet components are generally fastened together with nails applied by hopper-fed nailing machines. The two principal methods of nailing are illustrated in figure 5.4; the clenching being accomplished in the same operation as the nail driving by the use of a steel backing plate. Screws and staples (particularly those with divergent points so that the staple legs splay to increase withdrawal resistance) are sometimes used, but adhesives are rarely practical, partly because of the relatively high moisture content of the timber.

The moisture content (m.c.) of timber is important not only for pallets but for most timber applications including wooden cases. A short account

4.51

is therefore included here. The moisture content of wood is usually expressed in relation to the oven dry weight and is calculated as

$$\text{m.c.}\% = \left[\frac{\text{weight of sample} - \text{oven dry weight of sample}}{\text{oven dry weight of sample}}\right] \times 100$$

If therefore more than half the weight of a freshly felled log consists of water, the m.c. will be greater than 100%. Elm is an example of a hardwood used for pallet blocks; the m.c. green (freshly sawn) is about 135%, and the weight about 1040 kg/m^3 (65 lb/ft^3); seasoned elm weighs 544 kg/m^3 (34 lb/ft^3). Pine is a softwood used for pallet deckboards and stringers; the m.c. green is about 85% and the weight 800 kg/m^3 (50 lb/ft^3); seasoned pine weighs about 500 kg/m^3 (32 lb/ft^3). It may be noted that hardwoods come from deciduous trees and may be hard like oak or soft like balsa; softwoods come from coniferous trees and may be relatively soft like some pines or hard like yew.

After logs are sawn into boards, the latter are usually stacked with spacers (stickers) to season or dry out. Sawn timber can be kiln-dried in a controlled atmosphere to any desired moisture content: timber for furniture is often kiln-dried to 12% m.c., will then be approximately in equilibrium with the atmosphere of a heated house, and will not dry out further, shrinking and deforming as it does so. Kiln-drying is relatively costly, and most pallet and packaging timber is air-dried. In the British climate it is difficult to air-dry timber under cover to below 20% m.c.

(a) Two-way entry non-reversible (topside view).

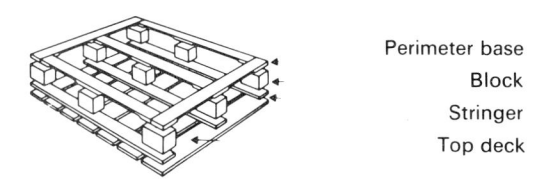

Top deck
Bearer
Chamfered edge (through or stop-chamfered)
Bottom deck

(b) Four-way entry (underside view).

Perimeter base
Block
Stringer
Top deck

Figure 5.4 The components of wooden pallets

PALLETS AND UNIT LOADS

(c) Annular ring nail used typically for nailing through the deck boards into the blocks or bearers. The withdrawal resistance is much higher than that of a plain shank nail and comparable with that of a screw.

(d) Section through a top deck nailed to the stringer with clenched nails. Again these have a high withdrawal resistance.

Note that in all pallet nailing the nail head should neither project (which may damage the load) nor be overdriven (when the board being fastened may pull off over the head). It should be flush with the surface as illustrated.

Figure 5.4

Much pallet and packaging timber is dried in the open, and the m.c. cannot be controlled, since it will depend on the time of year and incidence of rain, as well as on factors such as thickness and length of time in stack.

Wet timber has several limitations.

(1) It is relatively heavy.

(2) When a palletized load is stored in closed situations such as a freight container, the wood will dry out under warm dry conditions; when the temperature drops, condensation can easily occur and may affect the load. A pallet weighing say 30 kg may contain more than 5 kg of water.

(3) It may shrink and warp when it dries out.

(4) It may rot. Decay and rot are caused by fungi, and their activity is very dependent on moisture content. Below about 22% moisture, the water is held in the cell walls of the wood, which will not then be attacked by most fungi; above about 22% there is free water in the cell cavities, and the wood is susceptible to fungal attack.

A typical specification for a wooden pallet is given below and is shown in figure 5.5.

Example of a wooden pallet specification

The timber species, tolerances, nailing pattern, etc., will normally be arranged with the pallet supplier to accord with the particular application.

4.53

WOOD AND TEXTILE-BASED PACKAGING

Type
Four-way entry, perimeter base, 1000 × 1200 mm.

Materials
(a) Boards—the timber species, country of origin, grade or specification, and method of seasoning will be stated.
(b) Blocks—to be English Elm.

Dimensions
(a) Overall height to be subject to a tolerance of plus 5 mm minus 0 mm,
   i.e. maximum height 165 mm
       minimum height 160 mm
(b) Overall length and width to be subject to tolerances of plus 0 mm minus 5 mm.
(c) The pallet to be square and the pallet deck flat within 5 mm.

Timber components

Description	Material	Dimensions (mm)	Quantity
1 Top deckboards	softwood	125 × 19 × 1000	7
2 Base	softwood	100 × 19 × 1000	2
	(chamfered)	100 × 19 × 1200	3
3 Stringers	softwood	100 × 22 × 1200	3
4 Blocks	elm	138 × 100 × 100	6
		100 × 100 × 100	3

Construction
(a) The top deck boards will be evenly spaced with gaps of approximately 21 mm.
(b) Nails will be positioned not less than 20 mm from the edge of any block or board.
(c) Nail heads must not project above the boards nor be overdriven more than 4 mm. Clenched nails must be embedded in the underside of the stringers.
(d) The timber components will be fastened securely with the following nails:
   Top deckboards to stringers except over blocks—2 off 50 × 3·35 mm plain wire nails clenched.
   Top deckboards and stringers to blocks—3 off Tilgrip or similar annular ring nails 75 × 4 mm.
   Bottom deckboards to blocks—2 off Tilgrip or similar annular ring nails 75 × 4 mm.

Marking
Two diagonally opposite blocks stencilled with "User's name".

Inspection
Pallets may be subjected to inspection on receipt, and those not in accordance with the foregoing specification rejected.

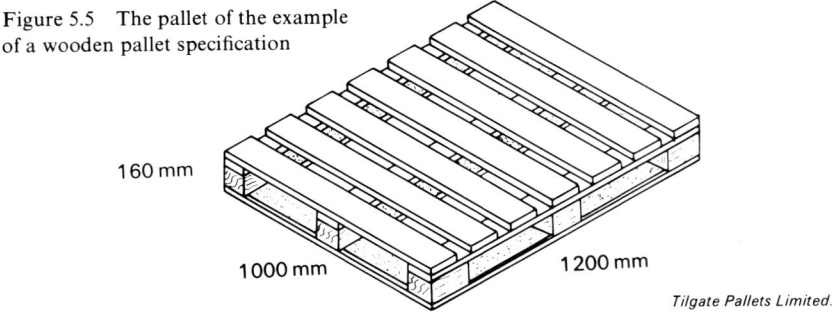

Figure 5.5 The pallet of the example of a wooden pallet specification

160 mm
1000 mm
1200 mm

Tilgate Pallets Limited.

Pallet performance

There are no generally agreed methods or standards for describing or measuring pallet performance. With the number of pallets in service, and the dangers to life and property should palletized loads collapse in high stacks or in transit, progress must be made in this area. The main factors which need to be taken into account to quantify the performance of pallets like those in figure 5.1*a–h* are summarized here, although additional factors may be important in certain situations, e.g. the ease of cleaning a pallet to be used for foodstuffs, and the vertical load resistance of the posts in the pallet shown in figure 5.1*k*.

Load capacity

The load capacity of a pallet is a prime consideration. Will the pallet safely transport $\frac{1}{2}$ tonne, 1 tonne or 5 tonnes? Unfortunately, a simple statement of capacity is meaningless unless other conditions are also specified. These conditions are discussed relative to the type of load, and methods of stacking and lifting.

Loads are classified into three types: self-supporting, divided and secured. As illustrated in figure 5.6*a*, a self-supporting load is one which exerts no bending forces on the pallet under the range of lifting and storage conditions. An example is a pallet load of wall boards, each one of approximately the same area as the pallet top deck.

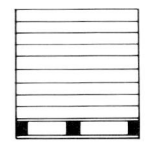

(*a*) Self-supporting load.

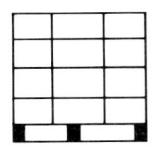

(*b*) Divided load.

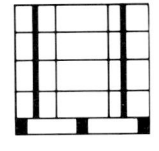

(*c*) Secured load.

(*d*) Pallet with divided load bending when lifted. The bending moment is accentuated when the tines are close-set.

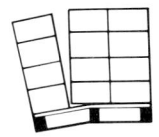

(*e*) Pallet with divided load bending when lifted either when the tines are offset or, at right angles, if the fork tines are shorter than the pallet.

Figure 5.6 Pallet load capacity

WOOD AND TEXTILE-BASED PACKAGING

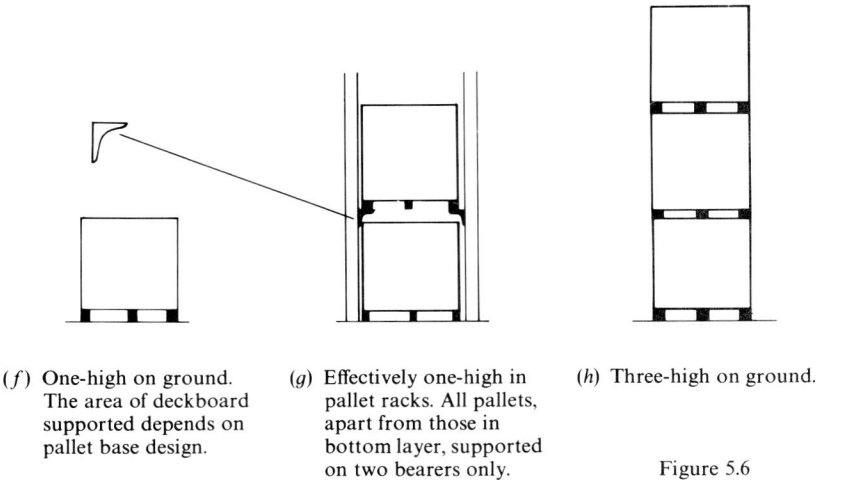

(f) One-high on ground. The area of deckboard supported depends on pallet base design.

(g) Effectively one-high in pallet racks. All pallets, apart from those in bottom layer, supported on two bearers only.

(h) Three-high on ground.

Figure 5.6

A divided load (figure 5.6b) is one which can exert bending forces during lifting and storage. An example is a pallet load of small filled cases which are not banded or shrink-wrapped onto the pallet. The effect of lifting a pallet load of this type with a fork truck is illustrated in figure 5.6d and e.

A secured load (figure 5.6c) is a divided load which has been banded in both directions, shrink-wrapped or similarly secured so that it behaves as a self-supporting load.

Three typical stacking conditions are illustrated in figure 5.6f, g, h.

The load capacity can thus be considered under two main sets of conditions:

(a) All loaded pallets are subjected to downward vertical forces taken by the blocks, bearers, feet or other spacers.
   Where these are of solid timber, failure is virtually unknown, since short lengths of wood have very high compressive strength. With self-supporting loads, therefore, wooden pallets have virtually unlimited capacity, and the stacking height is restricted by aspects such as stability and the compressive strength of the load, more than by the pallet strength.
   With spacers of material other than wood, such as plastics or fibreboard, the downward forces may limit both the pallet load capacity and the number of loaded pallets which may be stacked one on another. This is particularly so when the strength of the spacers is reduced by moisture or high temperature.
(b) All pallets with divided loads are subject to bending forces when lifted by fork-lift trucks. When pallet racking or sling handling is employed, such pallets are subjected to additional bending forces. The capacity of most types of pallet for divided loads will thus be limited by the pallet bending strength.

A third condition may be significant with very light timber pallets using a minimum of deck boards. As illustrated in figure 5.7a, some divided loads may apply bending forces to the deckboards between the

PALLETS AND UNIT LOADS

bearers, and the capacity may be limited in this way more than when the pallet is lifted.

Other pallet performance factors

The load-bearing capacity is the most important factor in pallet performance. There are a number of other factors, the importance of which will vary greatly with the type of pallet and the material with which it is made. These performance factors include:

(a) the resistance to fork tines, which may break, puncture or abrade many parts of the pallet.
(b) the ability to be put down on uneven ground.
(c) the resistance to impact when lowered rapidly onto level or uneven ground.
(d) the resistance to vibration during transit. Vibration tends to loosen fastenings and generally fatigue the material.
(e) the resistance to handling when not loaded, which typically occurs when multi-trip pallets are being returned empty.

Pallets often have to perform under conditions of high or low temperature and relative humidity. They are commonly stored outdoors and may be put down in puddles of rainwater. Conditions of use such as these should be borne in mind when selecting and specifying pallets.

Some common forms of damage to pallets, and their causes, are illustrated in figure 5.7.

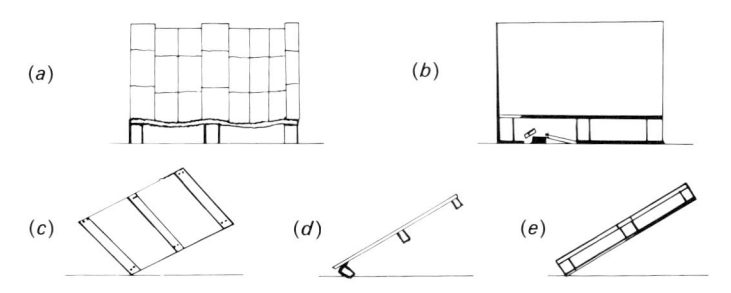

Figure 5.7 Some forms of damage to pallets

(a) 2-way-entry pallet with light deck boards bowing under a divided load.

(b) Lower deck boards of 4-way-entry pallet broken when the loaded pallet is lowered onto uneven ground or onto an obstruction such as a large stone.

(c) A wooden pallet changing from a rectangular shape to that of a parallelogram (diamonding) on being dropped empty onto a corner. Many wooden pallets are liable to wracking in this way, particularly if they have narrow deck boards providing only a short distance between fastenings.

(d) An outside bearer (or its equivalent lower deckboard unit in a 4-way-entry pallet) being knocked away from the top deck when the pallet is dropped onto one edge.

(e) A perimeter deckboard pallet is less likely to be damaged when dropped onto one edge.

4.57

Forming and securing unit loads

With unit loads such as box pallets or skid-based wooden cases, the methods of forming the load are inherent in the design. Internal locating and securing devices, such as checking and bracing engineering products in wooden cases, are often required—but these are outside the scope of this chapter.

Flat pallets can be loaded in various ways, but in general they should be loaded as evenly and squarely as possible. The factors affecting the arrangement of the load will depend largely on its nature; in this chapter it is discussed briefly in relation to pallet loads of small filled corrugated cases only. These may be block-stacked (with each case vertically over the one below) or bonded (by varying the arrangement in alternate layers in various patterns) so that more cases can be loaded per pallet or to restrict shifting. Where the cases themselves, and not the contents, provide most of the stacking resistance required, this will be reduced by bonded stacking. The cases may also be stacked within the pallet deck perimeter or overhang the edge; the latter arrangement further reduces stack resistance and may also make the cases more liable to damage from contact with other pallets, backs of vehicles, etc.

Flat pallets provide no built-in means of internal load restraint. Despite this, unit loads may be moved on them without restraint, particularly when, as with some sacks, there is little tendency for the load to shift on the chosen distribution system. When required, there are three main alternative or complementary methods for securing the load—adhesives, strapping and enveloping.

1. *Adhesives*

Low-shear adhesives are typically used, which bond cases together on the pallet, but enable them to be pulled loose without significant damage. The adhesives are often sprayed on during palletization.

2. *Strapping*

High-tensile steel, plastics or textile strapping is widely used, mainly applied with special hand or automatic tools. Relatively high forces are involved, and edge protectors may be necessary to prevent the straps cutting into the load. Strapping should be in continuous contact with the load and the underside of the pallet deck; with 4-way-entry pallets it should be parallel to the top deck stringers to avoid damage from forks.

3. Enveloping

Enveloping may take a number of forms:

(a) *Shrink-wrapping.*—A loose hood of plastics film dropped over the load and the pallet is then shrunk by heat. Shrinking can be performed in a tunnel, in an oven, by lowering a rectangular heating element over the load or by portable guns. Features of shrink-wrapping include:

    (i) relatively low forces over the whole load area, edge protectors unnecessary.
    (ii) effective with a wide range of loads, regular and irregular.
    (iii) weather protection provided.
    (iv) shrink-wrap below the pallet not affected by forks.
    (v) nature and condition of load can be observed through the film.

(b) *Stretch-wrapping* is similar to shrink-wrapping in result. An elastic film is used, applied with special equipment.

(c) The application of a fibreboard snood over the load, usually fastened to the pallet by nailing or strapping.

(d) *Vacuum packing* may be accomplished with a heat-sealed plastics envelope or a heavy-duty returnable envelope. The purpose is more usually to provide a high degree of climatic protection rather than to secure the load.

In addition to securing the load on the pallet, the unit loads themselves must be properly secured onto or into the vehicles or other means of conveyance. So far as road vehicles are concerned, guidance is given in the Code of Practice "Safety of Loads on Vehicles" published by the Department of the Environment and available from HMSO. This deals with a variety of loads and suggests the acceleration g levels against which their fastenings must restrain them. If palletized loads are stacked more than one high, each layer must be secured separately; sheeting is not generally adequate as a means of securing loads.

Pallet and unit load dimensions

The choice of pallet and unit load dimensions in any situation is related to:

(a) the dimensions, stacking pattern, etc., of the components of the load to be carried. For a returnable general-purpose pallet, the range of loads likely to be carried has to be considered.
(b) the limitations of the associated equipment, including fork-lift trucks, road vehicles, rail wagons, ships' holds, freight containers, etc.
(c) the dimensions of the existing pallets being used by the same company or on the same distribution system.

WOOD AND TEXTILE-BASED PACKAGING

BS 2629: Part 1:1967 recognizes that some industries need specialized sizes, but recommends the following for through-transit pallets:

    A   800 × 1200 mm
    B   1000 × 1200 mm
    C   1200 × 1200 mm
    D   1200 × 1800 mm

Sizes A, B and D are also recommended by ISO.

It should be noted, however, that the internal width of an ISO Freight Container is 2299 mm and none of the recommended pallet dimensions are modular to this.

A most important dimension is the free height for entry of the forks, which should be a minimum of 96 mm. Most pallets are made to this standard, but skids on wooden cases are often made more shallow; they may therefore not be handled by fork trucks as intended, or be damaged by forks being forced under the case.

Pallet utilization

A common way of classifying pallets is into single-trip (or expendable) and multi-trip (or returnable). To take two extreme situations, an expendable pallet will typically be used on an export journey where there is no prospect of it being returned economically. The single-trip pallet will normally be selected to give acceptable performance at minimum purchase price. In contrast, a closed-circuit distribution system, wholly within the pallet owner's control, will normally employ returnable pallets. The capital cost of the pallet can be balanced against the service life and repair cost (if a recovery and repair system is instituted) and a pallet with the lowest purchase price may not necessarily be the most economic choice. Higher first cost may be more than justified by longer life and lower maintenance.

On many distribution systems, however, pallets and their loads are delivered to a customer. It usually is not practicable to depalletize immediately, so that the delivery vehicle can collect and return the pallet. Various arrangements may be tried, an empty pallet from a previous delivery can be collected in exchange for a full one, or the pallets can be invoiced, and a credit given for return. Arrangements of this type work better in theory than in practice, and high pallet loss rates are common. The supplying company may therefore consider the use of an expendable pallet. The cost equation needed to make a comparison is straightforward so far as the expendable pallet is concerned, but requires an actual average service life figure to complete the returnable cost side. To date in the United Kingdom it is reckoned that returnable pallets usually offer economies, at least for general loads on most systems.

The difference between an expendable and returnable pallet is therefore one of utilization rather than one of type. By definition, a satisfactory expendable pallet should still be serviceable at the end of the trip—in which event it is serviceable for at least the start of a second trip.

Performance requirements are the same for expendable and returnable pallets, except that the latter have to meet them repetitively. The requirements are relatively severe, especially for heavy divided loads—which any general-purpose pallet may have to meet. For wide acceptance as an alternative, an expendable pallet probably has to cost only 25% of the cost of the returnable. From the nature of materials, it is difficult to achieve the bending and crushing properties needed within the tight dimensional and cost limitations.

As indicated earlier in this chapter, there are a wide variety of methods for palletization and unitization, some being specially suitable for certain types of load. While, therefore, a general-purpose expendable pallet may be unattainable at acceptable cost, it may be quite practical to design a low-cost pallet for specific loads and distribution systems.

Most returnable pallets in the United Kingdom are purchased and owned by individual users. An alternative system is the pallet pool. An example is the European Pallet Pool operated by the railways. This operates on one size of pallet, 800×1200 mm and two types, with or without sides. To participate, a company purchases the number of pallets they expect to use. For every loaded pallet delivered to the railway, they receive an equal number of empty pallets in return. There are a number of problems in the smooth operation of such a pool, and until recently in the United Kingdom pallet pools have been restricted to relatively small numbers of similar companies operating controlled distribution systems.

Economic factors

The economic advantages of unitization are both *direct* (in lower distribution costs) and *indirect*, e.g. in pack cost and damage rate reductions.

The **direct advantages** accrue not only to the consignor, but also to the transport agencies and often to the consignee as well. The consignor saves on handling and storage costs in his own warehouse, as well as on shipment charges. Unitization enables transport agencies to achieve larger and faster throughputs on very flexible systems. To encourage unitization, shipping companies, for example, may exclude the pallet in calculating the shipping volume, and may offer lower rates. A number of shipping companies formed the Unit Load Council* to promote the Unit Load

* Unit Load Council, Aslakveien, PO Box 52 R0A, Oslo 7 NORWAY.

Concept, which is based on the theory that cargoes should be packed so that they can be moved and handled by mechanical equipment on all links of the transport chain.

The consignee may receive a direct benefit from unitization in a similar way to the consignor, saving on handling and storage costs. Some UK companies have instituted systems with all their suppliers, so that all goods are sent unitized on pallets of agreed dimensions.

Prior to the 1939–45 war, virtually all freight was shipped break-bulk. The most publicized postwar change has been the introduction of containerization. This, however, demands major capital investment in ships, containers, handling depots, etc., and container lines still connect only a limited number of the world's ports. It is often assumed that containerization offers the ultimate in economy, but this is not necessarily so in comparison with unitization. Both certainly offer considerable economies compared with break-bulk. It should be noted, too, that a high proportion of containers are filled with unit loads.

The **indirect advantages** of unitization can also be considerable. They stem from the fact that large unit loads handled mechanically are subject to lower transit hazards, especially in regard to impact. One pack weighing, say, 10 kg and exported individually may have to withstand drops from heights of 900 mm onto edges and corners. Fifty such packs unitized onto a pallet may have to withstand impacts equivalent to less than 300 mm effectively onto the base only, and protected to some extent by the pallet. The unitized packs may therefore require much less protective packaging than those which are to travel individually. As in containerization, the limitations are not in the major portion of the export journey, but distribution to the final destination if the pallets are broken down near the port of entry.

Increasing attention is being paid to the safe transport of hazardous goods, such as corrosive or inflammable chemicals. Regulations are in force to ensure that the packs for such goods do not leak in transit. The reduced hazard inherent in through unitization may enable the regulations to be safely relaxed for unit loads, provided they are not broken down before arrival at final destination.

No statistics are available covering all the various types of pallet in service in the United Kingdom. It is estimated that wooden pallets hold the bulk of the UK market, and that the 1973 consumption was 12,000,000 units with an average value of £3·00 for a total wooden pallet market of £36,000,000. The ten largest producing companies manufacture some 60% of total output. Nearly half the market is for 1000 × 1200 mm four-way-entry open-deck perimeter-base nailed wooden pallets.

During the ten years prior to the start of 1973, the cost of a standard wooden pallet had actually gone down in cash terms. This was because

PALLETS AND UNIT LOADS

of the very competitive nature of the business, and the low capital required to make pallets, at least on a small scale. Over that period the standard pallet had changed, the volume and quality of timber being lowered, and nailing techniques improved in order to achieve high production rates of acceptable pallets. In passing, it is interesting to note that many users changed from North European softwood to Maritime Pine of lower mechanical properties and cost—the change was generally accomplished without problems, indicating that many pallets had been overspecified for years.

Timber was one of the first materials to show a dramatic price increase during early 1973. This was reflected in pallet prices, which more than doubled over the following year. Pallets which are non-standard, as probably an unnecessarily high number are, have increased even more in price and become more difficult to obtain. Other materials have also increased greatly in cost, so that timber still retains a dominant position.

Four trends can be expected. (1) The advantages and ease of unitization will be more widely recognized, resulting in more pallets and similar aids being used. (2) Greater recognition will be given to the vital part played by the pallet in modern distribution systems, and to the amount of capital which may be deployed in stocks of pallets. This may lead to more careful handling, possibly including such means as modifications of fork-lift trucks to minimize the risk of damage. (3) While the timber pallet is likely to remain dominant, and more attention will be paid to close specification, a larger number of alternatives to timber will be tried. (4) Further development of pallet pools will occur.

The Unit Load Concept is simple and fascinating in its ramifications. It seems to be one of those rare concepts with major applications giving advantages to all concerned and disadvantages to none. Essentially, it is based on the humble but invaluable pallet.

CHAPTER 6

Textile Sacks and Bags

W. G. Atkins

This chapter is a resumé of basic data which must necessarily be applied for the correct selection and use of jute fabrics and containers. Prospective users are, however, in their own interests, strongly urged to take advantage of the free technical advice and service available from most members of the trade.

Introduction

Jute, flax and hemp, when used for packaging purposes, are normally spun into yarns by dry-spinning processes, as distinct from the wet-spinning method used for finer flax yarns. Jute-substitute fibres are also used for sack production in certain areas of the world, although the quantities consumed are small in comparison with jute. Flax and hemp fabrics are normally too expensive to be used for other than special purposes, e.g. flax waterbags for use in hot countries, and tarred hemp sacks for the retail coal-carrying trade. Cotton is extensively used for the packaging of many commodities.

The larger part of the jute weaving industry employs plain looms of simple construction, although these are nowadays fitted with automatic weft replenishment and warp stop motions. In addition, some sections of the industry employ circular looms for the formation of tubular fabric and for the production of fully woven, i.e. seamless, sacks.

Jute sacks and bags are used for packing a wide range of materials, including powdered and crystalline substances, agricultural produce, textiles and small engineering parts. Apart from their intrinsic strength, they have a particular value in that they can be re-used in many instances.

Even where re-use for the same purpose is not possible, the sack usually has a value for other purposes.

Nomenclature

Jute fabric is specified by:

> The width and length of the piece required.
> The weight of a given area.
> The place of manufacture (e.g. Calcutta, Dundee).
> The type of weave (e.g. hessian, tarpaulin, bagging or sacking).
> The *porter*, or threads per unit length in the warp.
> The number of *shots* (threads per unit length in the weft).
> The finish.

Jute fabrics used for sack making are essentially simple in construction, being classifiable as follows:

Hessians: Plain-weave single yarns in both warp and weft.
Tarpaulins and baggings: Plain weave, warp threads laid double, i.e. double warp (D.W.).
Twilled sackings: Twilled 2:1 weave, usually with double warps, but sometimes single warps (S.W.).

Threads per unit length in the warp are commonly specified by the "porter" system, which specifies the number of splits in the loom reed. Thus a one-porter reed contains 20 splits in 37 in of reed width; ten porter contains 200 splits and so on. Hessians are usually drawn 2 warp threads per split, tarpaulins 4 threads, and twilled sackings 6 threads (D.W.) or three threads (S.W.) per split. If P = porter and N = threads per inch then for

Hessians	$N = (2 \times 20/37) P = 1·08 P$
D.W. Tarpaulins	$N = (4 \times 20/37) P = 2·16 P$
D.W. Twills	$N = (6 \times 20/37) P = 3·24 P$
S.W. Twills	$N = (3 \times 20/37) P = 1·62 P$

Since, however, fabric normally shrinks in width when removed from the loom, the actual warp threads per inch in Dundee jute fabrics are slightly higher than when calculated by these formulae. Calcutta fabric, on the other hand, uses *porter* in reference to finished cloth. *Shots* or *threads per inch* is the common designation for weft.

Hessians are generally specified by width, weight in ounces per yard linear at some fixed width and also by place of origin.

Example: "60 in, 6 oz/40 in Dundee" means that the cloth required will measure 60 in wide, and weigh approximately 6 oz per linear yard of 40 in material (i.e. 9 oz per linear yard of 60 in material) and originated in Dundee.

Tarpaulins are of double warp construction (D.W.), as also are baggings, but the latter are usually coarser fabric than tarpaulins. All are specified by the width required, the porter and the average weight per linear yard of whatever width is normal for that construction.

Example: "45 in 6/14/27 (or 45 in 6/27/14 oz) twilled sacking Dundee" means that the cloth required will measure 45 in wide, be of 6 porter construction and weigh approximately 14 oz per linear yard of material 27 in wide (i.e. 23·3 oz per linear yard of 45 in wide material).

Finishing treatments

A. *Mechanical*

For the most part jute fabrics receive only simple mechanical finishing treatments prior to being converted into sacks or bags. These may be classified thus:

1. Cropping and singeing.—The loose hairs on both faces are sheared off by rapidly rotating helical knives working against fixed blades, as in a lawnmower. Singeing is less common.

2. Damping.—The fabric is passed through machines which add water in the form of a fine spray produced either by jets or by long cylindrical soft-bristle brushes rapidly revolving in contact with rotating metal rollers semi-immersed in water. The amount of water added is controlled, and varies according to the fabric and the finish required.

3. Calendering.—Calendering is a process for flattening the threads of a cloth, and is normally achieved by running the cloth, which may be previously damped, between two or more alternate steel and compressed-paper cylinders, to which pressure is applied. The machine used is known as a *calender* and the cylinders, usually five in number, are referred to in the trade as *bowls*. The central bowl is of steam- or gas-heated steel with paper and steel bowls alternating above and below. Heat, water and pressure combine to flatten the threads and thereby reduce the size of the interstices, improving the appearance and ease of handling in later processes. The fabric face last in contact with steel is more polished than the other face. Speeds of up to 30 metres per minute are common.

4. Chesting.—The degree of flattening may be increased by allowing the fabric to wind up on the top or second bowl after passage through the lower nips. In contrast to calendering, this method is discontinuous and uses shorter lengths of fabric. The finish obtained lies between that given by calendering and that associated with mangling.

5. Mangling.—A process somewhat akin to chesting, but in this case the previously calendered cloth is rolled on a steel pin and rotated backwards and forwards between two steel bowls of the "mangle" under heavy pressure. The method closes the interstices more effectively than chesting.

According to the degree of damping and tension employed, calendering and chesting may tend to increase piece length and slightly reduce piece width; mangling on the other hand may tend to reduce piece length and to increase width. The effect is sufficiently well known to be taken into account when weaving, in order that the final width may be correct. Sackings and baggings are commonly calendered only, whereas hessian may be chested or mangled.

B. *Chemical*

The bulk of jute sacks and bags are used without chemical treatment, but a definite demand exists for specific treatments, more especially for imparting a more attractive appearance or resistance to certain adverse influences. In most instances these treatments are applied in the piece prior to mechanical finishing, but may also be applied at some earlier stage of processing or alternatively to the finished sacks. Such treatments include proofing against water, rotting, insects, rodents and fire.

Sacks and bags

The words *sack* and *bag* are often used synonymously, but the trade usually applies the first to those made from twilled sacking, and the latter to those made from hessian, bagging or tarpaulin. Sacks and bags, which are described by size, weight and the material from which they are made, are normally supplied with plain open mouths, but where necessary "valved" sacks can be specified. The two types of valves commonly in use are the Bates valve and the Sleeve valve. Although a little more costly, the latter allows a greater volume content and gives a better shape to the sack when filled.

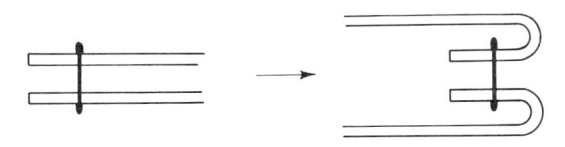

Figure 6.1 Plain seam

Sacks and bags may be hemmed and seamed in about a dozen different ways, and a variety of stitching is employed for these. It is not necessary within the compass of this chapter to describe all of these, but definitions of a few are as follows:

Plain seam.—The seam is made through two thicknesses of cloth and the bag is turned after sewing. One or two lines of stitches may be used, plain seams being used only on selvedge edges (figure 6.1).

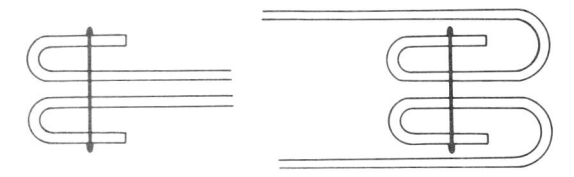

Figure 6.2 Counterlaid seam

Counterlaid seam.—Here the edges are turned outwards; the seam is formed by sewing through four thicknesses of cloth, and the bag is turned after sewing (figure 6.2). Again one or two lines of stitches may be used.

"M" seam.—This seam is formed by first turning in the edges to be sewn and then sewing through four thicknesses of cloth. The bag, which is not turned, is that commonly used for paper-lined bags.

Splay seam.—The selvedges of the cloth are laid together, one selvedge overlapping by about half an inch, and then sewn through two thicknesses of cloth, leaving the overlap protruding. The latter is then folded over, and again stitched through two thicknesses of cloth, the bag being finally turned.

Hemming.—Hemming is normally carried out in one of two ways, (a) using raw edges of the cloth, in which the fabric is turned in twice, and (b) on selvedges where the cloth is turned in once.

Sizes of sacks and bags

Sacks and bags can be made in any required size. A large variety of "standard goods" is, however, normally available from stock. A short selection of these is given in Tables 6.1, 6.2 and 6.3.

Standardization

Certain categories of standard Calcutta goods, when intended for export, have now necessarily to conform to the requirements of the relevant Indian Standards Institution specification in respect of cloth

TEXTILE SACKS AND BAGS

construction, strength, size, weight, etc., and all bales of such goods must carry a certification mark to that effect.

Failing such compliance/bale marking, shipment is not allowed. Goods at present included in this scheme are:

Cloth

I.S.I. No. 2818/1964	Hessian	10 oz/40 in	11 × 12
I.S.I. No. 2818/1964	Hessian	7½ oz/40 in	9 × 9

Sacks

I.S.I. No. 1943/1964	A twills	26 in × 44 in	8 × 9
I.S.I. No. 2874/1964	Heavy Cees	28 in × 40 in	8 × 9
I.S.I. No. 2875/1964	2¼ lb corn sacks	23 in × 41 in	8 × 9

Particulars and copies may be obtained from the Indian Standards Institution, Manak Bhavan, 9 Bahadur Shah Zafar Marg, New Delhi 1, India.

Sacks and bags of Dundee manufacture for delivery to home market are usually made up in uncovered bundles of 50 or less, but may be obtained in baled form if specified.

Table 6.1. Typical Calcutta standard sacks and bags (Hessian)

Name	Size	Weight	Porter and shot	Stripe	Sewing	Packing (yards)
Hessian bags	54 in × 27 in hd. w.i.p.*	7½ oz/40 in	9 × 9	plain	dry	1000
Hessian bags	56 in × 28 in hd. w.i.p.	7½ oz/40 in	9 × 9	plain	dry	1000
Hessian bags	56 in × 30 in hd. w.i.p.	7½ oz/40 in	9 × 9	plain	dry	1000
Hessian bags	50 in × 36 in hd. w.i.p.	7½ oz/40 in	9 × 9	plain	dry	1000
Hessian bags	47 in × 30 in hd. w.i.d.	10 oz/40 in	11 × 12	plain	dry	500
Hessian bags	47 in × 30 in hd. w.i.p.	10½ oz/40 in	11 × 12	plain	dry	500
Hessian bags	47 in × 30 in hd. w.i.p.	11 oz/40 in	11 × 12	plain	dry	500
Hessian bags	40 in × 22 in selv. Ex.	8 oz/40 in	9 × 10	plain	dry	1000
Hessian bags	40 in × 24 in selv. Ex.	8 oz/40 in	9 × 10	plain	dry	1000
Hessian bags	40 in × 22 in selv. Ex.	10 oz/40 in	11 × 12	plain	dry	1000
Hessian bags	40 in × 24 in selv. Ex.	10 oz/40 in	11 × 12	plain	dry	1000
Onion pockets	40 in × 22½ in hd.	12 oz	9 × 12	3 blue	tar	1000
Wheat bags	36 in × 22 in hd.	12 oz	11 × 12	plain	dry	1000
Australian bran bags	49 in × 30 in selv.	20 oz	11 × 12	plain	dry	600

* weight in proportion to

Table 6.2. Plain double warp bags

Name	Size	Weight (lb)	Porter and shot	Stripe	Sewing	Packing (yards)
Heavy Cee bags	40 in × 28 in hd.	2¼	8 × 9	plain or stripe	dry dry	400
Light Cee bags	40 in × 28 in hd.	2	8 × 8	plain or stripe	dry	400/500
E. bags	40 in × 28 in hd.	1¾	5 × 8	plain	dry	400/500
K. bags	40 in × 28 in hd.	1⅞	6 × 8	plain or stripe	dry	500
D.W. flour bags	56 in × 28 in hd.	2½	7 × 9 } 8 × 8 }	plain or stripe	dry	400
D.W. salt bags	45 in × 26 in hd.	1¾	6 × 8	3 blue	dry	500

4.69

WOOD AND TEXTILE-BASED PACKAGING

Table 6.3. Twill sacking bags

Name	Size	Weight (lb)	Porter and shot	Stripe	Sewing	Packing (yards)
A. twills	44 in × 26½ in hd	2⅜	8 × 9	3 in blue	dry	400
Liverpool twills	44 in × 26½ in hd.	2½	8 × 8	3 in blue	dry	250/300
B. twills	44 in × 26½ in hd.	2¼	6 × 8	3 in blue	dry	300/400
B. twills	44 in × 26½ in hd.	2	6 × 7	3 in blue	dry	300/400
Cuban sugar bags	48 in × 29 in hd.	2½	7 × 9/8 × 8	2 in blue	dry	400
Egyptian sugar bags	48 in × 28 in hd.	2½	6 × 8	{2 in blue 2 in magenta	dry tar	400
Australian cornsacks	41 in × 23 in hd.		8 × 9	plain	dry	300
N.Z. cornsacks	46 in × 23 in hd.	{w.i.p. 2¼ 41 × 23	× 9	plain	dry	250
Egyptian grainsacks	60 in × 30 in hd.	5	6 × 8	2 in magenta	tar	200
Egyptian grainsacks	60 in × 30 in hd.	3¼	6 × 8	2 in magenta	tar	250

Lined sacks and bags

Jute sacks and bags may be obtained with a variety of liners. A selection of these together with examples of their applications is set out in Table 6.4.

Where loose liner bags are used, the top of the liner should be folded over and closed before the jute bag or sack is closed. Loose paper liner bags should be not less than 3 in longer than the outer bags to allow for folding over, and not less than 1 in wider to prevent the inner bag bursting. Where air should not be excluded from the contents, as with certain glues, a double jute bag may be preferable.

Selection of correct size/type of sack

The selection of the correct type and size of sack for a given purpose is manifestly important, and in most instances the supplier is well qualified to advise on this. In many instances the particular application of a sack is apparent in the name by which it is normally known, for example Cuban sugar, D.W. Flour, Grain and Chaff, etc. Some typical specifications are given in Table 6.5.

In rare instances, it is necessary to determine mathematically the flat dimensions of a sack to hold a given weight of commodity. This may be achieved by the use of formulae devised by H. L. Parsons (1953).[1]

Methods of closing.—Sacks and bags may be closed by hand or machine stitching, or by tying with wire ties or twine.

Table 6.4. Jute bag liners

Liners	Usage
1. Jute fabrics combined with crepe or plain paper (united by bitumen or other water-resisting adhesives) generally known as "paper-lined".	For protection against moisture and sifting. This is used for packing such commodities as nitrates and other crystalline chemicals, pigments, fertilize. etc., provided the contents are not susceptible to contamination by the adhesive.
2. Loose crepe kraft paper liners (waxed if required).	For general protection against sifting where moisture is not detrimental.
3. Loose kraft or creped kraft paper liners (1 waxed, 1 plain).	For use under severe climatic or other conditions where jute fabric and paper united by bitumen or other adhesive is unsuitable.
4. Loose kraft paper liners. Each liner made from one or more plain kraft plies and one or more union kraft plies, the number and position of the union plies in the liner depending on the protection required.	(a) Liners with external ply (that ply in contact with the outer package) of union kraft when it is desired to prevent contamination of products by weather or contact with other materials in transit. (b) Liners with internal ply (that ply in contact with the contents) of union kraft when it is desired to prevent contents from contaminating other products during transit. (c) Liners with one or more plies of union kraft as the central plies, with the external and internal plies of plain kraft when the liner is used to reduce the permeability of the package.
5. Plastics in the form of liners, either loose or bonded to the sack, together with plastics-coated fabrics are nowadays finding increasing use, and it is reasonable to assume that this trend will continue. In general, such sacks are supplied to customers' particular requirements, and it is therefore not practicable here to define standard specifications.	

4.71

Machine stitching.—Can be accomplished by using either a stationary machine, using the normal conveyer belt feed, or by a portable stitching machine held in the hand. In either case, the line of stitching should be not less than 1 in from the edges of the fabric at the open end of the bag.

Hand stitching.—The bag should be rolled at the mouth to form two ears, and then oversewn with a packing needle. The filled bags should carry not less than 7 in of free cloth at the mouth, the mouth then being stitched to its full width, rolled down tight on its contents, twisting out an ear at each end. One ear is then securely knotted at its base, the rolled mouth overstitched, finishing with a loop knotted round the base of the other ear.

Wire tying.—The bag, which should have at least 7 in of free cloth, is bunched at the mouth and tied with a wire tie by means of a suitable tool. The wire tie should be made from not less than 16 S.W.G. wire and carry loops at each end. After tying, the end of the tie should be laid flat along the bag.

Tying with twine.—The 7 in of free cloth is "bunched" and the twine, knotted at one end is passed through both thicknesses of cloth by means of a packing needle, wound twice around the bunch, passed again through two thicknesses of cloth and securely knotted.

TEXTILE SACKS AND BAGS

Table 6.5. Jute bags and sacks

Commodity	Typical specification	
Bran	Hessian bag:	Capacity 112 lb Sizes: 27 × 54 in ex. 7½ oz/40 in; 28 × 56 in ex. 7½ oz/40 in; 30 × 56 in ex. 7½ oz/40 in Sewing: Overhead or herakles (dry jute), or union (cotton), hemmed or selvedge at mouth
Chaff	Hessian bag:	Capacity 8 bushels Size 33 × 58 in Weight of bag: Varying from 20 oz to 32 oz Sewing: Overhead or herakles (dry or tarred jute), or union (cotton), hemmed at mouth
Flour	Hessian bag:	Capacity 140 lb Size 24 × 40 in to 26 × 40 in ex. 10 oz/40 in to 10½ oz/40 in, various specifications; chested, cropped and chested, mangled, or cropped and mangled Sewing: Union (cotton), selvedge at mouth
	or tarpaulin bag:	Size: 25 × 44 in Weight of bag: 2 lb ex. 14 porter 15/16 shots Sewing: Splay seam, two sides (dry jute) hemmed at mouth
	or twilled sack:	Size: 25 × 44 in Weight of sack: 2½ lb ex. 10 porter twilled sacking 11/12 shots Sewing: Splay seam, two sides (dry jute), hemmed at mouth
Grain	Twilled sack:	Capacity 4 to 4½ bushels Size: 28 × 56 in, 10 or 12 porter sacking Weight of sack: 4 lb Sewing: Splay seam or overhead 2 sides (jute or hemp), hemmed at mouth, 3 in vents barred both sides in the case of splay seam
Nuts and bolts		Pockets made from good second-hand bagging or heavy twill sacking: Sewing: Union (flax or heavy cotton) or lockstitch Alternative— New twilled sack: ex. 8 porter 16 oz/27 in sacking Sewing: Herakles, overhead or single lockstitch (jute)

4.73

Table 6.5. Jute bags and sacks (*contd.*)

Commodity		Typical specification
Oats	Hessian bag:	Capacity 4 bushels Size: 27×54 in ex. 9 oz/40 in to 12 oz/40 in hessian Sewing: Union (cotton), herakles or overhead (dry jute), hemmed or selvedge at mouth
Potatoes	Hessian bag:	Capacity 56 lb Size: $19\frac{1}{2} \times 35$ in or 20×34 in ex. $7\frac{1}{2}$ oz/40 in to 10 oz/40 in hessian Sewing: Union (cotton), herakles or overhead (tarred jute), hemmed at mouth Capacity 112 lb Size: 22×40 in to 24×40 in ex. $7\frac{1}{2}$ oz/40 in to 10 oz/40 in hessian Sewing: Union (cotton), herakles or overhead (tarred jute), selvedge at mouth
	or twilled sack:	Capacity 112 lb Size: 23×41 in, $2\frac{1}{4}$ lb Australian corn sack
Powders	New laminated bag (paper-lined as described in Table 6.4)	Quality, weight, material, sizes and linings to meet requirements Sewing: Lockstitch

BIBLIOGRAPHY

1 Parsons, H. L., *J. Textile Inst.*, vol. **44**, 1953; or B.S. 1133 Section 9.

CHAPTER 7

Bales and Baling

W. G. Atkins

The process of baling is normally confined to compressible articles that are not likely to be damaged by compression. The press packing of bales reduces their volume, thereby effecting saving in shipping and warehouse space and costs.

A bale is a quantity of supplies, often highly compressed, normally of cubic shape, and formed in one of the following ways:

1. Unwrapped, and tied with rope or cord, or secured with wire or metal strapping.
2. Protected by one or more coverings and stitched on the outer fabric cover.
3. Protected by one or more coverings, stitched on the outer fabric cover and strapped with metal hoops.

A truss is similar to a bale, but the term is generally used to denote a small and less highly compressed unit package. Bales and trusses may, within reason, be of any size, dependent upon the nature of the materials to be packed, the forming pressure to be applied, and the handling facilities available at the points of assembly and destination.

The sequence of operations in the press packing of bales is:

(a) Where wrappings are used, the positioning of the lower portion of such wrappings on the press.
(b) Stacking of the materials to be packed on the wrappings.
(c) Positioning of the upper half of the wrappings on the goods or materials.
(d) Applying the pressure.
(e) Sewing the outer covering of the wrapper(s).
(f) Positioning and fastening the ropes, wires or metal straps.
(g) Releasing the pressure.
(h) Applying the bale markings.

In assembling the articles to be baled, note should be taken of the following points. Bulky parts should not be placed one on top of another, but positioned relatively to adjoining articles in such a manner as to compensate for any irregularity of form. Metal protuberances, such as buckles attached to the articles intended for baling, must be staggered, otherwise the application of pressure in the process of baling will most probably result in damage to both the metal (or similar part) and the article to which it is attached. Similarly, where the goods to be baled comprise a number of tied parcels, the knots must be at the ends of the parcels. It is essential that all articles to be baled are dry, otherwise mildew is likely to develop during transit or storage.

Covering

In general, bales are covered with jute fabrics, with or without a separate liner or liners of paper, with a view to protecting the contents from the effects of moisture, dirt, abrasion, insects, etc. The number and type of such protective wrappings will vary according to the length and hazards of the journey and the storage conditions obtaining at the destination.

Bales intended for dispatch to a destination in Great Britain might well be adequately protected by one covering of $9\frac{1}{2}$ oz/40 in hessian with or without an inner covering of kraft paper. Where oversea transport is involved, two coverings of hessian, one layer of waterproof creped paper and one layer of plain kraft paper are necessary. In the case of long sea voyages, particularly to countries in the Far East and for which purpose it is normal to form heavy bales, the outer hessian covering should not be lighter than $10\frac{1}{2}$ oz/40 in hessian. Where the hazards of the journey are expected to be exceptionally severe, it is common to use one layer of kraft paper, one of $6\frac{1}{2}$ oz/40 in hessian, one of "India tarpaulin" ($10\frac{1}{2}$ oz/40 in hessian coated with bitumen), and one outer of plain $10\frac{1}{2}$ oz/40 in hessian. It is usual in such instances to insert layers of greaseproof or other protective papers on either side of the bitumenized hessian in order to prevent softening of the bitumen by the oil content of the plain hessian.

It will be apparent that it is not possible to lay down any standard as to the type of fabric wrappings for use in baling, as these will be dependent on the degree of compression and the method of tying employed in forming the bale. In general, however, for heavy bales, where the maximum pressure is applied and steel hoops are used, the inside hessian should be $6\frac{1}{2}$ or 7 oz/40 in and the outside $10\frac{1}{2}$ oz/40 in hessian. Small bales formed under relatively light pressure and strapped with steel may be sufficiently protected by one layer of 9 oz/40 in hessian, though $10\frac{1}{2}$ oz/40 in would be preferable.

Forming of the bale

As previously indicated, the positioning of goods on the baling press is of great importance if damage from pressure is to be avoided. It is of equal importance when baling bundles or parcels to "break joint", otherwise it will be impossible to form a firm and compact bale, and cutting of the goods by the steel straps will be almost inevitable. In such instances, the pattern of piling of the goods should, as far as possible, resemble the positioning of bricks in a wall, i.e. the units in each course should lie at right angles to those in the course upon which they are placed, and joints should not lie one above another. When baling articles of a resilient nature, capable of accepting a high reduction in volume without damage, it is usual to build these up in a former or mould (box baling) in order to hold them in shape during the compression.

Compression

One of the objects of baling being the conservation of space, it is axiomatic that all bales should be of a maximum density consistent with the protection of the contents. Pressures employed in baling may vary from $600 \, lb/in^2$ to $4500 \, lb/in^2$ but the amount of compression desirable in a particular instance is entirely dependent on the nature of the goods being baled. Whilst, for example, textiles in piece-goods form will withstand a high degree of compression, stuffed and quilted textiles are preferably not compressed at all. It is obvious, therefore, that it is impossible to lay down a common standard for the degree of compression, and that this must be determined by the type of goods.

Strapping

Rope is not an entirely suitable material for the binding of bales in that it contracts in damp weather and stretches when exposed to a dry atmosphere. Its use should be confined to lightly pressed bales, the contents of which are of relatively low value.

Steel strapping is the most suitable tying material and is the most commonly employed. For small lightly-pressed bales, cold-rolled high-tensile steel $\frac{5}{8}$ in wide and not less than 26 B.G. should be employed. The strapping should be treated to prevent rusting. When applying strapping of this size, the outside straps should be positioned not less than 5 in from each end of the bale, the intermediate straps being placed equidistant one from another and from the end straps. Stretching should be achieved by the use of a mechanical stretching tool, and the degree of tension applied must be sufficient to ensure that the compression of the bale is fully held. Sealing should be carried out by the use of a metal sleeve or similar device, designed to be crimped or punched. The loose ends of straps should be either folded under or cut off so close to the seal that no projecting edges are left.

For bales of large size and weight, normally subjected to a high degree of compression, the straps should not be less than 1 in wide and no less than 18 B.G. thick. (In exceptional instances, e.g. for baling jute goods, 16 B.G. is specified.) These straps which are made from hot-rolled hoop

Table 7.1. Typical packing specifications

Contents		Gross weight (lb)	Approx. size of bale	Covering material	Approx. density ft³/cwt	Strapping or banding
Nature	Yards					
28 in khaki drill	1000	521	2'5" × 2'3" × 2'2" = 11·8 ft³		$2\frac{1}{2}$	3 hoops $1\frac{1}{4}$ in wide, 19 gauge
80 in bleached shirting	400	343	1'11" × 1'11" × 2'2" = 8·0 ft³		$2\frac{1}{2}$	3 hoops $1\frac{1}{4}$ in wide, 19 gauge
36 in rayon dress goods	1200	436	3'2" × 1'10" × 2'6" = 14·5 ft³	(i) Kraft paper inside (ii) Crepe union waterproof paper (iii) Double thickness of $9\frac{1}{2}$ oz or 10 oz 40 in hessian outside, boards top and bottom (optional)	$3\frac{3}{4}$	3 hoops $1\frac{1}{4}$ in wide, 19 gauge
36 in butter muslin	3600	355	3'2" × 2'2" × 2'2" = 14·6 ft³		$4\frac{1}{2}$	3 hoops $1\frac{1}{4}$ in wide, 19 gauge
46 in printed cotton furnishing	800	342	3'10" × 2'0" × 2'4" = 14·0 ft³		$4\frac{1}{2}$	3 hoops $1\frac{1}{4}$ in wide, 19 gauge
24 in terry towelling	1000	462	4'2" × 2'2" × 3'0" = 27·1 ft³		$6\frac{1}{2}$	3 hoops $1\frac{1}{4}$ in wide, 19 gauge
Turkish towelling	doz. 46	268	3'4" × 2'2" × 2'3" = 16·3 ft³		$6\frac{3}{4}$	3 hoops $1\frac{1}{4}$ in wide, 19 gauge

iron are referred to as "baling hoops" and should be lacquered or otherwise treated against rusting. Such hoops should be fastened by buckles or metal studs.

Marking

Bale marking is carried out by the use of stencils and a waterproof ink or paint. It is important to avoid the use of penetrating fluids. Tags or labels should not be used. Although, as has been implied above, it is not possible to lay down standards for the process of baling, the typical packing specifications given in Table 7.1 may prove useful as a guide.

PART FIVE

ANCILLARY MATERIALS

CHAPTER 1

Paper, Plastics and Fabric Sealing Tapes

A. O. D. Davies

Adhesive sealing tapes have long been accepted in packaging as a speedy, clean and efficient means of closing and sealing fibreboard cases, fibre drums, paperboard boxes and cartons, and paper-wrapped merchandise of all types. The correct selection of tape is of paramount importance, and careful consideration must be given to the specific pack on which it is to be used. Such matters as the width of tape, its strength, and its adhesive grip must be assessed with reference to the weight, bulk and value of the pack, and the transit hazards involved.

In addition, the means of application warrant careful thought on the grounds of both efficiency in sealing and economy in the use of labour. Efficient sealing necessitates a careful study of all these aspects, and consultation with the tape manufacturers in the preliminary stages is advisable.

Sealing tapes may be considered in three main divisions:

Glued or gummed tapes, paper or fabric-based, plain or reinforced—water-activated.
Self-adhesive or pressure-sensitive tapes, paper, plastics or fabric-based—already activated.
Heatfix or thermo-sensitive tapes, paper or fabric-based—heat-activated.

These divisions differ widely in many aspects, and each will be considered in some detail.

Gummed tapes

Adhesive compatibility in general usage

Gummed tapes will adhere satisfactorily to all packages or containers made from paper or paperboard, but are not satisfactory on metals or plastics. By virtue of their adhesive bond, they become, when properly applied, an integral part of the pack, providing considerable reinforcement to weak points. The added protection provided by mitred corners, as recommended for the traditional H taping of a fibreboard case, and the "gaitering" of large kraft-wrapped bales are examples of this. Tapes manufactured from hard-sized kraft paper and efficiently applied, provide reasonable moisture and water-resisting properties, adequate for normal transport hazards.

Materials

Base papers should consist of a kraft paper hard-sized to resist adhesive penetration, and flexible enough to conform with the shape of the package. Where exceptional strength is required, heavy-weight duplex glass-fibre-reinforced base papers may be used. The alignment of the glass-fibre strands may be varied according to requirements to the degree that they are virtually impossible to tear or rupture. Such tapes are used for single-strip case sealing and to produce the manufacturer's join on many fibreboard containers. Fabric and cloth glued tapes are available for specialized uses, mostly where reinforcement is required.

Adhesive coatings are mainly non-protein, based on modified starches, on which much progress has been made and which have now largely superseded animal glue.

Application and uses

Most gummed tapes are obtainable in different widths ranging from 24 mm up to 288 mm. They can be printed in several colours with a trade mark, slogan or other information relative to the contents of the package. Colour codes which can be used immediately to identify a batch or quality are also available.

The British Post Office accepts parcels sealed with tape alone for registration, providing that such tape bears the name or distinguishing mark of the sender. The Railway Clearing House of the British Transport Commission accept fibreboard containers at Railway Company risk when tape-sealed in accordance with their specifications.

The efficient application of gummed tape depends on correct moistening, either too much or too little moisture being detrimental to ultimate adhesion. Modern dispensing machines for gummed tape employ automatic moistening devices and will cut pre-set lengths of tape, correctly moistened for immediate application. Electrically operated and remote controls enable both hands of the packer to be free for the application of tape. Pre-determined lengths of tape are used when rectangular-shaped packages are being sealed by the H method and, for large-scale production, fully and semi-automatic case taping machines are available, both for single-strip or H-method applications (figure 1.1).

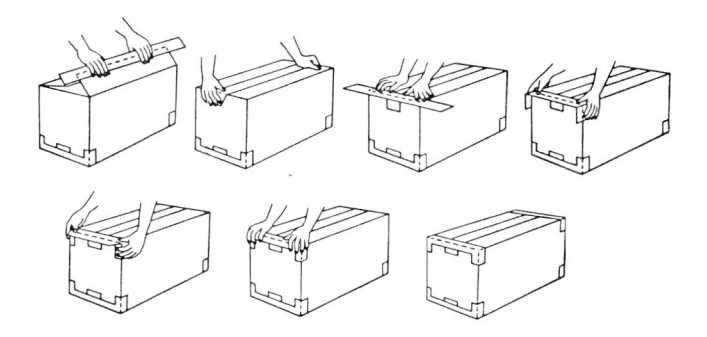

Figure 1.1 Method of applying gum strip to fibreboard cases

Storage

Gummed tapes should be kept in the original waterproof wrappers until required for use, and should be stored in comfortable warmth in a dry place. Storage in cold damp conditions, or on concrete floors in unheated store rooms, should be avoided.

Pressure-sensitive tapes

Adhesive compatibility in general use

Pressure-sensitive tapes will adhere satisfactorily to packages or containers made from paper, paperboard, metal, glass or plastics. Such tapes need only the application of pressure to cause them to adhere to almost any surface. Apart from seeing that the surface is free from dust, oil, grease or moisture, no special preparation before application is necessary.

Materials

Pressure-sensitive tapes are made from a very wide range of base materials, the adhesive coating being selected for its suitability to both the base material and the application. The adhesive is normally formulated from a rubber or synthetic elastomer and a resin. The elastomer is tackified to the required degree with the appropriate resin, thus providing

ANCILLARY MATERIALS

a method of varying the adhesive properties. The principal base materials used for tapes are:

Regenerated cellulose film.—Cellulose film in transparent or coloured form is used for making tapes for the general purposes of sealing, parcelling and bundling. Such tape is not waterproof but will resist oil, grease and many solvents.

Cellulose acetate film.—The properties of cellulose acetate film make it suitable for self-adhesive tape manufacture, particularly for special sealing duties in corrosive or damp conditions where cellulose tape would fail.

Polyvinyl chloride film.—The range of vinyl tapes covers many applications. The general packaging grade PVC is unplasticized and has a low water-vapour permeability. It has excellent mouldability to conform to uneven surfaces. Other advantages are high strength and ability to withstand deep freeze conditions. It is the most generally used form of filmic tape, both industrially and commercially.

Fabric.—Bleached cotton fabrics are the types principally employed for the manufacture of high-strength tape. They are frequently used for metal-canister and fibre-drum closures. Similar tapes with a low-adhesive-strength coating are used as protective and masking tapes.

Moisture-resistant paper.—A strong kraft paper coated on one side with a silicone resin to provide both moisture resistance and to permit very easy unwinding from the reel. It is a good general-purpose sealing tape of high strength, and is also very suitable for use in deep-freeze conditions.

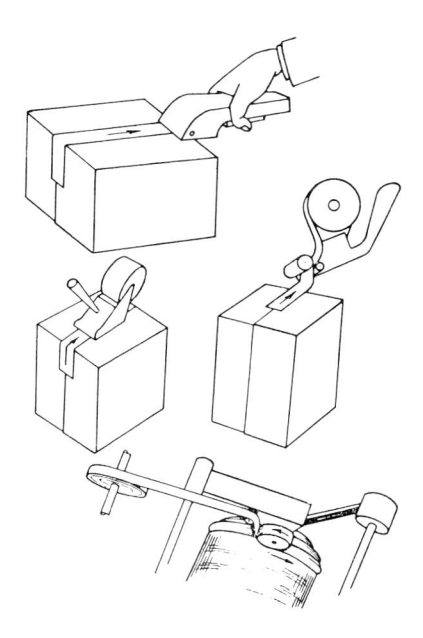

Figure 1.2 Applicators for pressure-sensitive tapes

Reinforced banding tapes.—A particularly strong type of pressure-sensitive tape, the backing usually of unplasticized PVC, is reinforced with rayon, nylon or glass-fibre filaments to give extremely high tensile strength. There is also a considerable range of self-adhesive tapes produced for specific industrial purposes, e.g. masking, insulating and other special protective requirements.

Availability

The greater proportion of all self-adhesive tapes can be printed and are obtainable in several colours. Most tapes are supplied in widths from 12 mm up to 96 mm and, as with gummed tape, there are many dispensing machines available, ranging from small hand applicators to fully automatic case-sealing machines (figure 1.2).

Storage

Bales of self-adhesive tape should be stored in their original wrappings until required for use, and should be protected from dust, solvent fumes and direct sunlight. High temperatures and humid conditions will also accelerate tape deterioration.

Heatfix tapes

Heatfix or thermo-sensitive tapes will adhere satisfactorily to all packages or containers made from paper or paperboard, glass and a wide range of plastics, provided the surface is clean and dry. Such tapes, once activated by heat, need only the application of pressure to cause them to adhere.

Materials

Heatfix tapes are in the main coated on to a kraft-base paper. The adhesive coating is normally formulated from a thermo-plastic emulsion which, once activated by heat, becomes immediately self-adhesive and retains this property for several hours, before finally "setting off" to give a dry permanent bond. This particular type of adhesive is referred to as a *delayed-action heatfix*. Other types of adhesive formulations are described as *immediate action* and require the tape to be activated by heat and applied to the packaging surface *in situ*.

Application and uses

Thermo-sensitive tapes are obtainable in different widths ranging from 36 mm to 72 mm. They can be printed in one or two colours. The equipment to apply heatfix tape correctly is available both in the form of a bench dispenser, where an operative can dispense a required length of tape by hand, as well as both semi- and fully-automatic case taping machines. The latter are designed to apply automatically a single strip of tape centrally to both the top and bottom of a carton or case.

Storage

Thermo-sensitive tapes should be kept in their original waterproof wrappers until required for use. They should be stored in a dry place and be protected from solvent fumes and direct sunlight. High temperatures must be avoided, as these could result in tape "blocking" if present for long periods.

Economics

As already stated, the correct selection of tape for any operation can be determined only by consideration of the specific packaging application.

Pressure-sensitive tapes are more expensive than gummed tapes but, since in packaging the overall cost of sealing rather than the cost of the material alone is important, it is necessary to take into account such factors as labour, productivity, and the relative lengths and widths of the possible tapes employed.

Heat-activated tapes offer many of the advantages of self-adhesive tapes and are priced midway between gummed and pressure-sensitive tapes.

Generally a detailed study of the problem will highlight an obvious choice. The manufacturers of sealing tapes are always ready to give advice, demonstrations and to co-operate in field tests in order that the efficiency of any tape to fulfil its function may be confirmed.

CHAPTER 2

Strapping and Stapling

V. Radcliffe

Steel or non-metallic strapping or wire can be used to perform one or more of the following packaging functions:

Bundling   Grouping and holding together several articles into a larger handling unit.
Palletizing   Securing one or more articles to a pallet.
Unitizing.   Bundling or palletizing with provision for pick-up by mechanical equipment.
Reinforcing.   Strengthening a shipping unit to withstand the hazards of transportation and handling.
Closing.   Securing the lid or top of a container, as with a telescope box or an interlocking flange container.
Baling.   Holding material together under compression, especially resilient items, to save space.

Steel is by far the most predominant strapping used, but non-metallic strapping, especially rayon, nylon and polypropylene are finding an increasing number of uses.

Steel costs less for equal length of equal strength. The ability of steel strapping to hold is not affected by age, heat or cold, sunlight or dampness, dirt or oil, and steel is not attacked by mould or insects. Some of the non-metallic strapping materials, however, are more convenient to dispose of (as by incineration), are more resilient and so may stay tight on a shrinking package, are light in weight and are non-staining.

Strapping was first applied by nailing it on to wooden containers. In 1913 tools capable of tensioning and sealing the strapping were introduced, and this greatly speeded the process and made possible its application to a wider variety of shipments, including those enclosed in paperboard, wrapped in paper and unwrapped. The increasing use of mechanical lift

ANCILLARY MATERIALS

trucks and other handling equipment has contributed to a very substantial growth in the use of strapping and similar binders in recent years.

Metal strapping materials are available in a variety of cross-sections (figure 2.1). Each of these has its own advantages and disadvantages. Wire can be twisted together at the ends to eliminate the cost of a separate seal. It can be draped diagonally across the face of the package if desired, but it tends to cut into the package at the corners, to stand away from the package, is harder to bend at the corners (especially in heavier wires) and is more difficult than flat strapping to pull tight around the package. Flat strapping can generally be applied faster with manual tools than can wire; it conforms better to the shape of the package, offers less friction for more effective tensional reinforcement, and is firm and straight, and so can easily be passed through a pallet or fed through chutes or tracks. Oval or flattened wires have characteristics intermediate between flat strapping and round wire.

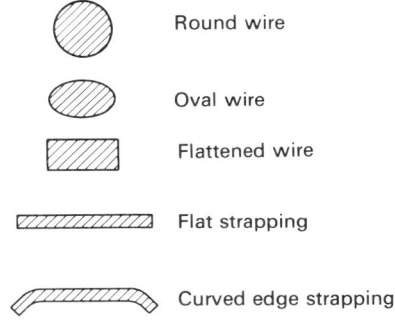

Figure 2.1 Cross-sections of metal strapping materials

Steel strappings and wires are available in a variety of finishes. A galvanized finish provides long-term protection against rust, roughly proportional to the amount of zinc deposited. Premium-grade galvanized strapping generally is protected by about $50\,\text{g/m}^2$ of zinc, while the more

Table 2.1. Typical test results to show the rust-free life of galvanized strapping

Type of service	Type of atmosphere	Amount of zinc/m²* 30 g	60 g
Indoor	clean rural	10 years	18 years
Indoor	urban	4 years	7½ years
Indoor	highly industrial	2 years	3¾ years
Outdoor	clean rural	2 years	3½ years
Outdoor	urban	1 year	1½ years
Outdoor	highly industrial	¼ year	¾ year

* Note: 1 oz per square foot is approximately $300\,\text{g/m}^2$.

usual commercial grade carries 25–30 g/m². The expected substantially rust-free life of galvanized strapping is shown in Table 2.1.

Steel strapping is also available in a choice of painted finishes. Some of the best painted finishes rival galvanized strapping in rust resistance. Both blue oxide and black oxide finishes have been popular in the past, while bright (untreated) metal finish is still manufactured, but rusts too rapidly for most packaging purposes.

Strapping for use in feed-wheel-type tensioning tools should have a finish coated with the proper type of wax or other lubricant to withstand the extreme pressures exerted by these tools. This lubrication also minimizes the loss of tension in pulling the strapping around the corners of a package. However, lubrication makes it more difficult to obtain a securely sealed joint; for this reason heavy-duty strapping that will be subject to heavy shock loads is often furnished "dry" (unlubricated).

Steel strapping

Strength

When strapping is bought, it is bought mainly for strength—strength to reinforce a package, strength to hold a bundle together, strength to hold compression in a bale, strength to hold a load in place. What happens when there is not enough strength? The package breaks open, the bundle comes apart, the bale expands, the load shifts and the purpose is not fulfilled. That is not economy. The selection of the right material for the purpose is essential.

Table 2.2. Tensile strength and weight of heat-treated and hot-rolled strappings

Width mm (in)	Thickness mm (in) hot-rolled	Thickness mm (in) heat-treated	Average tensile strength (N)	Weight (m/kg)
13 (½)		0·50 (0·02)	6450	19·6
13 (½)		0·50 (0·02)	8050	15·8
15 (⅝)		0·57 (0·023)	8900	13·6
18 (¾)		0·64 (0·025)	11650	10·5
18 (¾)		0·79 (0·031)	13790	8·5
31 (1¼)		0·79 (0·031)	24240	5·1
31 (1¼)	0·89 (0·035)		22680	4·4
31 (1¼)		1·12 (0·044)	33850	3·6
31 (1¼)	1·40 (0·050)		31150	3·0
50 (2)	1·40 (0·050)		48900	2·0

ANCILLARY MATERIALS

Adequate elongation

The *elongation* is a measure of the amount that strapping can stretch. It is a measure of its ability to absorb impact shock by "rolling with the punch". All hot-rolled medium-carbon strapping and cold-rolled heat-treated medium-carbon strappings should have an elongation of from 5 to 16%. On the other hand, elongation is an undesirable characteristic in those applications for which cold-rolled strapping is used. Too much elongation in such situations allows a strap to become too loose to reinforce packages adequately. The elongation of cold-rolled strapping is only between 1 and 3%, and this is ideal for package reinforcement.

Flat steel strapping is produced in four basic types with important differences in their physical characteristics:

Hot-rolled medium-carbon strapping has considerable elongation and is rather ductile, permitting heavy impact shocks to be absorbed by yielding (Table 2.2).

Cold-rolled low-carbon strapping has a high yield point and low elongation. The cold-rolling process permits an improved finish and closer control of tolerances, and this is the type of strapping most used in packaging (see Table 2.3).

Cold-rolled heat-treated medium-carbon strapping combines the improved finish and closer control of the cold-rolling process with the greater ductility of the hot-rolled product, together with a substantial gain in strength. It is a premium grade, justifying its extra cost through improved performance in many demanding applications.

Cold-rolled annealed strapping is a soft strapping capable of repeated bending and easily nailed through. It is widely used as a nailed-on reinforcement on cases for bottles.

Table 2.3. Tensile strength and weight of cold-rolled low-carbon strappings

Strap size (mm)	(inches)	Average tensile strength (N)	Weight (m/kg)
7½ × 0·30	5⁄16 × 0·012	1950	52·6
9 × 0·25	3⁄8 × 0·010	2000	52·6
9 × 0·38	3⁄8 × 0·015	2890	35·1
9 × 0·50	3⁄8 × 0·020	3690	26·3
13 × 0·25	½ × 0·010	2670	39·5
13 × 0·38	½ × 0·015	3825	26·3
13 × 0·50	½ × 0·020	4890	19·7
13 × 0·57	½ × 0·023	5515	17·2
15 × 0·38	5⁄8 × 0·015	4800	21·0
15 × 0·44	5⁄8 × 0·018	5600	17·6
15 × 0·50	5⁄8 × 0·020	6140	15·8
15 × 0·57	5⁄8 × 0·023	6890	13·7
18 × 0·38	¾ × 0·015	5740	17·6
18 × 0·50	¾ × 0·020	7340	13·1
18 × 0·57	¾ × 0·023	8270	11·5
18 × 0·70	¾ × 0·028	10000	9·4
18 × 0·89	¾ × 0·035	11560	7·5

Hand tools

(a) *Tensioners*

Tensioners are used to tighten a loop of strapping around the object to be strapped, to apply tension on two overlapping strap ends. Most tensioners have a base which goes under the strapping (figure 2.2) and which rests on a flat or nearly flat surface of the object. If this surface is narrow or if its girth is small, the freedom of the tensioner to move with take-up may be very limited. Some tension will be lost when the base is taken out from under the strapping. To avoid such loss, a push-type tensioner which has no base under the strap is required (figure 2.3).

Since it operates by pushing against the seal (by "tightening a slip knot") the push-type tensioner requires that seals be threaded on to the strapping, and the end of the strapping bent back under the seal. Push-type tensioners are preferred for small or irregular bundles and for narrow packages.

The majority of strapping applications utilize conventional tensioners with a base under the strapping. These are usually faster to use than the push type, and permit somewhat higher tensions to be attained, but they require a wide enough surface to permit the tool to move with take-up.

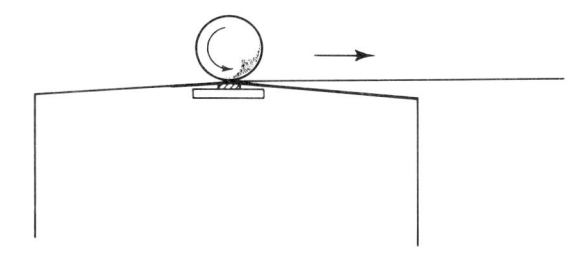

Figure 2.2 Steel-strap hand tensioner with base under strap

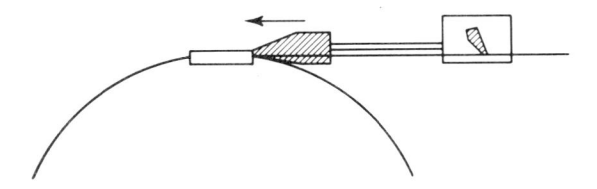

Figure 2.3 Push-type tensioner (no base under strap)

(b) *Sealers*

Sealers are used to join the overlapping strap ends with a metal seal. They mus. do this reliably, always producing a strong secure joint. A notch-type sealer cuts into the outer edges of the strapping and seal, and

ANCILLARY MATERIALS

turns the resulting tabs down (regular notch) or up (reverse notch). This type is generally used on waxed or coated strapping.

(c) *Combination tools*

The functions of both tensioner and sealer, and often cutter as well, may be provided in a single tool. It eliminates reaching for and then putting down these separate tools. A seal-feed combination tool carries a stack of seals in a magazine, and eliminates reaching for a seal and manually placing it on the strapping. For most efficient use on a production line, a combination tool should be mounted close to its operating position.

Power strapping machines

A power strapping machine can bring a variety of benefits. It increases an operator's productivity, so that he can perform additional functions such as inspection, stacking, labelling, marking and the like, as well as speeding the strapping operation. It improves the reliability and uniformity of the strapping job. It makes the strapping job easier to learn and more pleasant to perform.

There are three stages of automation in this equipment:

Semi-automatic machines leave to the operator the following functions: he positions the package in the machine, pushes the strap feed control, takes the end of the strapping from the upper chute, passes it over and down in front of the package and inserts it into a slot in the lower chute, pushes the cycling control, and removes the package. These machines save the cost and complication of the more automatic control systems, are extremely flexible as to package shapes and sizes, and are used where little would be gained by freeing the operator for other tasks.

Automatic machines require only three functions of the operator: positioning the package in the machine, pressing the foot switch, and removing the package. An operator with an automatic machine often applies more than a thousand straps per hour, in addition to performing other functions such as labelling and marking. Such speeds are, of course, limited to packages or bundles, light and small enough to be moved in and out of the machine at that rate.

Operator-less machines make use of a powered conveyor to bring the package in and out of the machine, a system of automatic controls to provide proper spacing and sequence, and a pusher cylinder or skewed conveyor rollers to position the package against the machine head. Operator-less machines have been installed which include the following position controls:

STRAPPING AND STAPLING

A photoelectric cell or feeler switch senses entry of the package. Straps are put on at predetermined distances from the front of the package.

The feeler switches locate cleats on a crate or box; a strap is applied over each cleat.

The carton or wrapper carries conductive or magnetic ink marks at each strap location. The head is sensitive to these marks, but to no other marking on the package.

Automatic machines have also been developed for specific industrial requirements, e.g. to secure steel coils, bricks, tubes, paper reels, compressed loads of corrugated containers and bagged products.

Non-metallic strapping

Non-metallic strapping in nylon, polypropylene or rayon cord has been introduced to overcome certain shortcomings found in all forms of steel strapping and wire. Table 2.4 gives some average figures for the various types of material used. In choosing a strapping material, a number of factors should be considered: price of strap, strength, elongation, elastic recovery, preferred method of application, danger of rusting and of damage to the packaged goods, safety hazards, disposability problems, etc. The first question is whether the strap must restrict or accommodate movement of the package. Basically there are three types of package:

1. Packages which tend to expand.
2. Packages which do not change in size.
3. Packages which tend to shrink.

Table 2.4. Comparison of strapping characteristics

	Steel strapping	Nylon and plastics strapping	Typical rayon cord strapping
Tensile strength (kN/m^2)	700,000–1,000,000	400,000–450,000	220,000–240,000
Tension transmitted around a 90° corner (%)	72%	72%	56%
Ease of feeding and threading	excellent	good	poor
Smoothness	excellent	excellent	fair
Stability under long-term load	excellent	good	fair
Weight (density kg/m^3)	7750	1100	1400
Resistance to abrasion	excellent	excellent	poor
Resistance to shearing (as on a sharp corner)	good	good	fair
Ease of disposal	good	excellent	excellent
Can be incinerated	no	yes	yes

ANCILLARY MATERIALS

Today mainly three types of strapping are used in Europe:
> steel
> nylon
> polypropylene

Figure 2.4 illustrates how the different types satisfy the service and package requirements.

Elongation and elastic return

When non-metallic strapping is placed under tension, it will elongate between 5% and 7½% as the tension is increased to load limit, according to the type of non-metallic strapping used. Elongation expresses how much the strap stretches when it is tensioned. Normally only the elongation at break is given, i.e. how far the strap will stretch before it breaks. This information is useful, but it is also very important to know how much the strap will elongate at any given tension. In other words to know the curve in figure 2.5 which compares steel strap, nylon strap and a typical polypropylene strap.

The first part of the curve (shaded) indicates which elongation of a non-metallic strap can be obtained in practice during a typical application. Generally there is a greater elongation with nylon and plastics strapping than with rayon type. When the tension is released, non-metallic strapping in most instances will return to its original length. The benefit of such elasticity is that, in the course of time, the tension on certain packages may be reduced because of fatigue (fibreboard cases), drying out (wooden boxes) or settling (bales).

Transmission of tension

As non-metallic strapping is brought under tension, it can be seen to glide smoothly and easily around the package. The low coefficient of friction prevents binding at the corners. The difference in strap tension between top, sides and bottom of the package is thereby reduced to a minimum.

Package conformation

The flexibility of non-metallic strapping enables it to "hug" the package and remain in position despite some irregularity in package shape. It can be successfully applied to some irregular shapes where a non-flexible binder, in order to remain in place, may require tension sufficient to damage the package. Good conformation also makes the strapping less subject to snagging.

STRAPPING AND STAPLING

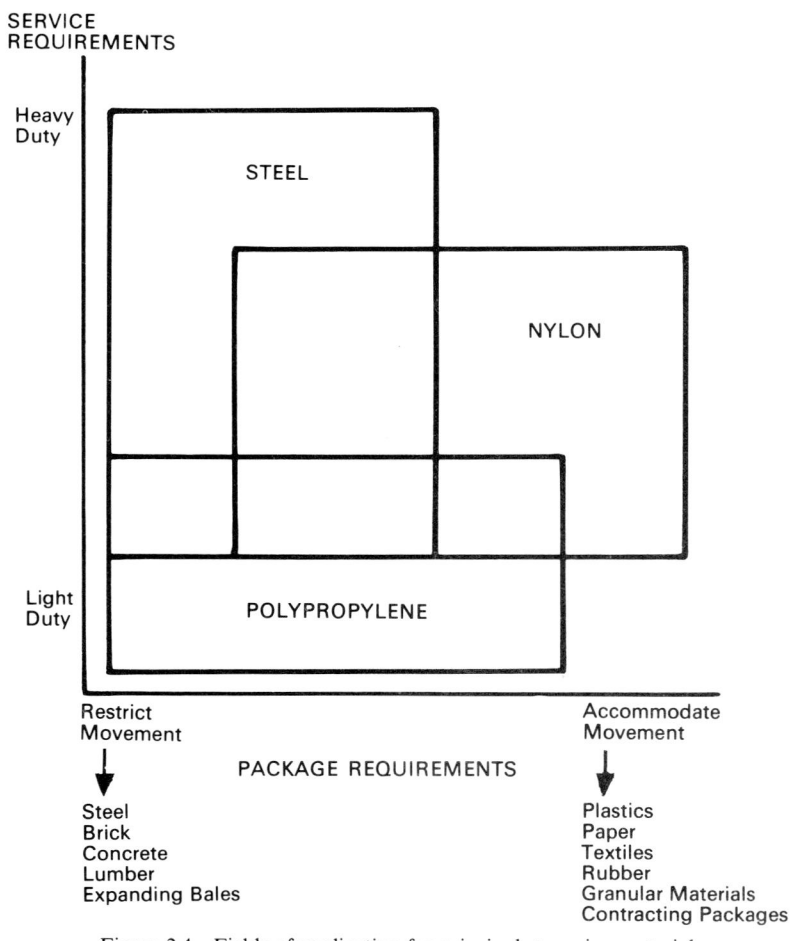

Figure 2.4 Fields of application for principal strapping materials

Moisture resistance

Moisture has no effect on non-metallic strapping for all practical purposes. Although shippers normally take precautions against damage to their packaged products, conditions cannot always be controlled. Where nylon or polypropylene strapping is used, high humidity conditions will not affect performance, but humidity can cause rayon strapping to change in length and lose tensile strength. Moreover, when rayon strapping becomes wet, it may disintegrate.

ANCILLARY MATERIALS

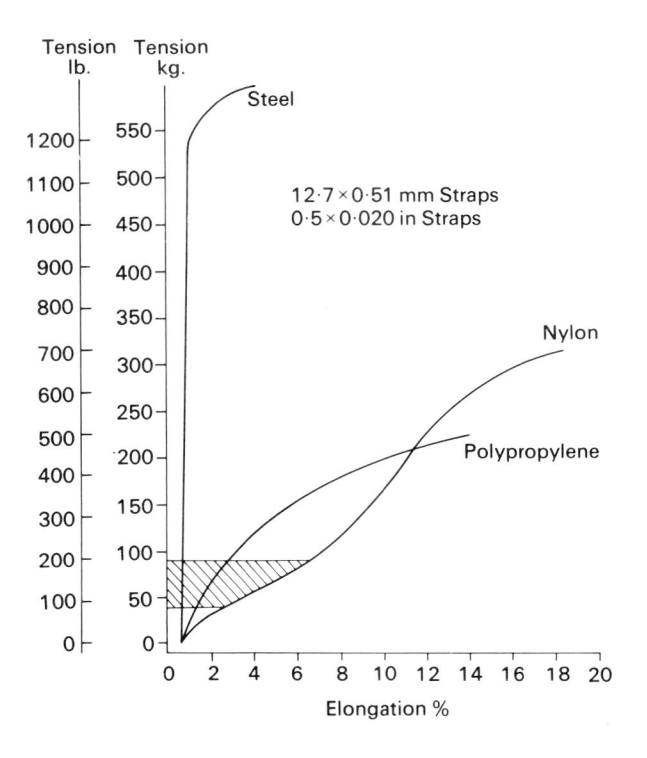

Figure 2.5 Elongation-tension curves for steel, nylon and polypropylene strap

Types of tool available

It is true to say that most non-metallic strapping can now be used with the same type of equipment as for steel strapping. In addition, machines which form the joint in non-metallic banding by a welding process have also been developed, the heat provided by electricity or by friction and impact welding, providing a completely non-metallic binding with no metallic seal. Light binders, i.e. automatic machines using strap sections of about 6 mm (¼ in), are now available with specially adapted models for wet fish packing and newspaper bundling. However, developments are so rapid in non-metallic strapping and tools that it is advisable to consult the suppliers. It is also suggested that comparisons should be made between each type of product.

5.18

Stapling

In recent years the King Size staple has gained favour as a corrugated fibreboard case closing method. These large preformed staples are clinched on a retractable anvil, thus enabling the case to be closed after it has been packed.

The main advantages are:

1. *Economy and speed of closure.*—The average case requires only four staples in the top, and the same number in the bottom, and can be closed in as little as five seconds.
2. *Security.*—It is almost impossible to pilfer a case closed in this way without detection, and the staples give an extremely strong closure which is not affected by atmospheric conditions.
3. *Efficiency.*—Closure can be made in the production area, and there is no problem with moving or storing pre-assembled cases. No drying time is involved as with some glues and paper tapes, and the case is thus ready for dispatch as soon as it is sealed. The strength of the closure readily lends itself to stacking, with little fear of the case flaps opening.

A wide range of equipment is available, including portable machines which can be either hand or air-operated. There are also production-line machines which will staple top and bottom, or either, or both ends of a case simultaneously, and there is also a completely automatic operatorless model available. The machines are adjustable, so that it is possible to control the depth to which the anvil hooks penetrate the board, and also the tightness with which the staple is clinched. There are several leg lengths of staple available, the length used being governed by the depth of fluting in the board, and also the case design. The most common is a 15 mm ($\frac{5}{8}$ in) leg which would, for example, be used on a centre seam of A-flute corrugated cases.

This type of closure does, of course, have limitations. In the first place it is not possible to drive the staples unless the case flaps are supported, i.e. unless the goods are packed very close to the top of the case, particularly with the B flute board. It is sometimes necessary to use a top pad of corrugated board to prevent damage to the contents.

King Size staples are most commonly used on corrugated fibreboard cases but, in ideal conditions of internal packing and with the use of an air-powered machine to drive and clinch the staple satisfactorily, they may be used with solid fibreboard cases.

CHAPTER 3

Labels and Labelling

J. U. Gooch and J. M. Montresor

Labelling is a means of performing the communication function of packaging, by which contents are identified, the customer is encouraged to buy, and the law with regard to consumer and carrier protection is satisfied.

A secondary function of a label is to close the package, as for example by the application of a circular heat-seal label to the end of a biscuit wrap.

Labelling is used in preference to pre-printing a container under several circumstances, principally the following:

1. When the required standard of graphics can be achieved more cheaply than by printing direct on to the container.
2. When the same basic container is to be used for a number of different products or purposes.
3. When the exact nature of the information to be displayed is not known until after the containers have been filled.

Types of labels

Glued-on labels are simply sheet material, usually paper, printed and cut to size. Adhesive may be applied, either at the time of application to the container or at the time of manufacture of the label, the adhesive being activated by moisture just before application to the container.

Thermosensitive labels (heat-activated) have a thermoplastic coating applied by the manufacturer of the label which is activated by heat at the time of application to the container.

In general these are more expensive than "glued-on" labels, but their application is simpler, and higher speeds can be attained. They have a special application when performing the dual role of a label and a means of closure, as on biscuit packs and bread-wraps.

Self-adhesive (pressure-sensitive) labels are supplied coated with an adhesive on the unprinted side, and mounted on a release paper, which is removed to expose the adhesive immediately before application.

Self-adhesive labels may be mounted individually and regularly spaced on the release paper, which can then be cut to facilitate their removal; or they may be mounted on a continuous web of release paper, perforated between each label so that the labels can be roll-fed to the application point.

Self-adhesive labels can be produced to adhere to a wide range of materials, either to give permanent adhesion or to provide the facility of easy removal. They are particularly useful when an addition such as "Special Offer" or price labels have to be added at short notice to an existing pack. They avoid several operations necessary in the application of glued-on labels so that, although more costly than glued-on labels, in many instances the overall operation may be cheaper.

Tie-on labels.—Tie-on labels are generally used for the utilitarian functions of addressing packages or marking items, the shape of which precludes other means of identification. The label itself will generally be made of strong paper, fabric or metal, and attached by twine or wire. There is, however, a use for tie-on labels on luxury consumer goods, for which an infinite variety of coloured threads and ribbons with seals, tags and labels of paper, plastics or metal attached have been devised.

Reversible labels required for returnable transit containers are generally affixed by insertion into a metal frame or a transparent plastics envelope, secured to the container.

Insert labels are inserted in transparent plastics packs and need no fixing, remaining visible to the customer through the pack.

Alternatives to labels

Many types of container may be stencilled, and a range of stencil-cutters is available.

Direct marking can also be achieved by type-holders and rollers, varying from hand-held to automatic conveyor-line marking equipment.

Glassware may be screen-printed with ceramic inks that are subsequently "fired" and fused to the surface of the glass. This is the most permanent method of labelling bottles; it is used on "returnables" in the soft-drinks industry to obviate the need for label removal and re-labelling

ANCILLARY MATERIALS

(and on glassware for pharmaceuticals where loss of label could endanger life).

Label forms and shapes

Labels may be supplied on a roll, from which each label must be cut before application; or may be supplied ready-cut (either in a guillotine, or by die-cutting which is more usual when the shape is other than rectangular).

Roll-cut labels are particularly used for pharmaceuticals, to avoid risk of wrong labelling.

Rectangular wrap-round labels are commonly used on cans. Labels on bottles may be wrap-round, or attached to the front, back, neck or shoulder of the bottle. Additional information indicating, for example, special offers are usually "spot" labels. With "combined" labels, a single label fulfills the purpose of two separate labels (see figure 3.1).

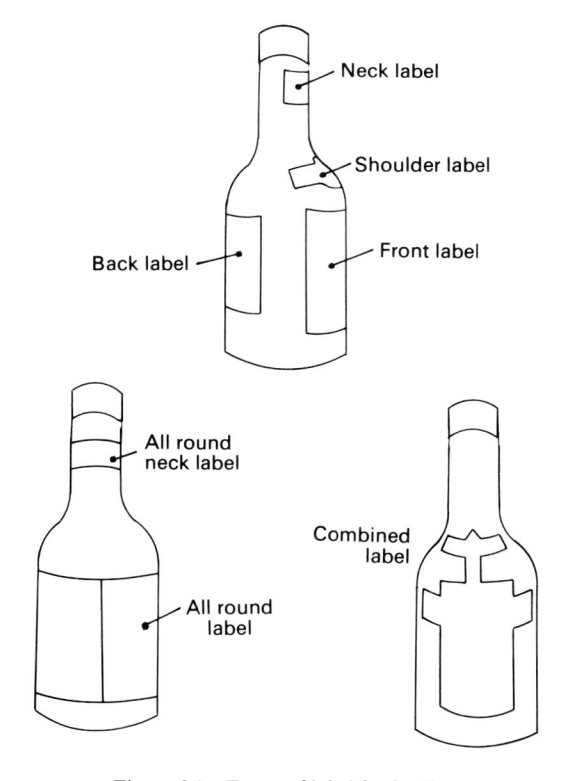

Figure 3.1 Types of label for bottles

5.22

Materials for labels

Papers

Uncoated, clay-coated, cast-coated and metallized papers are used in combination with all the commoner printing methods, and so one face of the label must have the characteristics needed by the printing method chosen to obtain the graphic effect required. After printing, good abrasion- and scuff-resistance are needed, which may involve over-varnishing.

The characteristics of the other, inner face of a label must ensure correct bonding by the adhesive used.

The label needs to be opaque and, when it is to adhere to a curved surface, must be sufficiently limp to conform, and be sufficiently free from curl to adhere without lifting (figure 3.2). Grammage is generally 80–100 g/m². The effects of curl can be limited to some extent by controlling the grain-direction of the paper, which is always parallel to the axis of curl. It is usual for the manufacturer of a labelling machine to specify the grain-direction of the labels to be used. Generally, but not always, it is parallel to the copy.

It is usual to specify that label papers should be flat within stated limits of relative humidity (e.g. between 35 and 80% r.h.).

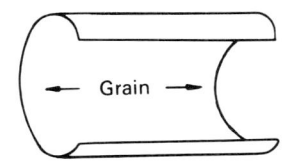

Figure 3.2 Finding the grain direction (curl). By moistening the back of the label paper, the direction of curl (and thus grain direction) can be determined

Foils and laminates

As well as producing special decorative effects, such as metallized appearance and deep embossing, foils or laminates with foil may be used where there are special environmental conditions, as when labels need still to adhere when bottles have been immersed in ice.

Plastics

The most common use of plastics labels is to give the effect of direct printing on to the container by use of a printed transparent plastics film.

Adhesion of labels

The initial action of the adhesive applied to the label is to bond sufficiently ("initial tack") to the surface to prevent the label assuming its natural shape when pressed to the container and so lifting at the edges or blistering. This bond must be maintained for sufficient time ("setting time") to enable the adhesive to set permanently ("final set"). Such a bond must then maintain adhesion in all conditions through the container's life, which may include standing in ice; but may also require to break down when a bottle is returned for re-use ("wet-off").

Achieving a bond between a label (which may be relatively absorbent) and a tinplate or glass container (which is non-absorbent and may have been surface-treated) requires careful formulation by the adhesive manufacturer to "match" the adhesive to label and container, within any constraints imposed by the labelling machine.

It follows that, having once achieved satisfactory performance, further supplies of label, adhesive and container should vary as little as possible from the original. This not only requires attention from the suppliers concerned, but from the user, to ensure that labels and adhesives are properly stored while awaiting use, and are used before they deteriorate.

Selection of adhesive

Usually a machine manufacturer, given samples of the label to be used and the container to be labelled, will make recommendations as to suitable adhesives. Such recommendations are sometimes too specific to permit buying in the cheapest market, or in times of short supply. Established adhesive manufacturers, provided with samples of labels and containers, and information on the machine to be used, can also advise on suitable formulations. Subsequent changes in label, adhesive or container, should be made only after a trial to check performance.

Further information on adhesives may be found in Chapter 4.

Labelling machinery and equipment

Labelling machinery can be categorized generally by the label/adhesive/container combination for which it is designed. As with other machinery, the higher the output required, the less versatility must be expected. If the production rate is low, most combinations will be satisfied with no more than a glue-pot and brush. A through-put of 60,000 bottles an hour will only be achieved, however, when label, adhesive and container are all carefully specified (figure 3.3).

LABELS AND LABELLING

Figure 3.3 Automatic labelling machine to apply body, shoulder, neck and back labels to 450 bottles per minute

Morgan Fairest Ltd.

In addition to the specific operation of applying the label, the machine may be required to code, count, weigh, price or overprint the label.

It follows that there is a great variation in the degree of automation offered to meet the requirements of different industries. The bottling industries are generally concerned with very high-speed lines, and every sub-operation needs to be fully automated. The pharmaceutical industry has a main concern for permanence of label, and absolute reliability that the correct label for the content has been used. In the self-service store there is a need for equipment that can be brought to the stock-in-hand, but first over-printed with the price, whereas a labelling machine for meat packs that are to be sold in this market will probably need to weigh, cost and over-print on a production line. The simplest equipment may do no more than dispense a label, ready for the operator to place it where she will on a container (figure 3.4).

The complexity of the machine is influenced by the design of both label and container. It must be recognized that the proper balance must be maintained between the marketing advantages of unusual and original designs, and the difficulties of fast reliable application.

Reliability is an important feature of any labelling machine, since a hold-up or breakdown at the point of labelling can cause stoppage of the whole line.

ANCILLARY MATERIALS

Figure 3.4   Hand-held hand-operated label printer/applicators
Norprint Ltd.

The principal operations performed by a labelling machine are four in number:

Feed label from magazine or roll.

Pick up label generally by suction cups, compressed air or by secondary adhesive.

Apply adhesive with either full coverage, vertical or horizontal stripes, generally from rollers on to either label or container.

Press label to container by pressure pads, compressed air, belt or brushes, during which process the container will be moved into position, held firmly while the label is applied, and then removed. This may be achieved by a rotary movement, the containers being held in a rotating turret: or by a "straight line" movement, by conveyor star-wheel or screw mechanism. "Straight-line" operation is generally more versatile, particularly where containers of unusual shapes are concerned. Most bottle-labelling machines hold the bottles vertically, but some machines, particularly for cans, hold the containers horizontally.

Purchasing, installation and operation of labelling machinery

The first step to successful labelling is to ensure that the machine supplier is aware of the exact requirement, and to this end he should be supplied with samples of the containers and labels to be used, throughput required and other details. An acceptance trial should be conducted, and a careful record made of the details of the label, adhesive and containers that were successfully used. The machine must be operated and maintained strictly in accordance with its Operating Manual, which should be readily available. Control should be kept of labels, adhesives and containers, to ensure they are as close as possible to those successfully used in the acceptance trial. Finally, when the same machine is to be used for a new

LABELS AND LABELLING

design of label and/or container, sufficient time must be allowed for trials to find a new, probably different, adhesive.

A check list of points that may need consideration is given in Table 3.1.

Table 3.1. Check list of details for consideration when selecting and ordering labelling machinery

Containers to be labelled
    Shape
    Sizes
    Dimensions
    Materials

Labels — User's requirements
    Number and type
    Dimensions
    Materials including treatments, e.g. whether embossed

Adhesive and application system
    Cold glueing   Full coverage
    Hot glueing   Stripe application
    Heat-activation   Machine maker's recommendation on
    Self-adhesion   adhesive supply

Payment and delivery
    Cost
    Method of payment
    Weight of machine
    Point of delivery (machine site? delivery bay?)

Installation
    Services needed:
        electricity (voltage, frequency, power, phase)
        compressed air (pressure and flow rate)
    Weight and dimensions of machine
    Heights of IN and OUT feeds for containers

Responsibilities; Supplier/User for:
    preparation of floor
    provision of electricity, air, etc.
    move machine to site
    instal machine
    conveyors to and from machine
    provision of safety devices (switches, guards, etc.)

Labelling machine capacity
    Maximum/minimum label widths
    Maximum/minimum label heights from base of bottle
    Maximum/minimum width and depth of rectangular bottle to be labelled
    Maximum/minimum diameter of cylindrical bottle

ANCILLARY MATERIALS

Attachments needed
 Coding
 Counting
 Wrong label detector
 Over-printing
 Special fittings for odd-shape bottles
 Logical sequence control to keep output speed at optimum, influenced by number of bottles on line before and after labelling machine

Operation
 Roll or magazine (hopper) feed
 Normal, maximum and minimum speeds required
 Left-to-right or right-to-left operation
 Labour skills and training arrangement for:
  operation
  maintenance
  changeover
 Changeover times
 Changeover parts:
  provision
  identification
 No bottle/no label, or no label/no bottle devices
 Re-load capacity during operation

Servicing
 Labour skill and training required
 Recommended spares holding by user
 Terms of machine manufacturers' repair service

Records
 Layout plan
 Electrical wiring diagram
 Operational manual
 Maintenance instructions
 Parts list

Security
 Label machinery manufacturer to maintain security on any information concerning the product to be bottled

Acceptance
 Bottle/label combinations to be used on acceptance trials
 Length of trial run for each bottle/label combination

LABELS AND LABELLING

Label printing

It is usual practice for the printer to decide the method used to print labels. Flexography is generally used when low-quality print is acceptable, as on simple labels giving warnings or with no other function than identification.

Quality labels may be printed by offset lithography, letterpress or gravure, the latter being particularly used for long runs. Silk-screen printing is widely used on plastics and for special applications. Lacquers and varnishes are applied to protect the printed surface from abrasion and dirt.

It is important that print and varnish are compatible with the content of the container and the adhesive used: and, in many instances free from undesirable odour.

Minimization of labelling faults and problems

Each foreseen combination of label, adhesive and container should be run successfully during the acceptance trials of the machine.

Samples of labels and containers used on the acceptance trial, and subsequent successful operation, should be retained for comparison in event of future breakdown.

The labelling machine must be run, maintained and adjusted strictly in accordance with the Machine Handbook, of which there should be at least one copy kept close to the machine (and a second copy in the Works Engineer's records). If either is lost, it should be replaced.

Suppliers of adhesives and labels should be asked to advise at once if, owing to shortages or other reasons, supplies cannot match previous consignments. This information must be passed by the Buyer to the shop floor.

All supplies must be kept in good storage conditions, with labels packed in moisture-proof wraps when necessary. Rapid changes in temperature before use should be avoided. Containers to be labelled should be clean and dry. When a new label or container is introduced, the labelling-machine manufacturer should be asked whether the particular machine is capable of the new requirement, and for recommendation as to adhesive, adjustment or modification to the machine. It follows that sufficient time must be allowed (and the machine made available) for trials prior to production.

A system of Quality Assurance on labels, adhesives, and containers should be operated and amended as experience grows of the particular characteristics required.

ANCILLARY MATERIALS

Correction of labelling faults

The commoner labelling faults are:

Tearing or creasing of the label
Inadequate adhesion, lifting of edges, blisters
Staining of labels
Scuffing of labels
Fading of labels
Incorrect positioning of labels
Failure to "wet-off"

Provided that the initial trial has been successful, failure will take place only when some change has occurred. Such changes may occur in the area of the machine (e.g. a new operator): of the label (e.g. new paper): of the adhesive (e.g. new supplier): of the container (e.g. new surface finish). The first step is therefore to narrow the likely source of trouble by inquiries as to what changes have taken place. The nature of the fault may also indicate the likely area of change.

Where the machine is the possible source of trouble, the most likely cause is that its operation or adjustment is not in accordance with the Manual. Where the label is the possible source, the approach should be to compare "good" and "bad" labels for such properties as grain-direction, curl, stiffness and absorbency. Most adhesive troubles come from using an adhesive different from that which has run successfully, e.g. by unauthorized adulteration with water or thinning agent.

In all investigations concerned with adhesion, it is worth attempting to make the label/container bond by hand since, if it cannot be done by hand, it is certain it will not be done by machine.

CHAPTER 4

Adhesives in Packaging

A. D. Brazier

Historical

The use of adhesives has been known for several thousand years. Prehistoric tribes placed their dead in tombs containing pieces of broken pottery which had been stuck together with a rosin. Statues over 6000 years old have been recovered from excavated Babylonian temples, on which parts had been glued together with a bituminous adhesive. There is a clear early reference to the use of animal glue recorded on stone in the city of Thebes, probably about 1500B.C. Tutankhamen's tomb provides additional historical evidence in the form of a glued wooden casket, now in a Cairo museum.

Still further evidence in history comes from Roman times, the Romans undoubtedly being knowledgeable in glues, as evidenced from specimens of veneering still to be seen. They almost certainly made up adhesives which were very similar in performance to our present-day animal glues.

The discovery of British gum and dextrine is believed to have been accidental, and resulted from an observation that some starch, which had been heated during a fire in a Manchester warehouse, yielded a sticky gummy solution when wet with water. In all probability the use of animal glue grew out of the fact that stews, especially those obtained from bones or skins, yielded a sticky solution which gelatinized when cooled. Violins and similar musical instruments made during the Middle Ages, especially in Italy, indicate that animal glue was used at that time. There are indications that early painters also used a glue size in preparing their canvases.

ANCILLARY MATERIALS

Chaucer (about 1386) writes in *Squire's Tale*:
The horse of brass that may not be remewed
I stant as it were to the ground yglewed.

Similarly in Lanfranc's *Chirurgeon* (about 1400)
as it were to bordis weren ioyned togidere with cole or with glu.

The importance of the manufacture of glue and gelatin was appreciated in Germany as a key industry. A German company, formed in 1895 with three plants, expanded until in 1912 it controlled the output of 17 plants, and also had factories in Austria, Russia, Belgium, Switzerland and France.

Further development since the turn of the century has shown an increase in animal glue and gelatin-type products, starch adhesives, dextrine and borated dextrine adhesives. In the last 40 years, polyvinyl-acetate-emulsion-based adhesives were developed, and a whole series of speciality adhesives, such as solvent-based adhesives and hot-melt adhesives.

In packaging today, polyvinyl-acetate-based adhesives are probably the most widely used, although there is still a considerable tonnage of dextrine and starch-based adhesives. In recent years hot-melt adhesives have become increasingly popular in packaging application.

Definition of terms

Polymer and polymerization

Polymerization is the process by which a polymer -A-A-A- is made from the relevant monomer A.

A homopolymer -A-A-A- is made up entirely from one monomer A. A copolymer -A-B-A-B- is made up from the monomers A and B.

Adhesion, adhesive, cohesive and adherend

Cohesive forces are the forces responsible for substances such as polyethylene or rubber having rigid shapes (i.e. are the forces acting between the molecules within a substance) whereas *adhesive* forces are similar forces acting between the molecules of dissimilar substances (e.g. between paint and metal). An *adhesive* is the substance used to bond together two *adherends* (substrates).

Thermoplastic, thermosetting

A *thermoplastic* material softens on heating and returns to its original state on cooling, the process being indefinitely repeatable. A *thermosetting* material undergoes a chemical change on heating, resulting in a solid material which does not revert to its original state on cooling. The change is irreversible.

Rheology

Rheology covers a multitude of physical properties; it is difficult to define exactly, but includes forces such as those opposing deformation and flow. Adhesives with the same viscosity but different rheological properties will behave differently on the same machine.

Thixotropy, dilatency

A *thixotropic* system is one that will thin out on stirring, but is capable of reverting to its original consistency on standing. A *dilatent* system has the opposite properties, i.e. it will thicken on agitation.

ADHESIVES IN PACKAGING

Viscosity

The property of internal forces within a liquid adhesive that tend to prevent the liquid from flowing.

Wetting

Wetting is the ability of an adhesive to flow out (wet) a surface by coming into intimate (molecular) contact with it.

Gel

A gel is a network of solid aggregates in which a liquid is firmly held. A gel can normally be disrupted by heat and/or mechanical forces, e.g. compression. Table jelly is a simple example.

Penetration

Penetration is the entry of the adhesive into the substrate(s). This invariably occurs to a greater extent with the substrate onto which the adhesive is first applied.

Shortness

Shortness refers to the lack of stringing, cobwebbing, and formation of threads during the separation of rollers.

Heat-set adhesive

A heat-set adhesive is one which forms a bond on the application of heat, and in which the water present is absorbed internally to form a gel—a patented process.

Heat-seal adhesive

A heat-seal adhesive is one in which a dry film is activated by heating immediately prior to bond formation.

Tack

The tack of an adhesive is the ability to form an initial bond of measurable strength immediately after the adhesive and adherend are brought into intimate contact, and generally while the adhesive is still liquid.

Blocking

Blocking is an undesirable adhesion between adjacent layers of a material, such as occurs under moderate pressure during storage causing them to stick together.

Solids content

The solids content of an adhesive is the weight of material expressed as a percentage of the total after all solvent has evaporated (by heat).

Setting time

Setting time is the time to form a bond under heat, pressure, etc., by means of a chemical or physical change. It gives a handling bond (initial bond).

Open time

Open time is the time between application of the adhesive to one or both of the substrates and the bringing together of their two surfaces.

Drying time

Drying time is the time to form the final bond.

Plasticizer

A plasticizer is a material added to an adhesive to render the dry film of adhesive more flexible.

ANCILLARY MATERIALS

An external plasticizer is incorporated in the adhesive as an addition after polymerization is complete, whereas an internal plasticizer is added during the polymerization process and forms an integral part of the polymer used.

Curing

Curing is a chemical reaction (cross-linking), usually brought about by increase in temperature, which results in irreversible physical change (hardening or setting).

Radio-frequency gluing (dielectric sealing)

This is a technique of bonding two substances together where the glue line is heated by radio waves. It is particularly suitable for substrates that would be damaged by the application of direct heat, as in radio-frequency welding; the substrate does not become heated.

Pressure-sensitive adhesive

A pressure-sensitive adhesive adheres to a surface at room temperature by briefly applying pressure alone. Such adhesives are permanently tacky.

Peel force

Peel force is the force used to measure the adhesive strength of a pressure-sensitive adhesive. Usually (180° peel) the substrate is pulled (peeled) back on itself and the force required recorded.

Consistency

Consistency is the property of an adhesive (paste) that causes it to resist deformation.

Paste

A paste is an adhesive with the consistency of thick cream. Starch and water pastes are typical.

Retrogradation

Retrogradation is a change from low to high consistency on ageing, that occurs in starch pastes, often referred to as setting back.

Sizing

Sizing is treatment with a liquid coating performed to fill the pores present in a surface.

Easi-clean

This is a term relating to certain PVA adhesives having the property that the dried or semi-dried film (on machine parts) can be cleaned easily by a wet rag. Such adhesives tend to dry slowly and do not clog nozzle applicators.

The terms defined above are largely self-explanatory, but three in common everyday use are of such importance as to be worth further comment.

Setting time/open time/tack

The *open time* as defined is the time elapsing between applying an adhesive to one or both surfaces, and the bringing together of those two surfaces. An alternative way of defining *setting time* would be the time between applying an adhesive to one or both paperboard surfaces and the time after bonding at which appreciable fibre tear occurs on pulling them

ADHESIVES IN PACKAGING

apart. In a normal carton sealing operation, the open time will be short with respect to the setting time, i.e. the time between the flaps coming together and fibre tear resulting. In the case of a hot melt, these times could be of the same order, i.e. two or three seconds only in view of the very fast setting speed of a hot melt.

Sometimes it is not possible to have a long compression time, but it may be necessary or desirable to have a fairly long open-time. In such an instance the adhesive would have to have high tack. This could be true with water-based adhesives when the tack concerned is "wet tack", or with a hot melt it would be "molten tack". The importance of tack when a long open-time, short compression-time system is being considered, is clear since, if the adhesive is applied to one surface, and then the two surfaces are brought together but not held firmly, it is essential that the surfaces do not separate. This can be a problem with water-based adhesives (not normally a problem with hot-melt adhesives) and therefore such water-based adhesives have to be formulated to give a high degree of wet tack in order that the flaps do not "pop open". It is probable that the wet-tack or molten-tack requirements of an adhesive will depend entirely on the open time/setting time/compression time characteristics of the operation in question, and it is therefore necessary to formulate a given tack requirement into an adhesive depending on the characteristics of a given packaging line. In this respect, tack is a relative rather than an absolute term.

Principles of adhesion

Mechanisms of adhesion

Even allegedly optically flat surfaces are in fact very rough when examined on a molecular scale, i.e. under high magnification. Therefore no two surfaces can ever be in 100% contact, and in fact two surfaces will seldom be in contact over more than 10% of their common area.

There are four basic theories of adhesion as follows:

(a) *Mechanical adhesion*

This theory is relevant only to absorbent materials. The bond strength between such materials is achieved by the polymer molecules between the surfaces interlocking and penetrating the crevices of each of the two surfaces to be bonded, e.g. a haystack is held together by purely mechanical forces involved in the intermingling of the straw threads. Mechanical adhesion is the prime contributing factor to bond strength in adhering paper-paper and rubber-textile systems. In bonding wood to wood, mechanical adhesion is not the major factor, as was thought until recent years. The figures in Table 4.1 illustrate this:

ANCILLARY MATERIALS

Table 4.1. Bond strengths in wood-to-wood adhesions

Wood surface	Bond strength
Planed	3210
Sanded	2360
Sawn	2690
Combed	2400

In shear tests it was found that the smoothest surfaces gave rise to the strongest bond. If the major factor in bond strength had been mechanical adhesion, it should have been the reverse, i.e. the combed specimens should have resulted in the highest strengths.

(b) *Chemical adhesion. Primary bonds*

Chemical adhesion due to primary bonds results from specific primary valency bonds. Experimental evidence to support this theory is lacking, although the strength of protein adhesives is almost certainly due to electrovalent (chemical) bonds between the polymer molecules.

Other examples are in fibre glass reinforcing, in which the primary bonds are covalent, and in welding and soldering, in which the primary bonds are similar to metallic bonds.

Secondary bonds (absorption theory)

This is the main theory of adhesion and involves secondary Van der Waals' surface forces, these forces being responsible for the attraction between molecules, i.e. cohesive forces in plastics materials, liquids, etc. Such forces act over only very short distances, and therefore for two substances to stick together by these forces they must be brought into very intimate contact—the importance for good wetting out of the adherend by the adhesive becomes apparent.

The best adhesives are therefore mobile liquids that readily wet out the substrate. By the same reasoning a flexible natural rubber would be a better adhesive than the less flexible SBR rubbers, or the still less-flexible styrene polymers.

(c) *Electrostatic adhesion*

This theory comes into play as a bond fractures. It has been shown that 90° peel tests using PVC bonded to glass in a dry inert atmosphere are accompanied by flashes of light due to electrostatic sparks. This effect is due to an electric double layer forming as the surfaces separate. The theory has been shown to be mathematically sound, although it has limitations since non-polar materials show good adhesion and not the poor adhesion predicted if the theory were the main contributing factor.

(d) *Diffusion theory (autohesion)*

This is applicable only to high-molecular-weight materials and is due to the inter-diffusion of adhesive and adherend. This process results in no clearly defined interface after diffusion has occurred.

Mutual solubilities of adherend and adhesive is an important factor, and experimental evidence from peel tests using rubber adhesives gives some support to the theory.

Having discussed the different methods by which adhesives form the final bond, let us now consider the development of bonds (figure 4.1).

Consider a 45% solids dextrine solution bonding an absorbent board to itself. It bonds by mechanical adhesion, and the rate-determining step is the loss of water (and adhesive solution) into the board surface. When the solids content of the film between the surfaces reaches 95%, the film

ADHESIVES IN PACKAGING

becomes coherent, develops tack and strength, and hence produces fibre tearing on separation.

For an emulsion, which is a stabilized suspension of discrete particles in water, the stabilization forces are sufficiently weak so that when the particle content reaches about 65 to 70%, these forces are overcome, and the emulsion particles coalesce to give a continuous film. Thus an emulsion produces initial fibre tear, while 20–25% water still remains in the film. Thus a PVC emulsion is about twice as fast in inter-bonding as a dextrine solution at the same solids/viscosity.

A hot melt forms a bond by solidification on cooling. In bonding, a little hot melt is placed on a large surface at room temperature, and hence it cools rapidly and gives fibre tear in a few seconds.

Thus hot melts form bonds fastest (by cooling), emulsions next fastest (by coalescence of the suspension), and solutions slowest (by penetration and loss of carrier).

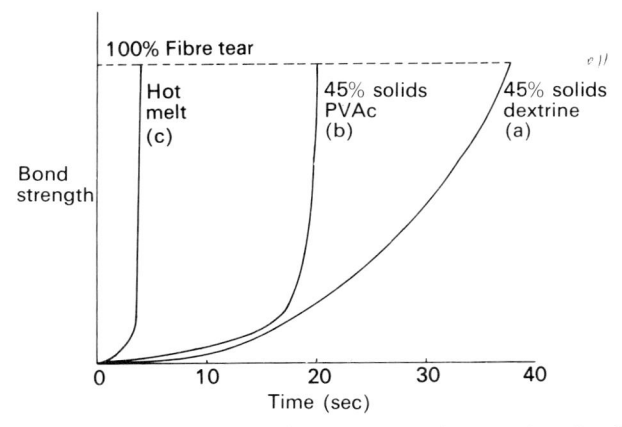

Figure 4.1 The development of bonds between absorbent boards using (a) a 45% solids dextrine solution (b) a 45% solids PVAc emulsion (c) a hot melt

Requirements for good bond strength

Many factors affect bond strength and can be related to both the adhesive and the substrates.

The basic requirements for good bonding are as follows:

(a) Good wetting-out (spreading) on the adherends by the adhesive. For this clean surfaces are essential.
(b) A thin film of adhesive.
(c) Increased pressure before curing (compression stage) to increase penetration of the adhesive into porous stock and/or to help air trapped at the interface to dissolve into the adhesive to attain better wetting-out.

5.37

ANCILLARY MATERIALS

(d) The thermal coefficients of expansion of the adhesive and adherends should be similar to prevent differential shrinkage on heating or cooling.

In addition to these points several factors concerning the adherends affect the final bond strength:

(a) Smoothness/roughness of the surface.
(b) Porosity of the substrate.
(c) Chemical nature of the substrate.
(d) Temperature of the substrate at time of bonding.

The smoothness of the surface will depend largely on the type of surface in question, and it is difficult to generalize. The more porous the substrate surface, the more adhesive will be required to make the bond since, for a given quantity of adhesive applied, the more highly porous surfaces will absorb more, which will therefore not be available for bonding. Ideally, the chemical nature of the surfaces should be as near as possible to that of the adhesive or, conversely, the adhesive recommended for a given bonding operation should be similar to the chemical nature of the surfaces. As a generalization, like bonds like and conversely.

A point often overlooked in commercial adhesive bonding operations is the temperature of the substrate. For ideal bonding conditions the substrate should be at the same temperature as the adhesive, i.e. in a commercial bonding operation, the substrates and the adhesive should be held for at least 24 hours adjacent to the area in which the bonding operation will be carried out. This is vital if the storage conditions for any of these materials is different to that of the environment where the bonding will be carried out.

Surface energies

For an adhesive (suitably chosen chemically) to bond two surfaces, it must have the correct surface tension in order to wet out the surface.

In order that a suitable adhesive with the appropriate surface tension may wet out a surface, it is generally found that the surface energy must be 45 dynes/cm or greater for good bond formation. Between 35 and 45 dynes/cm a good bond may or may not be formed, depending on the prevailing conditions. If the surface energy is 35 dynes/cm or lower, then a satisfactory bond will not be formed.

The standard test for evaluating surface energies is given in ASTM specification D2578-67 and uses test solutions of cellosolve (the methyl ether of ethylene glycol) and dimethylformamide in varying proportions).

Surface tension/contact angle

Mathematical relationships have been established relating the contact angle of an adhesive to a surface and the surface tension of the adhesive. The mathematical equations derived will not be detailed here but, when taken in conjunction with viscosity, it is evident that for maximum bond strength and maximum rate of bond strength development an adhesive will have

(a) A small contact angle to permit wetting of the surface.
(b) A suitably low surface tension with respect to the surfaces to be bonded to enhance the rate of wetting.
(c) A suitably low viscosity to ensure a maximum rate of dispersion across and into the surfaces to be bonded once the initial wetting has occurred.

Table 4.2, "A Guide to Adhesive Selection" typifies the adhesive types, their characteristics, and their main uses in packaging. It is intended primarily for use by people whose activities impinge on the application or purchasing of adhesives as part of a wider job function rather than for those whose work demands a knowledge of fundamental adhesive technology.

This being the main intention, it has concentrated on those adhesives that are used regularly and in significant quantities in packaging, as against attempting to provide an exhaustive survey of the entire product range now available.

It should also be remembered that many of today's adhesives have been formulated to meet a specific application which, in practical terms, means that a particular property or quality has been significantly modified. The notes below take account of the broad variations in properties within each adhesive type, that are possible through such formulations. Readers requiring more detailed information are recommended to contact their adhesives suppliers.

From this table it will be seen that all the packaging operations are covered, including some of the more sophisticated, such as pressure-sensitive applications (using modern "tailor-made" acrylic lacquers) as well as the more conventional starch and dextrine types. The choice of adhesive type is never based simply on technical criteria, but always in combination with economic factors, e.g. some carton-sealing operations with non-demanding end-use factors can quite satisfactorily perform with either dextrine or PVA adhesives, and the choice will either be made for low-cost criteria or perhaps for reasons like a preference for handling a given type of adhesive. The choice will, therefore, depend on several factors.

ANCILLARY MATERIALS

Table 4.2. A guide to adhesive selection

Adhesive type	Characteristics	Main uses
	ADHESIVES BASED ON SYNTHETIC RAW MATERIALS	
Emulsions	More accurately dispersions or suspensions of solid insoluble particles in water. When applied to substrate, water is released rapidly and hence more quick setting than solution adhesives. Adhesive film is normally strong and continuous and will vary in flexibility. Mostly based on polyvinyl acetate, and the polymer system can be designed to have a wide range of end properties (viscosities, adhesion to various surfaces, setting speeds, open time, solids/viscosity relationships, machining characteristics, resistance of dried film to water, transparency of dried film, initial tack, etc.). Frequently, the dried film is thermoplastic. "Heat Set" formulations available which give very rapid setting with minimum steam emission when subjected to R.F. or other heating.	Practically every kind of packaging operation where fast speed of set and specific adhesion to more difficult substrates is required. More particularly, case and carton sealing, tube winding, lamination (including plastics films), heat-seal coatings, bag making (including cellulose and plastics films), pressure-sensitive coatings, windowing, bar wrapping, glued lap (manufacturer's joint on shipping cases), adhesion to foil surfaces.
Formulated emulsions	Similar to emulsion adhesives but formulated to give a wider range of cost/performance relationships. Low-cost formulations give some of the properties of an emulsion adhesive, and higher-cost formulations extend the scope and performance range of emulsion adhesives.	Similar to emulsion adhesives.
Hot melts	"Solid Solutions" which need to be heated for use, and which soften as the temperature is increased and become fluid for application at the correct operating temperature. Contain no water or organic solvents. Good specific adhesion to difficult surfaces (dependent on formulation) and set very rapidly by cooling. Are available with a range of running characteristics (hot tack, viscosity, open time, setting speed) depending on the requirements of the application. Can be resistant to low and high temperatures, and are extremely water-resistant. Adhesive film can be very tough and flexible. Modern formulations give good stability in the glue pot (resistance to charring, discolouration and viscosity change) even when held heated for extended periods (several days). Temperature control in the glue pot important.	Most packaging applications dependent on availability of appropriate application equipment. Used extensively for case and carton sealing, carton manufacture, tube winding, heat seal and pressure-sensitive coatings, laminating, coatings, bag seams. Can be jetted, continuously extruded or applied by roller.

Table 4.2. A guide to adhesive selection (cont.)

Adhesive type	Characteristics	Main uses
Latex adhesives	Based on natural or synthetic latex. Can be very fast-setting and give good adhesion to a variety of surfaces, including printed, varnished, lacquered, foil and plastics. Dried films are usually very water-resistant and flexible. Some remain permanently tacky, and others (by coating both surfaces) give "self-seal" properties, i.e. the dried films combine under pressure. Tend to thin out in storage. In wet state, normally have ammonia or latex odour. In glue pots, avoid copper, brass, bronze, manganese, or as an adhesive for materials containing these substances.	Foil-to-paper lamination (can give good heat resistance), self-seal coatings, carton and case-sealing, carton manufacture, polythene bags, water-resistant labelling for drums, cans and bottles.
Solvent-based adhesives	Solids (which can be natural rubber or synthetic polymers) are dissolved in a solvent. Resins can be polyvinyl acetate, synthetic rubber, polyvinyl chloride, acrylics, etc. Setting and drying speeds vary according to the solvent used, but generally faster than water-based tacky flexible dry films. "Two-Part" Systems (i.e. adhesive plus catalyst) available, particularly for the lamination of impervious surfaces (e.g. cellulose and plastics films to each other in various combinations). The maximum degree of heat and fat resistance and non-yellowing when exposed to sunlight. Primarily polyurethane based. Some "one-part" polyurethane systems available where a lower standard of end performance in terms of heat and fat resistance is acceptable (e.g. normal "snack" foods packaging).	Pressure-sensitive adhesives for tapes and labels, film, foil and paper laminations (particularly in the case of "two-part" systems, where resistance to heat and fats, etc. is required), graphic arts lamination, coatings, cellulose film, coatings for skin and blister packaging.

ADHESIVES BASED ON NATURAL RAW MATERIALS

Pastes	Based on starch or white dextrine, normally pale in colour, with medium to low solids. Range in viscosity from very short buttery texture (viscosity reduces with agitation in some cases) to semi-fluid gelatinous and non-tacky types. Most are acid, but borated pastes (faster setting) are alkaline. Generally good resistance to humidity and sterilization conditions.	Seams and bottoms of paper bags. Lap paste in can labelling. Hand labelling of bottles and tins. Tube winding. Cup winding. Miscellaneous hand applications involving paper and board.

ANCILLARY MATERIALS

Table 4.2. A guide to adhesive selection (cont.)

Adhesive type	Characteristics	Main uses
Jelly gums	Starch based. Highly cohesive and tacky in a thin wet film. Excellent coverage. Noted for non-crystallizing characteristics in the dry film. Normally maintain good consistency in the glue pot under production conditions. Good final adhesion, humidity and sterilization resistance. Regarded as "semi-iceproof". Good instant tack to cold wet surfaces. Alkaline products available for oily or greasy bottles. Clean running and transparent dried film.	Fully or semi-automatic labelling of bottles, foil/paper lamination.
Unborated dextrines	Usually high solids combined with low or medium viscosities. Tacky and relatively fast-drying. Normally brown in colour but can range from off-white to dark brown. Usually can be spread in a thin film and formulated to machine cleanly on high-speed applicators. For envelopes and remoistenable coatings, can be formulated to have "layflat" or non-curl properties.	Envelopes, remoistenable coatings, carton and case sealing, bottle labelling.
Borated dextrines	Similar to unborated dextrines but normally higher tack on certain surfaces. Fluid, filmy and fairly fast-setting. Fairly good humidity resistance and good adhesion.	Carton and case sealing; tube winding; labelling; laminating.
Animal glues	Normally used hot and have high initial tack. Can be formulated to have a flexible dried film and to vary in setting speed. Can be liquefied so as to require little or no heating. Special "non-warp" formulations have good "layflat" properties.	Box covering, showcard manufacture, layflat mounting, tube-winding, gummed tapes and laminating.
Casein	Dried films are continuous and give moderate to high water resistance. Good adhesion to wood, plastics, glass, metal and a range of coated und uncoated papers.	Ice-proof adhesive for bottle labelling, tin labelling, labelling over coatings, case and carton sealing.
Latex	See Section on previous page.	

Table 4.2. A guide to adhesive selection (cont.)

ADHESIVES BASED ON COMBINATIONS OF NATURAL AND SYNTHETIC RAW MATERIALS

Vinyl dextrine	Combination of PVA and dextrine, primarily to give a remoistenable coating. High solids combined with low viscosity. Good compromise between end performance of remoistenable PVA and clean machining characteristics at high speed of a dextrine. Exceptional machine speeds and very fast drying with limited or no heat.	Envelopes, remoistenable coatings.
PVA/starch paste	Pasty texture, but better specific adhesion to more difficult surfaces than an ordinary starch paste. Range of viscosities and texture available. Some pastes can be highly formulated to cope with difficult substrates.	Bag making using difficult stocks.

CHAPTER 5

Package Cushioning Systems

D. C. Allen

The term *package cushioning systems* covers a very wide variety of techniques and materials for protecting goods from the effects of handling in transit. They range from the traditional application of straw and woodwool in the packaging of glassware, pottery and ceramics, to the highly sophisticated shock isolation systems, incorporating springs and hydraulic shock absorbers, used in packages for aero-engines and space satellites.

Cushioning systems fall into three main categories:

1 Resilient systems
2 Non-resilient or crushing systems
3 Space fillers

Most of the materials and devices used fall into one of these categories (figure 5.1).

Space fillers are used, as the name implies, to fill the space around an irregularly-shaped article placed in a regularly-shaped container. They permit the controlled movement which is essential to reduce the shock transmitted to the contents when the package receives an impact, but the level of shock protection is not easily predicted because of the non-homogeneous nature of these materials, and the difficulty of ensuring a uniform density throughout the space filled. This is not important in the bulk packaging of glassware and pottery, where the main function of a space filler is to prevent the pieces from making contact with one another, or with the walls of the container, but it is important in packaging certain engineering products. For this reason, space fillers are normally used only for fairly robust or low-value articles.

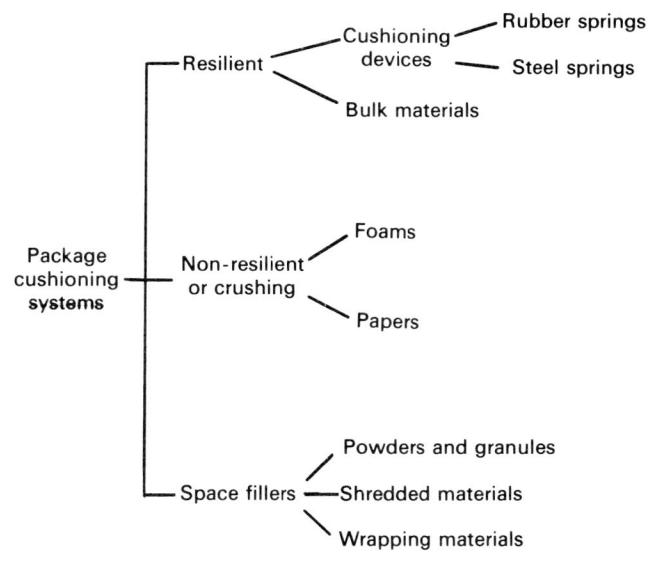

Figure 5.1 Package cushioning systems

Non-resilient or crushing systems are widely used in consumer packaging where low cost is the prime consideration, and only one journey is required—from factory to consumer. The most commonly used materials are corrugated fibreboard in various forms, and expanded polystyrene. The level of shock protection is predictable, although it deteriorates rapidly with repeated impact due to the crushing of the material structure.

There are two distinct types of resilient system: *bulk cushioning materials* and *cushioning devices*. Rubberized hair and plastics foams are typical of the first type. Their performance is predictable, and simple engineering-design techniques can be applied to the package design, provided suitable performance data are available. Such data can often be obtained from the material manufacturers. The term *cushioning device* includes a wide variety of rubber and steel springs. In general, resilient systems are expensive, and their use in consumer packaging tends to be restricted to high-value fragile goods, although they are sometimes employed in re-usable packages where their high cost can be justified. They are also widely used in the packaging of Service stores and equipment, where the long life required of the package (normally 5 years) again justifies the cost.

ANCILLARY MATERIALS

This chapter describes briefly the important properties required of a bulk cushioning material and the test methods used to assess these properties. It goes on to discuss in turn each of the more commonly used materials, and gives a simple example to illustrate the method of designing cushions. Some of the more commonly used cushioning devices are also described. The most important properties of all the materials discussed are summarized in Table 5.1.

Properties of cushioning materials

The properties required of a cushioning material depend on the nature of the article to be protected. As we have seen the range is very wide. The properties listed here do not apply in all instances. In general, only some of them will be relevant.

Dynamic performance

Dynamic performance describes the first and most important property —the ability to provide shock protection. The performance data are obtained on a drop test machine by dropping a number of weights several times[5] from selected heights on to test pieces of cushion of different thicknesses and measuring the maximum deceleration (peak g_n) experienced by the weight at each impact. The data are converted to peak deceleration (g_n)/static stress curves. Since most materials suffer some fatigue, performance curves are usually based on the decelerations recorded at the third impact. A typical set of curves is shown in figure 5.2. The parameter *static stress* is used because it is readily calculated from the design conditions (see Design Example page 5.57).

The term *cushion factor* is commonly used to describe the efficiency of a cushioning material[1] and reference figures are given for individual materials in Table 5.1. A low cushion factor denotes high efficiency, e.g. the thickness of a polyurethane foam with a cushion factor of 2·2 required for a given level of shock protection is only half that required if rubberized hair with a cushion factor of 4·5 is used. Thus an appreciably smaller package size is achieved with the polyurethane.

Fatigue and thickness loss

Few cushioning materials are completely resilient; their structure is damaged by impact, and this results in a progressive increase in the maximum deceleration recorded at each successive drop. This is known as *fatigue*. In addition, some materials do not recover to their original

Table 5.1. Properties of bulk cushioning materials

Material	Cushion factor	Density (lb/cu ft)	Thickness recovery	Fatigue resistance	Creep resistance	Temperature limits (°C) −	Temperature limits (°C) +	Moisture content (by weight) %	Water absorption	Corrosive effect	Mould resistance	Dusting
Wood wool	4·0–5·0	up to 5	poor	poor	fair	10	45	12–20	high	high when wet	fair	fair
Expanded polystyrene	3·0–3·5*	1·0–2·0	poor	poor	good	30	70	0·2†	low	low	good	low
Resilient expanded polystyrene	4·0–5·0	1·0–2·0	good	good	good	30	70	0·2†	low	low	good	low
Rubberized hair	4·0–4·5	4 & 6	fair	fair	fair	30	60	6– 8	high	low	good	low
Orientated rubberized hair	2·3–2·8	4 & 6	good	fair	fair		60	6– 8	high	low	good	low
Crimped rubberized hair	3·0–3·3	5	good	good	good	30	60	6– 8	high	low	good	low
Expanded rubber	3·5–4·5	11, 18 & 28	good	good	fair	0	60	up to 3	low	low	good	low
Polyurethane foam	2·0–3·0	1¼– 6	good	good	good	20	60	10–18	high	low	good	low
Bonded polyurethane chipfoam	3·0–3·5	3 –15	good	good	fair	—	—	—	high	low	good	low
Expanded polyethylene	3·0–3·3	2·5	good	poor	good	20	60	7– 8	low	nil	very good	nil
Expanded EVA	3·0–3·5	2·8	good	fair	good	—	—	—	low	nil	very good	nil
Bubbled polyethylene	4·0–5·0	—	good	good	good	—	—	very low	low	nil	very good	nil

* 1st drop
† By volume

ANCILLARY MATERIALS

thickness after each impact. This is sometimes undesirable, as it permits the cushioned article to rattle.

In general, space fillers and non-resilient cushioning materials suffer considerably both from fatigue and from thickness loss. With resilient materials, thickness loss is usually small and fatigue resistance is good, though materials vary considerably.

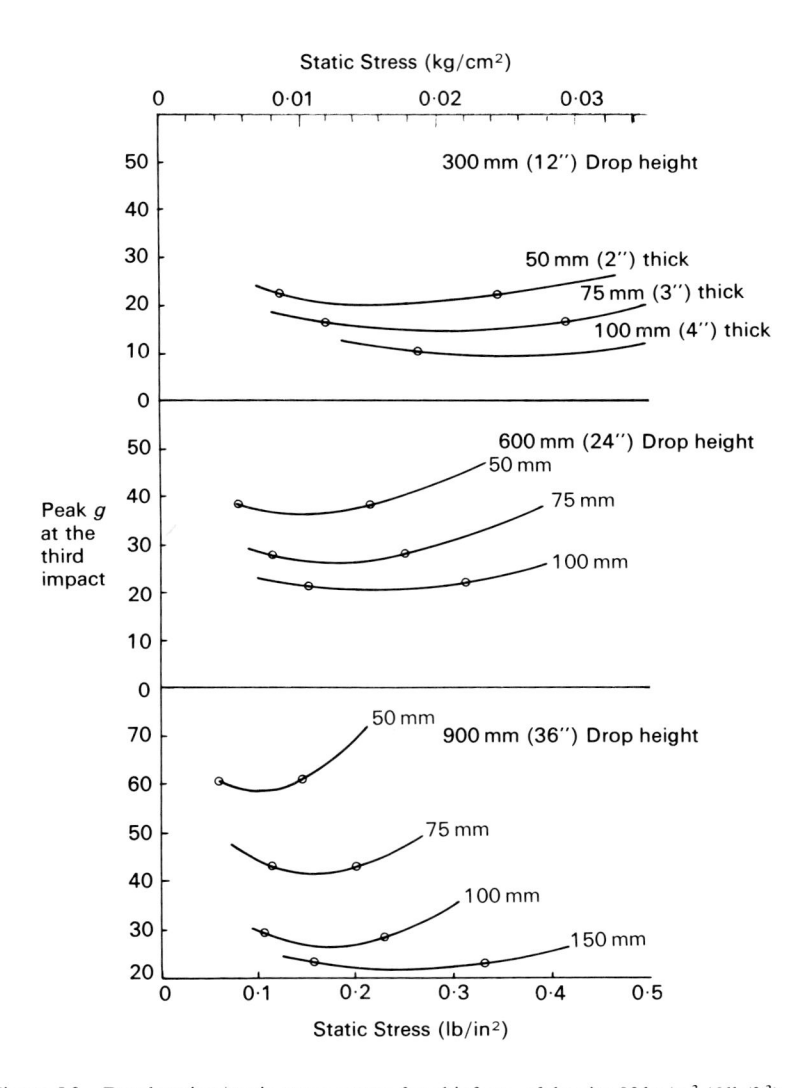

Figure 5.2 Deceleration/static stress curves for chipfoam of density 80 kg/m^3 (6 lb/ft^3)

5.48

PACKAGE CUSHIONING SYSTEMS

Creep

Most cushioning materials suffer a progressive loss of thickness under a static load, such as is applied by the item while the package is in storage. This is known as *creep*. Creep is measured by applying a static load to a test specimen of the cushion and measuring the loss of thickness at suitable intervals of time.[2]

Thickness loss due to creep usually decreases roughly logarithmically with time. Plotting the results on a log scale gives a straight line, from which it is possible to extrapolate to estimate the thickness loss over a longer time. A typical graph of creep characteristics is shown in figure 5.3. Such information is useful in comparing different materials but, if the data are used to calculate thickness loss over a long period, it should be borne in mind that the tests of short duration are carried out under constant climatic conditions. In practice, thickness loss due to creep will probably be smaller than the estimate derived from such curves.

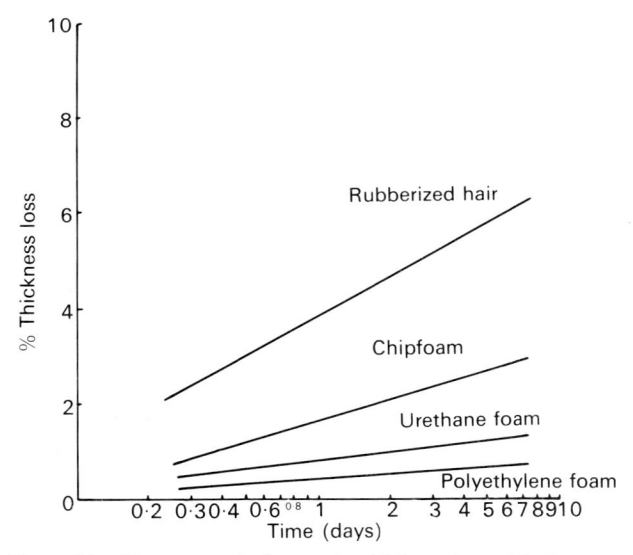

Figure 5.3   Measurement of creep, i.e. thickness loss with time

Moisture content

Many materials, and especially those derived from plants or animals, tend to be hygroscopic, i.e. if dry materials are exposed in a damp atmosphere, their moisture content will increase until it reaches equilibrium with the surrounding atmosphere. All the figures in Table 5.1 were obtained after equilibrating at 35°C and 95% relative humidity. In general,

5.49

a high moisture content has little effect on dynamic performance, even at low temperatures, but it does encourage mould growth.

Water absorption

Water absorption is usually measured by an immersion test. Clearly, materials of open structure, such as rubberized hair or polyurethane foam, will have a high water absorption under these conditions and, if there is any likelihood of a package being exposed to liquid water, it would be wise not to use such materials. Instead a closed-cell material, with a low moisture absorption would be preferred (i.e. expanded rubber or expanded polyethylene).

Corrosive effects

Corrosion will occur on metals if an electrolyte is present. In a packaging context, the electrolyte is usually contaminated water. Corrosion can be prevented in packages by controlling the internal relative humidity, and by ensuring that packaging materials are free from contaminants. Relative humidity may be controlled by enclosing the product in a water vapour barrier, often with a suitable desiccant included, and in special instances carrying out the packing operation in a controlled environment. Contaminants are controlled by prescribing acceptable levels in the packaging materials.

The standard test[3] for "freedom from corrosive impurities" is to boil a sample of the material in distilled water and determine the pH of the solution. Briefly, pH is a measure of the acidity or alkalinity, and refers to the concentration of hydrogen ions in the solution. Pure water is slightly ionized, and with a pH of 7 is considered neutral. All solutions with a pH below 7 are acid, and those with a pH above 7 are alkaline. Most cushioning materials are required to have a pH not less than 5 nor greater than 9.

With some materials, including those using rubber, corrosion may also be caused by the presence of reducible sulphur. This is checked by a silver staining test in which a small piece of the material is placed on a larger piece of silver foil in a covered dish and held at 70°C in an oven for 30 minutes. Non-silver-staining grades of rubberized hair and expanded rubber are obtainable.

Mould growth

Mould grows rapidly on damp materials, and it is, therefore, desirable to select a cushioning material which is resistant to fungal growth, especially if storage in warm damp climates is expected.

Dusting

When packing products such as optical instruments or those containing delicate mechanisms, it is important that the cushioning material used does not contain or generate dust. This is a difficult property to measure precisely, and a rough order of merit is given in Table 5.1.

Space fillers

Powders and granules

Cork is the outer bark of the holm oak, an evergreen species, *Quercus suber*, which grows mainly in Spain. It is tough, light and elastic, hence its use in granulated form as a space-filling cushioning system. No cushioning data published.

Kieselguhr is fine white siliceous powder containing the remains of diatoms. It is highly absorbent and non-inflammable, hence its main use as a space filler around tins or carboys of corrosive or other dangerous liquids.

Vermiculite.—The soft granular material is formed by heating thin flat flakes of crude vermiculite (a mineral that resembles mica) at temperatures up to 1100°C. Each flake grows to many times its original size by the heat expansion of both free and combined moisture in the flake. After heating it is called *exfoliated vermiculite*. It is normally used as a fireproof space filler and has good heat insulation properties. It is useful for keeping vaccines cool for several hours, and protecting aircraft flight recorders during a crash.

Wrapping materials

Cellulose wadding is made from a creped web or sheet of lightweight paper of open formation, made of cellulosic fibres and comprises two, three or more plies.

Single-faced corrugated board[4] consists of a corrugated sheet faced on one side with a flat liner. Both fluting and liner are usually made from straw pulp and waste pulps, as well as semi-chemical pulp. It is often rolled to form cushioning pads.

Bubbled polyethylene film is also used for wrapping.

ANCILLARY MATERIALS

Shredded materials

Wood wool is widely used as a space filler for packaging glassware, pottery, ceramics, etc. It also has many other uses.

Two forms of shredded paper are available—plain and waxed. Both have lower load-bearing properties than wood wool. The waxed form is to be preferred if damp conditions are anticipated.

Regenerated cellulose film is also shredded and is often used as an alternative to shredded paper.

Non-resilient or crushing materials

Wood wool[5]

Wood wool is a mass of tangled strands manufactured from sound well-seasoned softwood of low resin content. Cushion performance depends on packing density and moisture content, which is normally 12 to 20% by weight. The use of a waterproof lining in the outer case will help to keep moisture content low during transit. For engineering products a water-vapour-resistant barrier is advisable, since damp wood wool can cause corrosion.

When using loose material, care is required to ensure that the correct weight of wood wool is applied to each face, and that it is uniformly distributed. This can often be achieved more readily by using prefabricated pads or bolsters.

Wood wool can also be bonded, using resin or latex, into conforming moulds for squaring off irregular shapes. A range of densities is available.

Wood wool is primarily a space filler but, used with care, it qualifies as a non-resilient cushion and some performance curves are published.

Expanded polystyrene

This is a closed-cell material, made from beads of polymerized styrene, which are expanded by pentane and air, and moulded in a steam heating process. It is available as slab stock in sheets up to 8 ft × 4 ft in area and of various thicknesses. The standard density is a nominal 1 lb/ft^3 but other densities are available. Performance curves are published for 1 lb/ft^3 slab stock material.

Expandable polystyrene beads can be moulded very cheaply in a range of densities and colours to make fitted furniture for products such as cameras, TV sets, radios, etc. It is also available in spaghetti-like strands for use as a space filling material, although used in this way its performance is claimed to be comparable with that of some resilient slab materials.

Its main advantages are low cost, light weight and good load-bearing capacity; the main disadvantages are flammability (though fire-retardant grades are available) and poor resistance to attack by solvent vapours, including hydrocarbons.

Slab stock can be made more resilient by compressing it to about 20% of its original thickness and allowing it to recover partially.

Moulded fittings

Moulded fittings can be used to square off irregular shapes. Using a suitable bonding resin, paper pulp and wood-wool mouldings are available in a range of densities. They are being replaced by expanded polystyrene mouldings which are cleaner, lighter and often cheaper.

Corrugated fibreboard

Corrugated fibreboard consists of a fluted sheet of paper faced on both sides with a flat liner sheet, often of kraft paper. This is known as double-faced or single wallboard. Double and triple wallboards are also made. The main use of these materials is in the fabrication of containers, but they are also frequently used to make cushioning pads and fittings (see chapter 6, Part One).

Resilient materials

Rubberized hair

Rubberized hair is made from animal fibres, usually consisting of about 80% hog's hair and 20% horse mane, horse tail or cow tail bonded with rubber latex and heat-vulcanized. It is available in four different forms.

(i) Oriented rubberized hair

In the plain variety, the hair lies parallel to the surface. By cutting the sheet into strips, and bonding the strips together at 90° to their original position, the hairs are re-orientated so that they are at 90° to the plane of the sheet. In this position they are loaded as struts. This improves efficiency, reducing the cushion factor from about $4\frac{1}{2}$ to $2\frac{1}{2}$.

(ii) Plain rubberized hair

Plain rubberized hair is most commonly used.

ANCILLARY MATERIALS

(iii) *Crimped rubberized hair*

In this process curled hair is first carded into a delicate uniform layer of low density and then gathered by a system of needles, by which it can be condensed to any desired degree. In this way, not only is the original curl of the hair preserved, but in addition a crimped structure develops as a result of the needling action. Resilience, fatigue resistance and cushion factor are all improved.

(iv) *Atlas modules*

Each module is 75 mm (3 in) square with a central hole about 21 mm (1 in) square. They are supplied in slab form, each slab measuring 450 mm × 450 mm (18 in × 18 in), comprising 36 modules in 6 rows of 6, each module being joined to its neighbours by an integrally moulded flash about 3 mm ($\frac{1}{8}$ in) thick. This flash serves as a hinge and permits the slab to be folded or curved to suit the contours of a cargo. The flash can easily be cut with a knife or scissors, enabling the modules to be used singly or in any desired combination. For example, by cutting out three modules in an L-formation and folding up the arms of the "L" a corner block can be made.

Expanded rubber

A closed-cell material made from natural rubber, it is made by moulding under pressure in an autoclave, which produces a sheet with a surface skin. If this skin is cut, the cells are exposed which, while not affecting cushioning performance, may be undesirable for other reasons. In such instances the exact requirements must be specified to the supplier. The main advantage of expanded rubber is its high load-carrying capacity, but it has a number of disadvantages. The most important is that it is seriously degraded by exposure to sunlight. The visible signs are extensive cracking and crazing of the surface. This is accompanied by loss of resilience and creep resistance, and an increase in cushion factor. Expanded rubber is also relatively heavy, and becomes stiffer rapidly as temperatures are lowered. Expanded rubbers are beginning to be phased out and replaced by materials based on expanded polyethylene, which do not have these disadvantages.

Polyurethane foams (flexible)

Polyurethane foams are of open-cell construction and are formed by the polymerization and simultaneous expansion of an isocyanate and a hydroxyl compound. There are two types of foam—polyesters and polyethers. Polyesters tend to be unstable and are degraded by exposure to high humidities and temperatures. Polyethers are, therefore, preferred for most applications, although even these break down if exposed for long periods to sunlight. By altering the chemical constituents, an extremely wide range of foam stiffness and hence cushioning properties can be produced.

A method of grading according to static stiffness is given in BS 3667[6] but this is of limited value for package cushioning because, unlike many other cushioning materials, stiffness is not related to density. Care is, therefore, needed in specifying requirements. The safest way is to specify a particular product, allowing alternatives known to be similar.

Polyurethane foams are amongst the most efficient of cushioning materials, with a cushion factor in the range 2 to 2·5. At impact, the air contained in the open-cell structure must be expelled in a very short time, thus doing work additional to that done by the material itself. Hence, it is important when using these materials in a closed container to leave some free space.

The material is available in sheets in a range of densities from 24.0 kg/m^3 to 96.0 kg/m^3. It can be readily and cheaply machined to make conforming shapes.

Polyurethane foam can also be moulded in the conventional way, although mould costs are high. Alternatively, the foam can be moulded *in situ*. The article to be cushioned is suspended in its container, and the mixed ingredients are poured in and allowed to foam so as to completely envelop it.

Bonded polyurethane chipfoam

Bonded polyurethane chipfoam consists of small pieces or flakes of polyurethane foam of the polyether type, bonded with a binder, and cured to form a homogeneous network of interconnecting cells. It is available in a range of densities between 32.0 and 240.0 kg/m^3. A binder of a different colour is used for each density to aid identification. It is supplied in the form of sheet, roll, strips, corner blocks and mouldings.

As a bulk cushioning material it is more efficient than rubberized hair, and comparable with expanded rubber.

Expanded polyethylene

Polyethylene is expanded about thirty times from the solid state to form a lightweight closed-cell material. It is available in two forms—cross-linked and uncross-linked. The cross-linked form has superior thickness

ANCILLARY MATERIALS

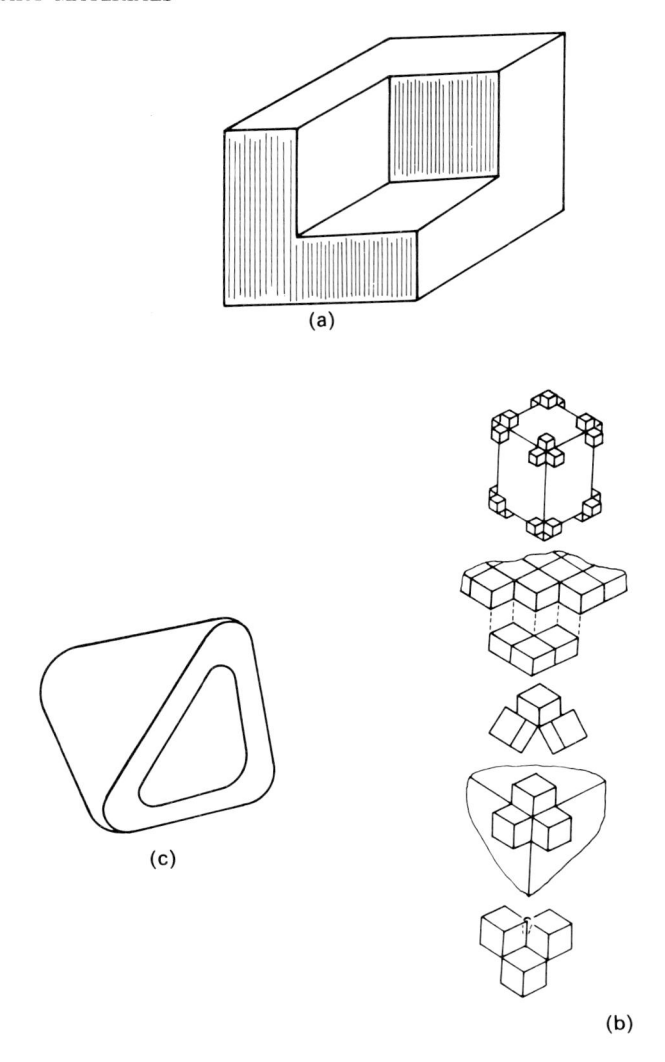

Figure 5.4 Corner pieces
(a) expanded polystyrene mouldings
(b) fabrication of corner blocks from sheets of expanded polyethylene
(c) blow-moulded from polyethylene sheet

loss and creep characteristics, especially at elevated temperatures. As no chemical additives are used, expanded polyethylene is free from sulphur staining compounds or other ingredients liable to tarnish silver, cellulose, or lacquers. It can be sawn or cut by a sharp knife or a hot wire. It can also be shaped by moulding, hot hobbing or cold forming.

5.56

Expanded polyethylene is a cushioning material of moderate efficiency (cushion factor about 3) and light weight—about one quarter the weight of the lightest expanded rubber, which it is now superseding.

Ethylene vinyl acetate (EVA) foam is a co-polymer of polyethylene and vinyl acetate. It is much more rubberlike in its behaviour, but is less efficient than expanded polyethylene, having a cushion factor of about 4·0. Its performance is, therefore, very similar to that of the expanded rubbers.

Bubbled polyethylene

Bubbled polyethylene is made from two sheets of polyethylene film, one of which is embossed to entrap air bubbles. Several grades are available with bubbles of different sizes. Small bubbles are recommended for protecting small parts, and large bubbles for heavier items. The material can also be used for wrapping, or in rolls or pads as cushions. Used as a heat-sealed envelope, it also provides a barrier to water vapour. It is supplied in rolls 1·25 m (4·10 ft) wide and up to 240 m (787·40 ft) long.

Corner pieces

A very simple method of packing articles which are generally rectilinear in shape is to use corner pieces. These are easily applied by unskilled labour and are effective and inexpensive. However, care is needed to ensure that the total loaded area on each face gives a static stress in the optimum range for the material employed.

Corner pieces can be made in most of the materials described here. Expanded polystyrene mouldings are commonly used (figure 5.4a) and expanded polyethylene lends itself to the fabrication of corner blocks from sheet (figure 5.4b). Other designs are also available, including corner pieces blow-moulded from polyethylene sheet (figure 5.4c).

Selection of bulk cushions

To select a package cushion we need to know

about the product   (i) its overall dimensions
　　　　　　　　　　(ii) its weight
　　　　　　　　　　(iii) its fragility factor
about possible cushioning materials
　　　　　　　　　　(iv) their peak g_n/static stress curve
　　　　　　　　　　(v) their cost

ANCILLARY MATERIALS

Finally, we need to know the height from which the package is likely to be dropped in transit. Trials with instrumented packages[7] have confirmed the general correctness of the traditional assumption that the lighter the package the greater the height from which it is likely to be dropped. The measured drop heights agree reasonably well with those derived from observation and judgment, and show that drop height is inversely proportional to the logarithm of gross weight (figure 5.5).

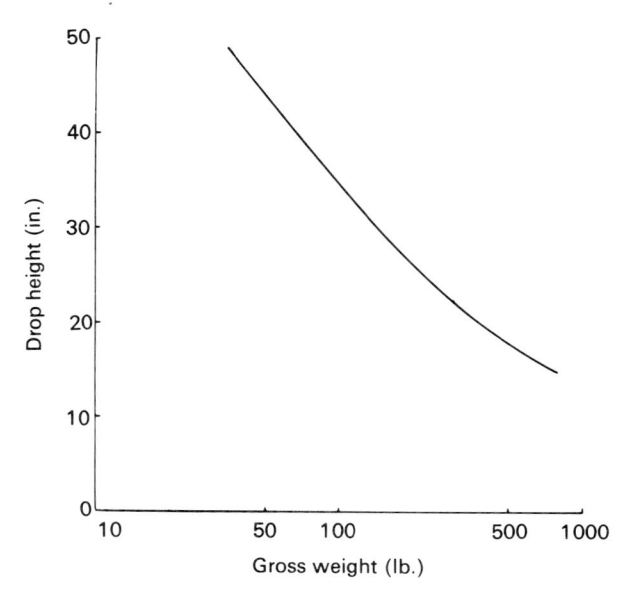

Figure 5.5 Drop height is inversely proportional to the logarithm of gross weight (approximately)

The size and weight of an item are readily obtainable, but its fragility factor is more difficult to determine. The best way is to carry out a fragility test by dropping the article, fitted with an accelerometer, from increasing heights until damage occurs; but to obtain a reliable figure, this will probably entail damaging a number of articles. Another method is to make an estimate based on a careful examination of the article. Experienced packaging engineers can get remarkably close to the true fragility of an article by this method, but it must be stressed that experience with similar articles is the key to success.

Peak g_n/static stress curves for many materials are now widely available. The following simple example is given to illustrate the basic design method. It assumes that the article is either rectilinear or can be made so by the addition of suitable fittings.

Consider an article contained within the volume $0.5 \times 0.4 \text{ m} \times 0.3 \text{ m}$

(20 in × 16 in × 12 in), weighing 20 kg (45 lb) and having a fragility factor of 40. We must first calculate the static stress on each face as follows:

$$\text{Sides} \quad \frac{40}{20 \times 12} = 0.17 \text{ lb/in}^2 \quad \frac{20}{0.5 \times 0.3} = 133 \text{ kg/m}^2$$

$$\text{Ends} \quad \frac{40}{16 \times 12} = 0.21 \text{ lb/in}^2 \quad \frac{20}{0.4 \times 0.3} = 167 \text{ kg/m}^2$$

$$\text{Top \& bottom} \quad \frac{40}{20 \times 16} = 0.13 \text{ lb/in}^2 \quad \frac{20}{0.5 \times 0.4} = 100 \text{ kg/m}^2$$

Next we must specify the drop height, and this requires an estimate of the gross weight of the finished package. Assuming, for the purpose of this example, that gross weight will be twice the cargo weight (i.e. 40 kg (90 lb)) we see from figure 5.5 that drop height should be about 0·9 metre (36 in).

Referring now to figure 5.2 we see that, for a 0·9 m (36 in) drop, between 75 and 100 mm (3 and 4 inches), say 90 mm (3½ in) of 6 lb/ft³ Chipfoam is required. About 40 g_n will be achieved at static stresses between 75 and 150 kg/m² (0·1 and 0·2 lb/in²), which spans the actual loading range.

If the article weighed 10 kg (22 lb), then the actual static stresses would have been halved. In such instances, the area of the cushions must also be halved in order to achieve the static stress at which the cushion is most efficient (i.e. gives the lowest peak g_n).

If the article weighed 40 kg (90 lb), it would be necessary to use a stiffer material—possibly a high-density Chipfoam.

Cushioning devices

Bulk cushioning materials cannot provide the answer to all cushioning problems, and it is sometimes necessary or advantageous to use cushioning devices. This is particularly so when a product is already provided with mounting or attachment points, which can be used to secure it to a shock mounted platform. For cargoes up to say 120 kg weight, the benefits are probably most evident with articles of irregular shape, since there is no need to "square-off" to provide faces for the cushioning material, and this results in a simpler, smaller, and often lighter package. With larger products, such as aero-engines, where size and weight are such that the volume of bulk cushioning material required results in a considerable increase in package size and weight, the cost of the cushioning material may well exceed the total cost of shock mounts and mounting platform. The greatest advantage of shock mounts for this type of product is that the cargo is secured only to the base of the case and, if required, the top sides and ends can be removed in one piece, sewing-machine fashion, to give instant access to the contents.

In designing a cushioning system with shock mounts, several problems arise which do not occur with bulk cushioning materials. (1) The design procedure is more complex since, in most packages, the system must provide roughly equal protection against shocks from all directions. (2) Shock mounts have less natural damping than bulk cushioning materials, and care is needed to ensure that the spring/mass system is not likely to be excited at resonance in transit. On larger packages, hydraulic dampers are sometimes used to control resonance, and these have the double advantage that they greatly increase the efficiency of the shock isolators.[8] With smaller packages space and cost does not allow this, but experience shows that satisfactory systems can be designed without dampers. (3) Special containers must be designed, with local reinforcement, to take the high concentrated loads.

For all these reasons it is advisable, except perhaps for the simplest designs, to seek the advice either of the mount manufacturer, or of a specialist firm of packaging engineers.

Bonded-rubber shock mounts

Shear mountings

A shear mount is a block of rubber bonded between two parallel steel plates, which incorporate holes or studs for attachment. Occasional shock deflections up to twice the rubber thickness are possible in the shear plane. In compression, the deflections available are too small to be useful and various geometrical arrangements can be adopted to give the necessary all-round protection.[8] Mountings are available with shear stiffnesses in the range 2000–30,000 kg/m (100 lb/in to 1400 lb/in), and capable of shear deflections up to 60 mm. There are several manufacturers of shear mounts, but other types of shock mount are often proprietary designs.

Conical mounts

Conical mountings are designed to overcome the limitation of the shear mount, by providing flexibility in three planes. Some examples are illustrated in figure 5.6. Generally, the axial stiffness of these mounts is much lower than the lateral stiffness. This is because they are designed to be mounted beneath a heavy equipment where, in simple drops, they are loaded axially in compression. In side impacts, the fact that the centre of gravity is well above the plane of the mountings means that a rolling movement occurs, and the mountings are subjected to both axial and lateral loads. Mountings are available to support static loads in the range 170 kg to 800 kg.

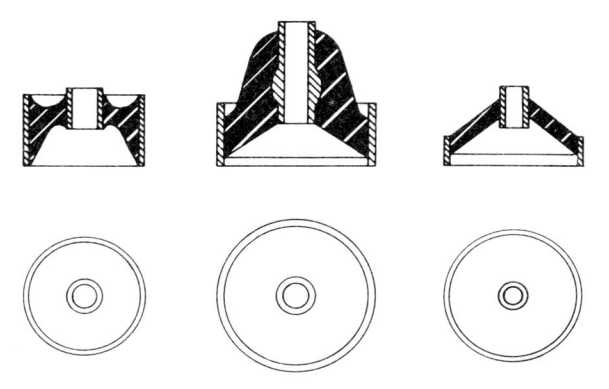

Figure 5.6 Conical mounts

Metalastik buckling mounts

The buckling mount is a development of the conical mount. Stiffness is similar in tension, compression and lateral shear (figure 5.7). Two sizes are available. The smaller gives up to 55 mm deflection in all directions and takes static loads between 30 and 70 kg, depending upon rubber hardness. The larger gives up to 100 mm deflection in all directions, and takes loads from 45 to 100 kg according to hardness.

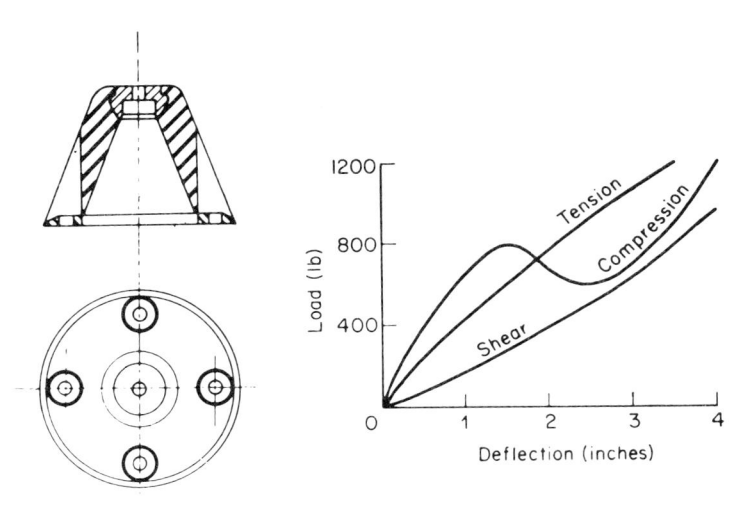

Figure 5.7 Buckling mount

5.61

ANCILLARY MATERIALS

Tubular mountings (BTR)

The tubular mounting (figure 5.8) is an alternative to the conical mounting and is used in a similar way, i.e. mounted beneath the load. It is available in a range of hardness to take loads from 200 to 2200 kg per mount.

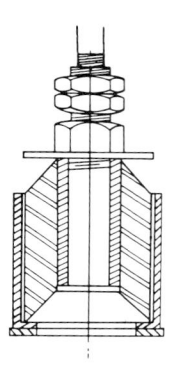

Figure 5.8 Tubular mount (BTR)

Delta mountings (Silentbloc)

The Delta Mounting (figure 5.9) represents a novel approach to shock mount design. It is available in a range of sizes. The smallest is 92 mm high and gives a maximum deflection of about 35 mm. The largest is 550 mm high with a maximum deflection of 225 to 250 mm.

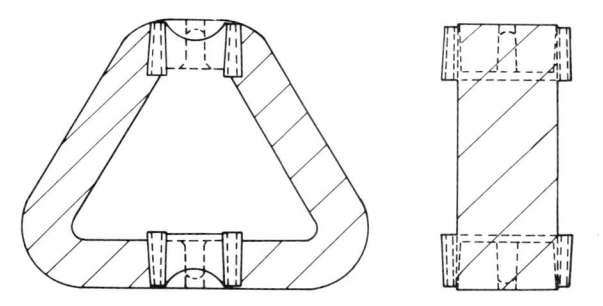

Figure 5.9 Silentbloc delta mounting

Steel spring systems

Coil springs

When deflections in excess of about 280 mm are needed, bulk cushions

5.62

PACKAGE CUSHIONING SYSTEMS

and the conventional cushioning devices are seldom practical or economic. Deflections of this order are sometimes necessary for products of low fragility and light weight. Many years ago a tension spring system was evolved to deal with this problem, and it was adopted by the Ministry of Defence as a standard method of packaging electronic valves. A design procedure was published by Mindlin in 1945.[9] A typical package is illustrated in figure 5.10. Damping is low in the system and may sometimes be a problem.

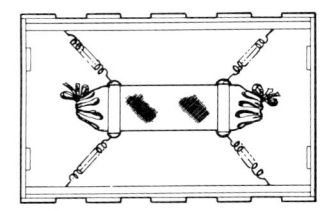

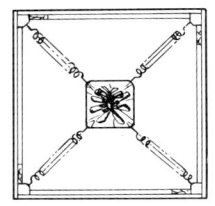

Figure 5.10 The use of coil springs

Leaflote springs

This is a laminated leaf spring system suitable for heavy fragile loads requiring large deflections. The springs are curved to form a semicircle, and the ends are curled over so that they slide easily over the surface of a steel plate secured inside the container. The semicircular shape gives a very high initial stiffness (figure 5.11) and hence high efficiency. The chief disadvantage of the system is its great weight.

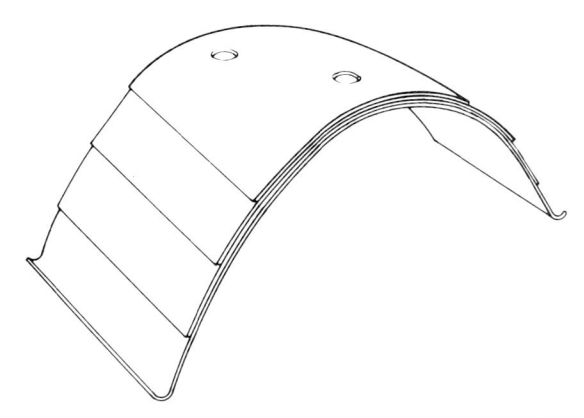

Figure 5.11 Leaflote springs—laminated steel plates of semicircular form giving a very high initial stiffness and hence high efficiency

5.63

ANCILLARY MATERIALS

"*Pnucush*" mounts (*Wilmot packaging*)

The Pnucush mount is an alternative to the bonded rubber conical mount. It is especially useful in meeting extreme low-temperature requirements. Each mount comprises a number of individual torsion springs secured between two metal end plates (figure 5.12). They are designed to provide deflection by both axial compression and lateral shear, thus giving protection from both drops and horizontal impacts. Mountings are available to cover the weight range 14 to 140 kg per mount and to give impact load factors up to 30.

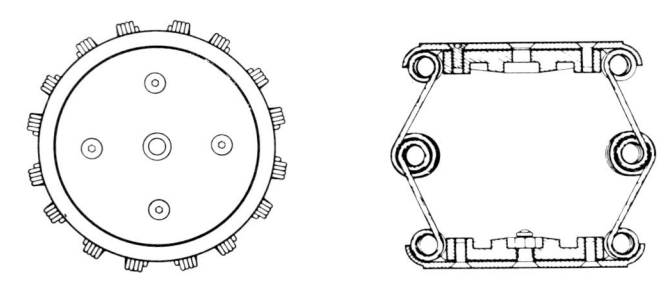

Figure 5.12 The Pnucush shock mount

REFERENCES

1. *Fundamentals of Packaging*, edited by F. A. Paine (Blackie).
2. BS 1133—Packing Code. Section 12—Cushion Materials, Appendix "E".
3. BS 903—Method of Testing Vulcanized Rubber. Part F.9.
4. BS 1133—Packaging Code—Section 7.
5. BS 2548—Wood Wool for Packaging.
6. BS 3667—Methods of Testing Flexible Polyurethane Foam.
7. "Maximum Drops experienced by Packages in Transit," by D. C. Allen, *J.S.E.E.*, June 1972.
8. *Rubber in Engineering Practice*, by A. B. Davey and A. R. Payne (McLaren).
9. *Dynamics of Package Cushioning*, by R. D. Mindlin, 1945.

EPILOGUE

During the last twenty-five years, the public awareness of the use of materials in the distribution of goods of all kinds has increased considerably. In the last five years, awareness of the utilization of fuels of all types for the production of energy has moved from an attitude of almost complete indifference to one of concern. The solutions to questions of material resources, energy usage and conservation are all dependent on the type of society in which we desire to live.

The so-called "throw-away society" has developed in response to society's desire to save labour and time, both of which are priced at a level above that of most materials and of energy. Consequently, packaging designers have produced packages tailored to carry out all their functions effectively at the lowest overall cost in money terms. Packaging has therefore become a techno-economic function in the distribution of products, with the objective of minimizing overall costs whilst maximizing sales and profits. Society determines its needs, and the package-making industries, the producers of goods, the transport organizations and the retail outlets try to meet them as efficiently as possible.

Packaging is, in fact, a very small user of energy (Table 1); it is one of the major forces for preserving and protecting goods and for keeping waste to a minimum; and continually strives to use less material to carry out its very specific functions. All concerned with making and using packaging continually seek to understand better why they operate as they do and, where inefficiencies exist, to rectify them. Because packaging materials and techniques are continually being improved, 100% efficiency is never likely to be possible and, no doubt, at any given time, some 25–30% of all packaging could be improved. It is unlikely, however, that

EPILOGUE

Table 1. Energy used in container production (tonnes of oil equivalent per tonne)

	Aluminium	Plastics	Paper	Tinplate	Glass
Raw material production	6·00 (95·3%)	2·30 (78·7%)	1·45 (91·2%)	1·00 (86·3%)	0·35 (92·5%)
Conversion to containers	0·20 (3·2%)	0·40 (13·7%)	0·05 (3·2%)	0·10 (8·6%)	
Heating and lighting factories	0·08 (1·2%)	0·16 (5·5%)	0·07 (4·4%)	0·04 (3·4%)	0·02 (5·0%)
Transport to user	0·02 (0·3%)	0·06 (2·1%)	0·02 (1·2%)	0·02 (1·7%)	0·01 (2·5%)
Total	6·30	2·92	1·59	1·16	0·38

this would result in more than, say, a 5% saving overall—but even 1% of the £1600 million per annum (1973) packaging materials bill is worth saving (Table 2). For this very reason the package-using industries will attempt to save it, and their suppliers will assist them.

Table 2. Estimated expenditure on packaging 1973 (£millions)

Paper sacks	52
Paper bags and carrier bags	39
Paperboard cartons	176
Fibreboard cases	204
Rigid boxes	36
Cellulose film	46
All plastics	200
Glass containers	123
Metal cans, etc.	240
Aerosols	47
Steel drums	45
Aluminium foils	48
Collapsible tubes	8
Jute sacks	8
Wooden containers and pallets	70
Miscellaneous	250
Total	ca 1600

The package-making industries are growing in importance. As consumers demand "convenience", the need for quality packaging will increase. There is likely, therefore, to be more packaging, not less. This need will, however, be questioned all along the line in relation to the total cost of distribution. The package must be effective.

At its most fundamental, packaging is designed to prevent damage and hence reduce waste—not only of goods and materials, but also of energy and labour. Its major functions are directly concerned with this. When the package has done its job, it can be used for something else and, when it finally becomes a waste material to be disposed of, it often has already reduced the waste disposal problem by reducing the amount of trim (husks, bone, etc.) appearing in urban waste. Even where the primary function of packaging is not that of conservation, it is designed either to improve sales or to provide convenience—both will lead to better profits and will contribute to improvements in the quality of life.

Packaging will, therefore, continue in the future to serve mankind by containing, protecting, preserving and identifying the products that are needed. Thus it will contribute to improvements in the quality of life in all countries, from the developing to the most sophisticated, by getting goods in the right quantity, at the right time and in prime condition, to the people who need them, at the minimum overall cost. It will continue to do this, although the manner in which it does so may well change in a world where the costs of raw materials and of energy are increasing disproportionately.

There are not likely to be any fundamental changes in the way packaging performs its tasks, and the three trends that have been taking place over the last twenty-five years will probably continue for the next twenty-five. These are:

1 Packaging designers will continue to develop lighter-weight more-economical and more-convenient packages for the purposes for which they are needed.

2 The functions of containment, protection and preservation, together with identification, will be necessary for all types of packaging. The amount that will be spent on the selling function and the convenience and service aspects of packaging may, however, change in the course of the next decade. In general, the overall cost of the operation of delivering products to consumers will still be a major criterion in the decision on how such changes will occur.

3 Whenever packaging can contribute to a reduction in labour costs, particularly when these are high, then the interaction between packaging materials and filling, forming and closing machines will play a big part in determining which packaging material will be used for any particular purpose. This will be particularly true where flexible materials are concerned in the packaging of food products and other household goods.

The challenge in packaging today is to continue to serve people in a way that meets their needs and interests, and to make the facts known in

EPILOGUE

ways that can be understood by the general public. This book is only a beginning to that end, and an understanding of the packaging function is essential to the optimum use of packaging materials.

INDEX

A

accelerometer, 5.58
adhesive, 1.112, 2.97, 3.104, 4.58, 5.3, 5.24, 5.27, 5.29, 5.31
adhesive selection, 5.39–5.43
aerosols, 2.73
AFD foods, 2.111, 2.112, 3.68
aluminium, 2.3, 2.5, 2.14, 2.44, 2.46, 2.98
aluminium foil, 1.112, 2.6, 2.9, 2.57, 2.102, 5.23, 5.41
ammunition, 2.54
analytical tests, 1.38
animal feed, 3.75, 3.78, 3.89
annealing, 3.10
artwork, 1.58

B

bacon, 3.70
bags, 1.24, 3.49, 4.16, 4.64, 4.67, 5.41
baler bags, 1.51
bales, 4.75, 5.7, 5.9
barrels, 3.94, 3.101, 4.21
barriers, 1.23, 1.30, 1.42, 1.50, 1.57, 1.112, 2.55, 2.104, 2.107, 3.58, 3.62, 3.68, 3.78, 4.16
beating, 1.8
beer, 2.40, 3.104, 4.23
beetles, 2.107
beverages, 2.40, 3.3, 3.104, 3.139
biscuits, 3.58, 3.63, 3.68
bitumen, 1.112, 2.59, 4.76, 5.31
blackplate, 2.5, 2.44, 2.45
blister packs, 2.7, 3.67
blow moulding, 3.9, 3.25, 3.32, 3.95, 3.101
board making, 1.12, 1.55
board properties, 1.60, 1.62, 1.65, 1.90, 1.97, 1.98
boil in bag, 3.70
bond strength, 5.35, 5.37
bottles, 3.3, 3.107, 5.22, 5.24
boxboard, 1.54, 1.77, 2.53, 2.56
boxes, 1.77, 1.85, 2.101, 4.8, 4.37, 5.3
bran, 4.73
bread, 3.52, 3.59, 3.63, 3.68, 3.106
break-bulk, 4.62
breathing wares, 3.59, 3.66, 3.68
Brikpak, 1.74
bubbled polyethylene, 5.57
building products, 3.104
bulk handling, 1.51
bundling, 5.9
burst test, 2.71

C

cable reels, 4.40
cakes, 3.64
cans, 2.5, 2.13, 5.22
caps, 2.74, 3.108, 3.120
cartoning systems, 1.72, 1.74
cartons, 1.54, 2.101, 5.3
cases, 1.99, 4.5, 4.35
casks, 2.84, 4.20, 4.22
Cekatainer, 1.74
cellulose acetate, 3.65, 5.6
cellulose film, 3.64, 5.6
cements, 2.18
chaff, 4.73
cheese, 2.111, 3.66, 3.69, 3.70
chemical pulping, 1.6

1

INDEX

chemicals, 1.96, 1.112, 3.78, 3.104
child resistance, 3.119, 3.130
chipboard, 1.15, 1.55, 1.78, 1.91
cigarettes, 1.70, 2.6, 2.13, 3.58, 3.64, 3.66
clay-coated board, 1.15, 1.55
closure, 1.43, 1.95, 2.41, 2.59, 2.65, 2.93, 2.101, 3.36, 3.86, 3.91, 3.107, 4.17, 4.24, 4.70, 5.5, 5.9, 5.19, 5.40
Clupak, 1.26, 1.42
coated boards, 1.55
coated papers, 1.29, 2.57, 3.63, 3.68
coating, 1.31, 2.99
Cobb test, 2.71
cocoa, 3.92
coffee, 3.92
cohesive, 5.32
cold forming, 3.42
compatibility, 3.129, 5.3
composite containers, 2.52, 2.97
compression strength, 1.98, 2.72, 4.30
Concora, 1.105
confectionery, 2.111, 3.53, 3.58
containers, retail, 1.54, 1.77, 2.13, 2.52, 2.76, 2.97, 2.108, 3.3, 3.25, 3.107, 5.24, 5.27
containers, shipping, 1.3, 1.40, 1.88, 1.106, 2.84, 3.94, 4.10
convenience, E2, E3
convolute (straight) winding, 1.107, 2.52, 2.54
cooperage, 4.20
cork, 5.51
corrosion, 2.8, 2.10, 3.59, 5.47, 5.50
corrugated fibreboard, 1.88, 1.90, 4.30, 4.49, 5.51, 5.53
cosmetics, 2.97, 3.39, 3.104
costs, 1.3, 2.6, 4.12, 4.17, 4.61, 5.8, 5.9, 5.26, 5.57, 5.59, E1, E2
cover papers, 1.78, 1.81, 1.84
crates, 3.94, 3.104, 4.8, 4.12
creasing, 1.61, 1.65, 4.28
creep, 5.47, 5.49
creped paper, 1.25, 1.42
cullet, 3.5
cushion factor, 5.47
cushioning, 1.114, 5.44
cutting, 1.58, 1.61, 1.78, 1.94, 2.37, 4.27

D

deceleration, 5.48
decoration, 2.46, 2.94, 2.103, 2.109
deep-drawn tins, 2.31
definitions, 1.54, 1.116, 2.73, 3.3, 3.18, 3.25, 3.62, 3.94, 4.20, 4.25, 4.45, 4.75, 5.32
design, 1.57, 1.75, 3.19, 3.132
detergents, 1.70, 3.38, 3.69, 3.104
diotite, 1.73
dispensing, 3.107, 3.119, 3.135, 5.4
domestic appliances, 1.96

double seam, 2.14, 2.15, 2.64, 2.87
drawn and wall ironed (DWI) cans, 2.21, 2.32, 2.46
drop number, 1.49
drop test, 1.49, 3.87, 5.58
drums, 1.106, 2.84, 2.86, 2.91, 3.94, 3.98, 3.99, 4.38, 5.3
dry goods (wares), 1.112, 3.50
dry offset, 2.50
duplex boards, 1.15, 1.55
dyestuffs, 3.104

E

economics of tins, 2.34
egg trays, 1.118
electrical components, 1.118
electromotive series, 2.9
emulsions, 5.40
energy, E1, E2
engineering components, 1.118
envelopes, 1.24
enveloping, 4.59
Erichsen test, 2.71
essences, 1.118
European Pallet Pool, 4.61
expanded polyethylene, 5.55
expanded polystyrene, 3.31, 5.47, 5.52

F

fabric sealing tapes, 5.3, 5.6
facing materials, 3.130
fertilizer, 3.74, 3.92
fibreboard containers, 1.88, 3.78, 5.3, 5.5, 5.19
fibre building board, 4.25
filling cement, 2.80
filling sacks, 1.42, 1.44, 1.45, 3.81
finishing, 1.18, 1.64
fish, 2.6
fish meal, 3.92
flax, 4.64
flexible barriers, 3.44
flexography, 1.60, 3.81
flour, 3.92, 4.70, 4.73
fluting medium, 1.71, 1.77
foams, 2.74
foil-lined board, 1.55
food, 1.23, 1.96, 1.112, 2.11, 2.26, 2.40, 2.107, 3.3, 3.38, 3.78, 3.104
footwear, 1.71, 1.77
forestry, 1.3, 4.3
fork lift equipment, 4.46, 5.10
FPCMA, 1.105
fragility factor, 5.57
freight containers, 5, 4.60
frozen food, 1.55, 2.108, 5.6
fruit, 3.60, 3.65, 4.33
furniture, 1.96

2

G

gas packing, 3.72
general-line tins, 2.22
glass, 1, 1.118, 2.13, 2.76, 3.3, 5.21
glassine, 1.14, 1.28, 2.58, 2.107
glued labels, 5.20
grain, 3.92, 4.70, 4.73
gravity packer, 1.44
gravure, 1.60
greaseproof paper, 1.14, 1.24, 1.28, 2.109, 4.76
guidelines for good retail packages, 6
gummed tape, 5.3

H

handling paper sacks, 1.47
handwrapping, 3.47, 3.66
hardboard, 4.28
heat seals, 2.106
heat-sensitive tape, 5.3, 5.7
hemp, 4.64
Hermetet, 1.74
hessian, 4.65, 4.76
hosiery, 1.77
hot melt coatings, 1.33, 2.105
hot melts, 5.40
household goods, 1.96, 3.40

I

imitation krafts, 1.27
industrial packaging, 1
inhibitors, 2.10
injection moulding, 3.28, 3.33, 3.95, 3.97, 3.99
inks (printing), 1.112, 2.47, 3.104
insects, 2.107, 4.76, 5.9
insert labels, 5.21
inspection, 3.12
interleaving, 1.24
iron, 2.9

J

jars, 3.3, 3.107
Jenkins test, 2.71
Jerrycans, 3.94
jute, 3.91, 4.64

K

kegs, 2.84, 2.89, 3.94, 4.20, 4.49
Kieselguhr, 5.51
Kliklok, 1.73
kraft paper, 1.7, 1.14, 1.25, 1.42, 1.91, 1.112, 2.56, 4.16, 4.76, 5.4, 5.7

L

labelling, 2.51, 2.67, 2.77, 5.20, 5.40
laboratory transport trial, 1.98
lacquer coating, 1.30, 1.33, 2.7, 2.11, 2.96, 2.99, 5.28, 5.39

laminates, 2.57, 2.103, 3.52, 3.63, 3.71, 3.84, 5.23
lard, 1.112
latex, 5.41
lead, 2.18
letterpress, 1.59
lever lid, 2.8, 2.25, 2.62
lightweighting, 3.20
lignin, 1.6
liners, 1.91, 1.112, 2.57, 2.95, 3.90, 3.108, 3.126, 4.12, 4.16, 4.70, 4.71
linoleum, 1.112
lithography, 1.59
loads, 4.55
locked corner, 2.28

M

machine tools, 4.36
marking, 4.79
meat, 3.60, 3.64
measuring composites, 2.68
measuring tins, 2.38
mechanical pulp, 1.6
mechanical wrapping, 3.50
metal packaging, 1, 2.3, 2.72, 2.97
mild steel plate, 2.3
milk, 2.7, 2.109, 3.68, 3.104, 4.26
moist wares, 3.58
moisture content, 2.71, 4.51
mould, 5.9, 5.47, 5.50
moulded pulp, 1.116
multiwall paper sacks, 1.40, 3.76

N

nails, 1.96
nesting, 2.92
non-metallic strapping, 5.15
non-resilient cushioning, 5.45
nozzles, 2.100
nuts and bolts, 4.73
nylon, 3.70, 5.15

O

oats, 4.73
oblong pourer tins, 2.27
odour, 2.11
offset lithography, 1.59, 2.47
oils, 2.36, 3.104
open time, 5.34
open-mouth sacks, 1.42
open-top cans, 2.14
opening devices, 1.66
optical properties, 1.22, 1.39
oxygen, 2.8

INDEX

P

package types, 2
packaging boards, 1.15
  material, 2
  papers, 1.14, 1.20
pH, 2.9, 5.50
pails, 2.90, 3.94, 3.98
paint, 2.13, 2.25
pallets, 4.41, 4.48
palletization, 5, 4.46, 5.9
paper, 1, 2.103, 3.62, 5.3, 5.4, 5.23, 5.40
paper board, 1
paper making, 1.3, 1.9, 1.116
paper properties, 1.22, 1.48, 1.86
parcel wrap, 3.46, 3.51
parison, 3.7
passivity, 2.10
pastes, 5.41
peas, 2.12
peat, horticultural, 3.78
Pemplex, 1.74
perfumery, 3.3
Perga, 1.74
pesticides, 3.89
pharmaceutical, 2.11, 2.97, 2.105, 3.3, 3.40
pickles, 1.113
pilfer proofing, 3.118, 3.128
pinholes, 2.102
plastics, 1, 1.112, 2.6, 2.13, 2.44, 2.52, 2.65, 2.66, 2.76, 2.97, 2.100, 3.25, 3.74, 3.94, 3.107, 4.48, 4.71, 5.3, 5.23
plywood, 4.7, 4.10, 4.20
polarization, 2.8
polyethylene, 3.66, 3.74, 3.103, 5.47, 5.51
polymer, definition, 5.32
polypropylene, 3.103, 5.15
polyurethane foam, 5.47, 5.54
potatoes, 4.73
pouches, 2.110, 3.20
poultry, 3.68
pouring devices, 3.119, 3.138
prebreaking, 1.64
pressure-sensitive material, 5.3, 5.5, 5.25, 5.41
Pridham decision, 1.89
printing, 1.58, 1.93, 1.114, 2.95, 3.67, 3.80, 5.4, 5.21, 5.28
propellant, 2.74, 2.77, 2.79
puffer pack, 2.55
pulping, 1.6
puncture, 2.71
Pure Pak, 1.74
pure pulp boards, 1.56
PVC, 1.114, 3.65, 3.74, 5.6

Q

quality control, 1.66, 1.105, 3.15, 5.29

R

refuse, 1.51
reinforcement, 4.30, 5.4, 5.7, 5.9
release coatings, 1.112, 2.58
resilient cushioning, 5.45
retail package, 6, 1.54, 1.57, 1.72, 1.75, 1.87, 1.116
reversible label, 5.21
roller coating, 2.50
rope, 4.77
rotational casting, 3.31, 3.94, 3.97, 3.98
rubber, expanded, 5.54
rubberized hair, 5.47, 5.53
rub resistance, 1.61

S

sachet, 1.24
sacks, 1.40, 3.74, 4.64, 4.67
sampling, 3.17
scoring, 1.78
screw cap, 3.36
sealing tape, 5.3, 5.41
self-adhesive, 5.3
semi-chemical, 1.7, 1.91
semi-liquid, 1.112, 1.114
setting time, 5.34
shaving creams, 2.7
shelf-life, 2.11
sherry, 4.23
shipping containers, 3, 5, 1.88, 1.106, 3.74
shirts, 1.72
shock, 1.24, 5.48, 5.59, 5.60
shrinkwrapping, 3.75, 4.59
sift proof, 1.73
silk screen, 2.51
skin pack, 3.54
slip lids, 2.23
softboard, 4.25
solid fibreboard, 1.88, 1.89
solvent, 2.36
space filler, 1.24, 5.44, 5.51
specification, 1.66, 1.67, 1.115, 3.15, 3.21, 4.51, 4.53, 4.68, 4.78
spiral winding, 1.107, 2.52, 5.40
spirits, 4.23
stainless steel, 2.85
stapling, 5.18
staying, 1.80
steel drum, 1.106
steel spring, 5.62
steel strapping, 5.10, 5.11
stitching, 4.28, 4.72
storage of material, 1.35, 1.46
strapping, 4.58, 4.77, 5.9, 5.15, 5.17, 5.18
strawboard, 1.77, 1.91
strength properties, 1.37
stretch forming, 3.43

INDEX

stretchable kraft papers, 1.25
strip packaging, 2.109, 3.54
sugar, 3, 92, 4.70
sulphite papers, 1.7, 1.28
surface energy, 5.38
surface properties, 1.38, 1.61
surface tension, 5.39
surface treatment, 1.18, 3.23
syrups, 2.74

T

tack, 5.24, 5.34
tamperproof devices, 3.118, 3.121, 3.142
tea, 4.26
thermoforming, 3.30, 3.35, 3.65, 3.68
thermosensitive label, 5.20
thread, 3.110
tie-on labels, 5.20
tray, 4.33
truss, 4.75

U

ullage, 2.92, 3.112
union kraft, 1.42, 1.50
unit load, 4.45, 4.58, 4.62

V

vacuity, 3.112
vacuum packing, 3.72, 4.59
vacuum tobacco tin, 2.30
Valeron, 3.84
value of bulk packaging, 1
value of packaging purchased, 4
valve, aerosol, 2.73, 2.74, 2.76
valve, sacks, 1.44, 3.77, 3.82
VCI, 2.59, 3.59, 4.16
vegetable parchment, 1.14, 1.23, 1.29, 1.112, 2.57, 2.109
vegetables, 3.60, 3.65, 4.23
veneers, 4.7
venting closures, 3.123
vermiculite, 5.51
vitamins, 2.109, 3.78

W

wads, 3.108
waterproofing, 1.118
water-soluble wrapping, 3.69
waxed board, 1.55
waxing, 1.18, 1.34
weatherproof board, 1.96
welding, 2.85
wet-strength paper, 1.27, 1.42, 1.50
whisky, 1.118, 3.112
wine, 4.23
wood, 1, 1.5, 4.3
wood pulp, 1.3
wood wool, 5.47, 5.52
wooden cases, 4.10, 5.9
wooden casks, 1.106
woven plastics sacks, 3.77
wrappings, 1.20, 3.44, 5.3, 5.51

Z

Zupak, 1.74

5